Subcellular Biochemistry

Volume 80

Series editor

Robin Harris, Northumberland, UK

For further volumes:
http://www.springer.com/series/6515

Gregor Anderluh · Robert Gilbert
Editors

MACPF/CDC Proteins - Agents of Defence, Attack and Invasion

Springer

Editors
Gregor Anderluh
Laboratory for Molecular Biology
and Nanobiotechnology
National Institute of Chemistry
Ljubljana
Slovenia
Department of Biology
Biotechnical Faculty
University of Ljubljana
Ljubljana
Slovenia

Robert Gilbert
Division of Structural Biology
Wellcome Trust Centre
for Human Genetics
University of Oxford
Oxford
UK

ISSN 0306-0225
ISBN 978-94-024-0339-8 ISBN 978-94-017-8881-6 (eBook)
DOI 10.1007/978-94-017-8881-6
Springer Dordrecht Heidelberg New York London

Printed on acid-free paper

Springer is part of Springer Science+Business Media (www.springer.com)

Preface

Cellular membranes are crucial for the survival of organisms and maintaining the lipid membrane structure is therefore paramount for the proper functioning of cells. Many efficient strategies for damaging membranes have evolved and pore formation is one of the most intensely studied and best understood. Pore formation is frequently used in toxic attack on cells, as it can lead to efficient disruption of cell metabolism or even cell death. There are several different ways in which proteins efficiently attach to lipid membranes, oligomerize, and punch holes in them. This book describes the functioning of two important pore-forming protein families, cholesterol-dependent cytolysins (CDC), and proteins of the membrane attack complex/perforin (MACPF) family, which are nowadays understood to belong together. Cholesterol-dependent cytolysins are key virulence factors of many pathogenic bacteria, while proteins of the membrane attack complex and perforin family enable removal of unwanted cells from organisms. CDC and MACPF proteins represent perhaps the best studied pore forming protein families and in the last couple of years we have witnessed a significant increase in our understanding of how they function. Perhaps the most surprising discovery was the realization that MACPFs and CDCs are evolutionarily linked through a common domain that forms the basis for transmembrane pore formation, and the two protein families are now collectively referred to as the MACPF/CDC superfamily. Many new members have been described and functionally characterized in recent years and the different contributors to this book provide a comparative analysis of the structure and function of a diverse range of MACPF/CDC superfamily members.

The introductory two chapters by Gregor Anderluh, Robert Gilbert and colleagues, and Masaru Nonaka set the stage for the rest of this book by providing an overview of the MACPF/CDC members and their evolutionary relationships. Next, the book deals with the structural properties of the cholesterol-dependent cytolysins (Robert Gilbert) and MACPF proteins (chapter by Andreas Sonnen and Philipp Henneke). The chapter by Benjamin Johnson and Alejandro Heuck provides a mechanistic explanation of membrane interaction and pore formation by perfringolysin O, one of the best-studied MACPF/CDCs. The next part of the book describes functional properties of MACPF/CDC proteins. A general chapter by Gregor Anderluh and colleagues provides an overview and general information about membrane interactions, pore formation, and effects on cells. Functional properties and the biological role of several MACPF/CDC members are then described in more detail: pneumolysin (by Tim Mitchell and Catherine Dalziel), listeriolysin O (by Stephanie Seveau), perforin (by Judy Lieberman and Jerome Thiery, and Apolonija Bedina Zavec and colleagues), MACPF proteins of apicomplexan parasites (by Robert Menard and colleagues), and chlamydial MACPF proteins (by Lacey Taylor and David Nelson). Next, an interesting group of MACPF proteins from fungi that require a second, accessory protein for efficient pore formation is described by Peter Maček and colleagues. This book finishes with a chapter on fluorescence imaging of MACPF/CDC proteins by Michael Senior and Mark Wallace. This will be an important biophysical approach for study of MACPF/CDC interactions with the membrane, their assembly mechanism, and the properties of the final pore.

We hope that this book will set the stage for future studies on the evolution, structure, and function of MACPF/CDC members, particularly those that are as yet not so well known. There is certainly a lot more we need to learn and understand about these proteins, and it will be especially interesting to see what is discovered in the coming years about their roles in development, as part of the toxic arsenal of marine organisms and in the pathogenic mechanisms of bacteria and apicomplexan parasites. We enjoyed editing this volume and would like to thank our contributors very much for their involvement in the project, and for providing excellent chapters and enabling this book to offer fresh insights into the subject. We thank them for their engagement and hard work toward their contributions and are also grateful to Gisou van der Goot for her historical overview of pore-forming proteins. We would also like to express our sincere gratitude to the production staff at Springer, particularly to Thijs van Vlijmen for his patience and answers to our endless inquiries.

Ljubljana, December 2013 Gregor Anderluh
Oxford Robert Gilbert

Contents

Part I Overview of MACPF/CDCs

1 **Introduction: Brief Historical Overview** 3
Gisou F. van der Goot

2 **Distribution of MACPF/CDC Proteins** 7
Gregor Anderluh, Matic Kisovec, Nada Kraševec
and Robert J. C. Gilbert

3 **Evolution of the Complement System** 31
Masaru Nonaka

Part II Structures of MACPF/CDCs

4 **Structural Features of Cholesterol Dependent Cytolysins and Comparison to Other MACPF-Domain Containing Proteins** 47
Robert Gilbert

5 **Perfringolysin O Structure and Mechanism of Pore Formation as a Paradigm for Cholesterol-Dependent Cytolysins** 63
Benjamin B. Johnson and Alejandro P. Heuck

6 **Structural Biology of the Membrane Attack Complex** 83
Andreas F.-P. Sonnen and Philipp Henneke

Part III Functional Aspects of MACPF/CDCs

7 Membrane Interactions and Cellular Effects of MACPF/CDC Proteins 119
Miša Mojca Cajnko, Miha Mikelj, Tom Turk, Marjetka Podobnik and Gregor Anderluh

8 The Biology of Pneumolysin 145
Tim J. Mitchell and Catherine E. Dalziel

9 Multifaceted Activity of Listeriolysin O, the Cholesterol-Dependent Cytolysin of *Listeria monocytogenes* 161
Stephanie Seveau

10 Perforin: A Key Pore-Forming Protein for Immune Control of Viruses and Cancer 197
Jerome Thiery and Judy Lieberman

11 Perforin and Human Diseases 221
Omar Naneh, Tadej Avčin and Apolonija Bedina Zavec

12 The Role of MACPF Proteins in the Biology of Malaria and Other Apicomplexan Parasites 241
Joana Tavares, Rogerio Amino and Robert Ménard

13 Chlamydial MACPF Protein CT153 255
Lacey D. Taylor and David E. Nelson

14 Fungal MACPF-Like Proteins and Aegerolysins: Bi-component Pore-Forming Proteins? 271
Katja Ota, Matej Butala, Gabriella Viero, Mauro Dalla Serra, Kristina Sepčić and Peter Maček

15 Fluorescence Imaging of MACPF/CDC Proteins: New Techniques and Their Application 293
Michael J. Senior and Mark I. Wallace

About the Editors 321

Index 323

Contributors

Rogerio Amino Département de Parasitologie et Mycologie, Unité de Biologie et Génétique du Paludisme, Institut Pasteur, Paris, France

Gregor Anderluh Laboratory for Molecular Biology and Nanobiotechnology, National Institute of Chemistry, Ljubljana, Slovenia; Department of Biology, Biotechnical Faculty, University of Ljubljana, Ljubljana, Slovenia

Tadej Avčin Department of Allergology Rheumatology and Clinical Immunology Children's Hospital, University Medical Center Ljubljana, Ljubljana, Slovenia

Apolonija Bedina Zavec Laboratory for Molecular Biology and Nanobiotechnology, National Institute of Chemistry, Ljubljana, Slovenia

Matej Butala Biotechnical Faculty, Department of Biology, University of Ljubljana, Ljubljana, Slovenia

Miša Mojca Cajnko Laboratory for Molecular Biology and Nanobiotechnology, National Institute of Chemistry, Ljubljana, Slovenia

Mauro Dalla Serra Istituto di Biofisica, Consiglio Nazionale delle Ricerche and Fondazione Bruno Kessler, Trento, Italy

Catherine E. Dalziel Institute of Infection, Immunity and Inflammation, University of Glasgow, Glasgow, UK

Robert J. C. Gilbert Division of Structural Biology, Wellcome Trust Centre for Human Genetics, University of Oxford, Oxford, UK

Philipp Henneke Center for Chronic Immunodeficiency, University Medical Center Freiburg, Freiburg, Germany

Alejandro P. Heuck Department of Biochemistry and Molecular Biology, University of Massachusetts, Amherst, MA, USA

Benjamin B. Johnson Department of Biochemistry and Molecular Biology, University of Massachusetts, Amherst, MA, USA

Matic Kisovec Laboratory for Molecular Biology and Nanobiotechnology, National Institute of Chemistry, Ljubljana, Slovenia

Nada Kraševec Laboratory for Molecular Biology and Nanobiotechnology, National Institute of Chemistry, Ljubljana, Slovenia

Judy Lieberman Program in Cellular and Molecular Medicine, Boston Children's Hospital and Harvard Medical School, Boston, MA, USA

Peter Maček Biotechnical Faculty, Department of Biology, University of Ljubljana, Ljubljana, Slovenia

Robert Ménard Département de Parasitologie et Mycologie, Unité de Biologie et Génétique du Paludisme, Institut Pasteur, Paris, France

Miha Mikelj Biotechnical Faculty, Department of Biology, University of Ljubljana, Ljubljana, Slovenia

Tim J. Mitchell Institute of Microbiology and Infection, University of Birmingham, Birmingham, UK; School of Immunity and Infection, College of Medical and Dental Sciences, University of Birmingham, Birmingham, UK

Omar Naneh Laboratory for Molecular Biology and Nanobiotechnology, National Institute of Chemistry, Ljubljana, Slovenia

David E. Nelson Department of Microbiology and Immunology, Indiana University School of Medicine MS411B, Indianapolis, USA

Masaru Nonaka Department of Biological Sciences, Graduate School of Science, The University of Tokyo, Tokyo, Japan

Katja Ota Biotechnical Faculty, Department of Biology, University of Ljubljana, Ljubljana, Slovenia

Marjetka Podobnik Laboratory for Molecular Biology and Nanobiotechnology, National Institute of Chemistry, Ljubljana, Slovenia

Michael J. Senior Department of Chemistry, Oxford University, Oxford, OX1 3TA, UK

Kristina Sepčić Biotechnical Faculty, Department of Biology, University of Ljubljana, Ljubljana, Slovenia

Stephanie Seveau Department of Microbiology, Department of Microbial Infection and Immunity, The Ohio State University, Columbus, OH, USA

Andreas F.-P. Sonnen Center for Chronic Immunodeficiency, University Medical Center Freiburg, Freiburg, Germany

Joana Tavares Département de Parasitologie et Mycologie, Unité de Biologie et Génétique du Paludisme, Institut Pasteur, Paris, France

Lacey D. Taylor Laboratory of Intracellular Parasites, Rocky Mountain Laboratories, National Institute of Allergy and Infectious Disease, National Institutes of Health, Hamilton, USA

Jerome Thiery INSERM U753, University Paris Sud and Gustave Roussy Cancer Campus, Villejuif, France

Tom Turk Biotechnical Faculty, Department of Biology, University of Ljubljana, Ljubljana, Slovenia

Gisou F. van der Goot Global Health Institute, Ecole Polytechnique Fédérale de Lausanne (EPFL), Lausanne, Switzerland

Gabriella Viero Istituto di Biofisica, Consiglio Nazionale delle Ricerche and Fondazione Bruno Kessler, Trento, Italy

Mark I. Wallace Department of Chemistry, Oxford University, Oxford, UK

Part I
Overview of MACPF/CDCs

Chapter 1
Introduction: Brief Historical Overview

Gisou F. van der Goot

Abstract Membranes are essential in defining the border and ensuring function of all living cells. As such they are vulnerable and have been a preferred target of attack throughout evolution. The most powerful way of damaging a membrane is through the insertion of pore-forming proteins. Research over the last decades shows that such proteins are produced by bacteria to attack bacterial or eukaryotic cells, vertebrates to kill invading organisms or infected cells, and by eukaryotic cells to "kill" mitochondria and trigger apoptosis. The breadth of effect of these proteins is bringing together, in a very exciting way, research communities that used to be unaware of each other.

Keywords Pore-forming · Toxins · Bcl · Perforin · Aerolysin

Abbreviations

CDC Cholesterol-Dependent Cytolysins
PFO Perfringolysin O
PFPs Pore-forming proteins
PFTs Pore-forming toxins

Whether life has originated from an RNA-world [7] or a lipid-world [17], there is no doubt that ancestral cells possessed a lipid bilayer separating their inside environment from the outside. By defining the limits and the integrity of the cell, the plasma membrane emerges as an ideal target for attack from others cells. Not surprisingly many organisms have evolved the ability to produce molecules that have the capacity to affect cellular membranes. These include peptides as well as proteins with diverse activities such as hydrolysis or clustering of lipids. Arguably

G. F. van der Goot (✉)
Global Health Institute, Ecole Polytechnique Fédérale de Lausanne (EPFL), Station 19, 1015 Lausanne, Switzerland
e-mail: gisou.vandergoot@epfl.ch

G. Anderluh and R. Gilbert (eds.), *MACPF/CDC Proteins - Agents of Defence, Attack and Invasion*, Subcellular Biochemistry 80, DOI: 10.1007/978-94-017-8881-6_1,

the most powerful of the membrane affecting molecules are pore-forming toxins (PFTs) [2, 9]. These are synthesized as rather large soluble proteins by the producing cell but have the intrinsic capacity to change their conformation, often following multimerization, to convert to a transmembrane pore structure. While these pores are often meant to permeabilize membranes to ions of varying sizes, some have evolved to translocate proteins and are used by bacteria to introduce proteins with enzymatic activity into the cytosol of targets cells. Examples include the translocation domain of diphtheria toxin [4] and the protective antigen of anthrax toxin [5].

PFTs and toxin translocation domains have been studied in the bacterial world for many decades [1], initially based on the observation that many pathogenic bacteria have the capacity to lyse red blood cells, hence the name "hemolysins". This has led to the discovery of a variety of protein folds that allow the conversion of soluble proteins to transmembrane pores. The first to be solved was the pore-forming domain of colicin A, an alpha helical bundle which shelters a central hydropbophic helical hairpin involved in membrane insertion [14], a fold that was subsequently also found in the translocation domain of diphtheria toxin [4]. In the early 1990s, it was however appreciated that hydrophobic helices might not be the only means to cross a lipid bilayers and that certain PFTs might cross the membrane as β-barrels, as do outer-membrane bacterial porins [12]. The X-ray structure of the soluble form of the pore-forming toxin aerolysin indeed soon after revealed that some PFTs are largely composed of β-strands [13]. The structure of the soluble form of PFTs is however not always predictive of the secondary structural elements, α-helix or β-strand, with which the toxin pore will cross the membrane. Indeed when the structure of the cholesterol-dependent hemolysin Pefringolysin O (PFO) was solved it was unclear which segment of the protein was responsible for membrane perforation [16]. Using cysteine-scanning mutagenesis combined with fluorescence quenching techniques, Tweten and colleagues made the surprising finding that the membrane spanning segments adopt an alpha-helical fold in the soluble form, but convert to a β-conformation upon membrane insertion [18, 19]. Although still rare, such a conversion had at that time only been reported for the transition between the native and the pathological, scrapie, form of Prion protein [3].

The production of pore-forming proteins (PFPs) is however by no means restricted to bacteria, and in particular the mammalian immune system produces at least two pore forming systems, perforin and the membrane attack complex. While in particular the early work of late Jurg Tschopp and his group [21] had not escaped researchers studying bacterial pore forming toxins, the fields of immunology and bacterial pathogenesis largely ignored each other until 2007 when the structure of a bacterial MACPF protein, Plu-MACPF from Photorhabdus luminescens [15] revealed the similarity in fold between the MACPF domain and members of the cholesterol-dependent-cytolysin (CDC) family of which PFO has been, in terms of structural biology, the leading member [8].

Although possibly the most charismatic example, the MACPF-CDC similarity is not the first between bacteria and vertebrate pore-forming proteins. Based on the structure of the membrane-spanning domain, PFTs have classically been divided into two large families, the α- and the β-PFTs. Since their ability to be transmembrane can be detected using bioinformatic means, i.e. identification of hydrophobic stretches, α-PFTs were initially better characterized and indeed the first PFT structure to be solved was that of the pore-forming domain of colicin A. The big surprise came, when in the field of mammalian apoptosis, the structure of Bcl-xL was solved and revealed that is had a fold similar to that of the diphtheria toxin translocation domain [10]. The mechanism of action of Bcl2 family members was at that time unknown, but based on years of research in the PFT field, it was then rapidly demonstrated that these proteins could, in a regulated manner, insert into membranes in order to favor or inhibit apoptosis.

More is however to come. Indeed the β-PFT family contains at least 3 subfamilies based on their structure: the CDCs, the *S. aureus* hemolysin like proteins and the aerolysin superfamily [9]. While eukaryotic members of the second family have so far not been found, bioinformatics analyses have revealed that proteins with an aerolysin fold can be found in all kingdoms of life [20]. Interestingly, the existence of aerolysin family members in vertebrates was recently confirmed. It was indeed found that the frog species, *Bombina maxima*, expresses a protein complex, named betagamma-CAT, which contains an aerolysin-like protein subunit and two trefoil factor subunits [6]. It has been proposed that betagamma-CAT might be a component of the *B. maxima* immune system, which is very similar to that of mammals [22], and would help the frog live in harsh environments, such as microorganism-rich mud pools.

Maybe the reverse will also happen, i.e. that a pore forming protein found in vertebrates is later found in bacteria. Will the pore-forming antibacterial human intestinal C-type lectins of the RegIII family [11] have a bacterial counterpart that affects mammalian cells?

References

1. Bernheimer AW (1996) Some aspects of the history of membrane-damaging toxins. Med Microbiol Immunol 185:59–63
2. Bischofberger M, Iacovache I, van der Goot FG (2012) Pathogenic pore-forming proteins: function and host response. Cell Host Microbe 12:266–275
3. Carrell RW, Lomas DA (1997) Conformational disease. Lancet 350:134–138
4. Choe S, Bennett MJ, Fujii G, Curmi PM, Kantardjieff KA, Collier RJ, Eisenberg D (1992) The crystal structure of diphtheria toxin. Nature 357:216–222
5. Collier RJ (2009) Membrane translocation by anthrax toxin. Mol Aspects Med 30:413–422
6. Gao Q, Xiang Y, Zeng L, Ma XT, Lee WH, Zhang Y (2011) Characterization of the betagamma-crystallin domains of betagamma-CAT, a non-lens betagamma-crystallin and trefoil factor complex, from the skin of the toad Bombina maxima. Biochimie 93:1865–1872
7. Gilbert W (1986) Origin of life: The RNA world. Nature 319:618
8. Hotze EM, Tweten RK (2012) Membrane assembly of the cholesterol-dependent cytolysin pore complex. Biochim Biophys Acta 1818:1028–1038

9. Iacovache I, Bischofberger M, van der Goot FG (2010) Structure and assembly of pore-forming proteins. Curr Opin Struct Biol 20:241–246
10. Muchmore SW, Sattler M, Liang H, Meadows RP, Harlan JE, Yoon HS, Nettesheim D, Chang BS, Thompson CB, Wong SL, Ng SL, Fesik SW (1996) X-ray and NMR structure of human Bcl-xL, an inhibitor of programmed cell death. Nature 381:335–341
11. Mukherjee S, Zheng H, Derebe MG, Callenberg KM, Partch CL, Rollins D, Propheter DC, Rizo J, Grabe M, Jiang QX, Hooper LV (2014) Antibacterial membrane attack by a pore-forming intestinal C-type lectin. Nature 505:103–107
12. Nabedryk E, Gravito RM, Breton J (1988) The orientation of b-sheets in porin. A polarized fourier transform infrared spectroscopic investigation. Biophys J 53:671–676
13. Parker MW, Buckley JT, Postma JPM, Tucker AD, Leonard K, Pattus F, Tsernoglou D (1994) Structure of the *Aeromonas* toxin proaerolysin in its water-soluble and membrane-channel states. Nature 367:292–295
14. Parker MW, Pattus F, Tucker AD, Tsernoglou D (1989) Structure of the membrane-pore-forming fragment of colicin A. Nature 337:93–96
15. Rosado CJ, Buckle AM, Law RH, Butcher RE, Kan WT, Bird CH, Ung K, Browne KA, Baran K, Bashtannyk-Puhalovich TA, Faux NG, Wong W, Porter CJ, Pike RN, Ellisdon AM, Pearce MC, Bottomley SP, Emsley J, Smith AI, Rossjohn J, Hartland EL, Voskoboinik I, Trapani JA, Bird PI, Dunstone MA, Whisstock JC (2007) A common fold mediates vertebrate defense and bacterial attack. Science 317:1548–1551
16. Rossjohn J, Feil SC, McKinstry WJ, Tweten RK, Parker MW (1997) Structure of a cholesterol-binding, thiol-activated cytolysin and a model of its membrane form. Cell 89:685–692
17. Segre D, Ben-Eli D, Deamer DW, Lancet D (2001) The lipid world. Origins Life Evol 31:119–145
18. Shatursky O, Heuck AP, Shepard LA, Rossjohn J, Parker MW, Johnson AE, Tweten RK (1999) The mechanism of membrane insertion for a cholesterol-dependent cytolysin: a novel paradigm for pore-forming toxins. Cell 99:293–299
19. Shepard LA, Heuck AP, Hamman BD, Rossjohn J, Parker MW, Ryan KR, Johnson AE, Tweten RK (1998) Identification of a membrane-spanning domain of the thiol-activated pore-forming toxin Clostridium perfringens perfringolysin O: an alpha- helical to beta-sheet transition identified by fluorescence spectroscopy. Biochemistry 37:14563–14574
20. Szczesny P, Iacovache I, Muszewska A, Ginalski K, van der Goot FG, Grynberg M (2011) Extending the aerolysin family: from bacteria to vertebrates. PLoS ONE 6:e20349
21. Tschopp J, Masson D, Stanley KK (1986) Structural/functional similarity between proteins involved in complement- and cytotoxic T-lymphocyte-mediated cytolysis. Nature 322:831–834
22. Zhao F, Yan C, Wang X, Yang Y, Wang G, Lee W, Xiang Y, Zhang Y (2013) Comprehensive transcriptome profiling and functional analysis of the frog (Bombina maxima) immune system. DNA Res (in press). doi:10.1093/dnares/dst035

Chapter 2
Distribution of MACPF/CDC Proteins

Gregor Anderluh, Matic Kisovec, Nada Kraševec and Robert J. C. Gilbert

Abstract Membrane Attack Complex/Perforin (MACPF) and Cholesterol-Dependent Cytolysins (CDC) form the MACPF/CDC superfamily of important effector proteins widespread in nature. MACPFs and CDCs were discovered separately with no sequence similarity at that stage being apparent between the two protein families such that they were not, until recently, considered evolutionary related. The breakthrough showing they are came with recent structural work that also shed light on the molecular mechanism of action of various MACPF proteins. Similarity in structural properties and conserved functional features indicate that both protein families have the same evolutionary origin. We will describe the distribution of MACPF/CDC proteins in nature and discuss briefly their similarity and functional role in different biological processes.

Keywords Cholesterol-dependent cytolysins · MACPF/CDC protein superfamily · Membrane attack complex · Perforin · Protein evolution

---

G. Anderluh (✉) · M. Kisovec · N. Kraševec
Laboratory for Molecular Biology and Nanobiotechnology, National Institute of Chemistry, Hajdrihova 19, 1000 Ljubljana, Slovenia
e-mail: gregor.anderluh@ki.si

G. Anderluh
Department of Biology, Biotechnical Faculty, University of Ljubljana, Večna pot 111, 1000 Ljubljana, Slovenia

R. J. C. Gilbert (✉)
Division of Structural Biology, Wellcome Trust Centre for Human Genetics, University of Oxford, Roosevelt Drive, Oxford OX3 7BN, UK
e-mail: gilbert@strubi.ox.ac.uk

G. Anderluh and R. Gilbert (eds.), *MACPF/CDC Proteins - Agents of Defence, Attack and Invasion*, Subcellular Biochemistry 80, DOI: 10.1007/978-94-017-8881-6_2,

Abbreviations

APC-β	Apicomplexan perforin-like proteins C-terminal β-pleated domain
ASTN	Astrotactin
BRINP	Bone morphogenic protein/retinoic acid inducible neural-specific protein
CDC	Cholesterol-dependent cytolysins
MAC	Membrane attack complex
MACPF	Membrane attack complex/perforin
Mpg-1	Macrophage-expressed gene 1
PFN	Perforin

General Overview of Structural Similarity of MACPF/CDC Protein Families

MACPF and CDC proteins are important effectors in the immune system and bacterial pathogenesis, where their observed effects are primarily mediated through membrane interactions and the formation of transmembrane pores. Their activity is, however, not restricted to membrane interactions and it will be important to determine the exact role and molecular mechanism of action for many representatives. MACPF proteins were initially defined by the sequence similarity among proteins of the Membrane Attack Complex (*MAC*) of the complement system and perforin (*PF*) [59, 83, 93]. These proteins have crucial roles in immune defence against viral and bacterial pathogens and in removal of cancer cells. For example, perforin is the only molecule of the human immune system that allows entry of granzyme into the target cells and initiation of apoptosis. The MAC is the terminal stage of the complement system of innate immunity. On the other hand, CDCs are predominately found in Gram-positive bacteria [29, 94], although a recent report indicates that this protein family may be more widespread as they are at least present in Gram-negative bacteria too [41] and diatoms and plants (see below). Because they were discovered separately, in quite separate niches, there was originally no sequence similarity apparent between the proteins of MACPF and CDC families. The realisation that these two families are evolutionary related was the result of intense structural work on MACPF proteins; in particular crystal structures of Plu-MACPF from bacteria *Photorhabdus luminescens* [77], Bth-MACPF from bacteria *Bacteroides thetaiotaomicron* [104], the MACPF domain of complement components C6 [5] and C8 [35, 62, 85] and mouse perforin (PFN) [57] revealed their similarity to the CDC protein family. Both protein families are now collectively referred to as the MACPF/CDC pore forming proteins [30, 78]. Structure-based alignment allows the identification of very limited sequence conservation of glycines present in all family members [30], which in fact

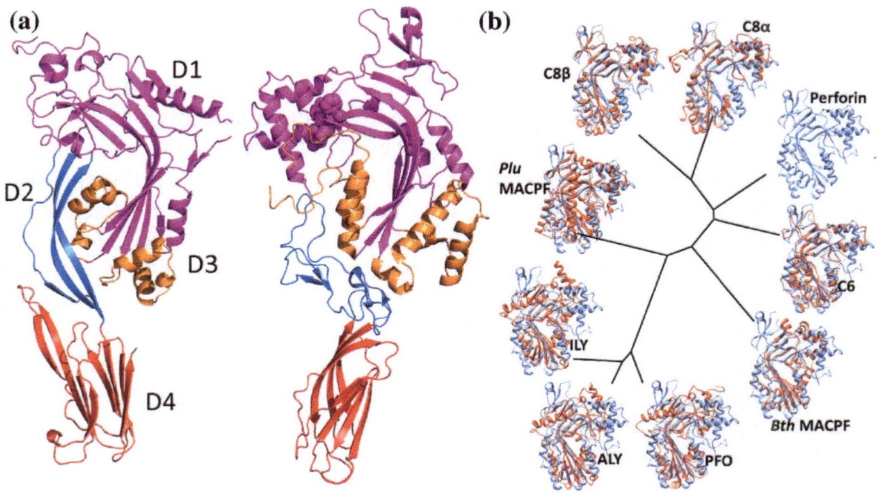

Fig. 2.1 Structure and similarities of CDC and MACPF proteins **a** Crystal structure of perfringolysin O (PFO), the first CDC to have its structure determined (*left*) [79] and mouse PFN (*right*) [57]. The original domains of the CDC are labelled D1–D4. The MACPF/CDC domain is shown in *magenta*, with helical regions that refold to form the transmembrane β-barrel coloured in *orange*. The membrane interacting domains in both proteins, immunoglobulin-like domain in PFO and C2 domain in PFN, are labelled *red*. The conserved MACPF residues in PFN are labelled with *spheres* (*see below*). **b** Structural phylogenetic tree of determined MACPF and CDC structures. The PFN structure is shown in *blue* and each MACPF/CDC structure in *orange*. The figure was adapted from [30]

represent a continuum in sequence space (see below) and several other shared features can be identified (see also Chap. 4).

The structure of CDCs is traditionally divided into four domains, each with particular functions during the pore-forming process (Fig. 2.1a) [29, 94]. The C-terminal domain 4 is used for the attachment of the protein to the lipid membrane, domain 1 is primarily used for contacts between the monomers, domain 3 is used for the formation of the final transmembrane β-barrel pore and domain 2 is a linking structure that provides flexibility needed during the process of pore formation (Fig. 2.1a) [29, 42]. See Chaps. 4 and 5 in this volume for description of a detailed understanding of the CDC mechanism of action. Structural similarity between MACPFs and CDCs is centred on the central bent β-sheet composed of four strands and flanked on the sides by clusters of α-helices (the MACPF/CDC domain) [35, 77] (Fig. 2.1). From the structural comparisons it seems that domains 1 and 3 of CDCs could be seen as a single functional unit with a fold that is conserved throughout the tree of life (Fig. 2.1b). The universally-conserved glycines permit conformational rearrangements of this region of the protein to allow insertion of part of it into the lipid membrane, forming a transmembrane pore.

Distribution of MACPF Proteins

The MACPF protein family (Pfam PF01823) is the biggest family of eukaryotic pore-forming proteins and the current version of the Pfam database (version 27.0) [75] contains 1,329 unique sequences. The MACPF-containing proteins are widespread in nature and may be found in all domains of life [30]. Proteins with the MACPF domain are most abundant in metazoans and many representatives may be found in higher eukaryotes. Some organisms possess large numbers of genes containing MACPF domains—for example amphioxus (*Branchiostoma floridae*) possesses 29 such genes [43]. In this organism approximately 10 % of all genes are defence-related and since MACPF proteins represent an important part of the immune system such a high number is not surprising. MACPF proteins are also abundantly represented in Apicomplexans, where they have a crucial role in tissue and cell invasion and egress [47]. MACPF proteins were even found in anguilid herpesvirus 1 [96].

Prokaryotic representatives, primarily from the taxonomic groupings Chlamydia, Proteobacteria and Bacteroidia, may be found in various habitats and come from the following genera: bacteria of the human microbiome (*Alistipes*, *Bacteroides*), bacteria from the sea (*Dokdonia*, *Leeuwenhoekiella*, *Plesiocystis*, *Zunongwangia*), from land waters (*Beggiatoa*), plant pathogens (*Calvibacter*, *Ralstonia*), bioluminescent pathogens of insects (*Photorhabdus*), human pathogens (*Chlamydia*, *Chlamydophila*, *Chryseobacterium*, *Filifactor*, *Porphyromonas*), cyanobacteria (*Trichodesmium*), etc.

MACPF proteins are abundantly represented in mammalian genomes. In the human genome, for example, there are at least 12 genes that code for a MACPF domain (Fig. 2.2). Complement proteins C6, C7, C8α, C8β and C9, perforin and macrophage-expressed gene 1 (Mpg-1) have important roles in the immune system, while astrotactins (ASTN) 1 and ASTN2 and bone morphogenic protein/retinoic acid inducible neural-specific protein (BRINP) 1, BRINP2, and BRINP3 have important roles in the neuronal system (see below). Some of these MACPF-containing proteins have relatively simple domain structures, for example Mpg-1 and the BRINPs have a MACPF domain at the N-terminus and a C-terminal region of unknown structure and function. Mpg-1 is an interesting example of a MACPF protein that is permanently tethered to the membrane in contrast to other well-known MACPF proteins such as complement proteins and perforin. Some of these have a complex structure containing many different associated domains (see Chaps. 6 and 10 for a detailed domain organisation of complement proteins and perforin, respectively), while ASTN1 and ASTN2 are the largest human MACPF proteins with multiple domains. They are composed of two N-terminal transmembrane helices, three epidermal growth factor-like domains in the central part followed by the MACPF domain and finally a fibronectin type-III domain on the C-terminus.

Although family membership is simply identified, there is in fact only limited sequence conservation in the MACPF family, as the sequences are around 20 %

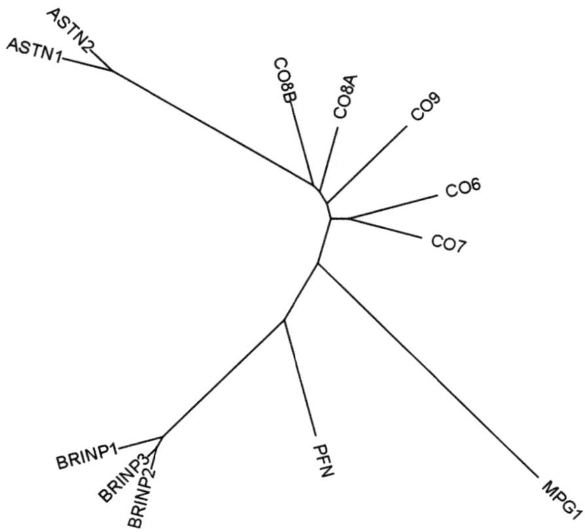

Fig. 2.2 Maximum-likelihood phylogenetic tree of human MACPF proteins. Protein sequences from UniProtKB (www.uniprot.org) were aligned using PSI-Coffee on T-Coffee server [51] and the phylogenetic tree was constructed with RAxML BlackBox [87]. The tree was depicted using *Archaeopteryx* [37]. *Sequence names* ASTN1 (Astrotactin 1, O14525), ASTN2 (Astrotactin 2, O75129), CO8B (complement component C8β, P07358), CO8A (complement component C8α, P07357), CO9 (complement component C9, P02748), CO6 (complement component C6, P13671), CO7 (complement component C7, P10643), MPG1 (macrophage-expressed gene 1 protein, Q2M385), PFN (P14222), BRINP1 (BMP/retinoic acid-inducible neural-specific protein 1, O60477), BRINP2 (Q9C0B6), BRINP3 (Q76B58)

identical (Fig. 2.3a). The most conserved amino acids are within the so-called MACPF motif [73] (Fig. 2.3b). This motif may be found in the structure at the top of the twisted β-sheet and particularly conserved are two glycine residues at the hinge between the upper and lower segment of the β-sheet (Fig. 2.1a). Other amino acid residues are not so well conserved (Fig. 2.3b). Thus the low sequence conservation presents a great challenge in discovering novel MACPF-containing proteins, as is highlighted by the fact that the original discovery of MACPFs in mammals and of CDCs in Gram-positive bacteria hid their deep evolutionary relationship until their structures were solved. This is demonstrated nicely by our recent analysis of MACPF proteins in fungi detailed in the next section.

MACPF Proteins in Fungi

Proteins with a MACPF/CDC domain are present in fungi as well. It is believed that more than 1.5×10^6 fungal species exist, with diverse lifestyles. They are important pathogens, saprophytes and symbionts. A huge expansion of available fungal

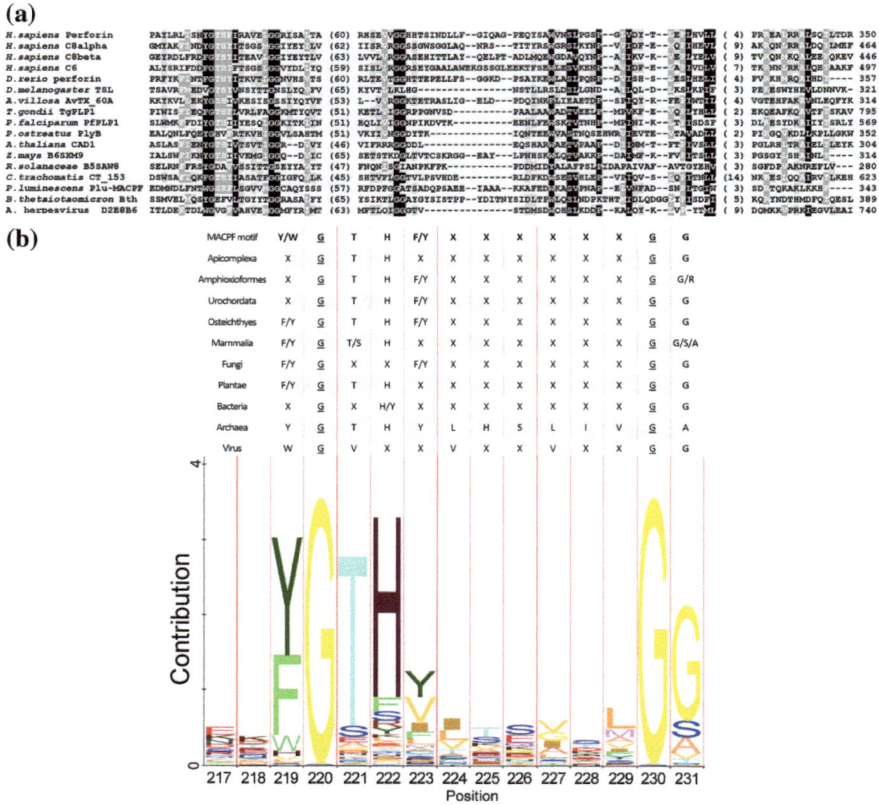

Fig. 2.3 Sequence conservation of MACPF domain **a** Alignment of the most conserved region in MACPF domain of various representatives. The numbers in brackets denote the length of the segments that are not conserved between the sequences. The numbers on the *right* are according to mature proteins. Adapted from [30]. **b** The MACPF motif. The *table* shows the motif as found in sequences of different taxonomical groups in Pfam database. The canonical MACPF motif is shown in the *first row* [73]. *Below* is the HMM logo representation of MACPF motif [81]. The data for generation of HMM logo was downloaded from Pfam website. The numbering below the motif is according to PFN (Uniprot P14222)

genome data will enable identification of proteins with major impacts for health and biotechnological applications. BLAST or PSI-BLAST algorithms were used to search databases at NCBI, JGI Mycocosm [32] and the *Aspergillus* genome database AspGD [13] with perforin or pleurotolysin B [92] sequences. Due to low sequence conservation of the MACPF domain hits with low significance (E = 2) were included in the further analysis. The hits were analysed for the prediction of the MACPF/CDC domain fold by using the PHYRE2 server [50]. Hits with the predicted MACPF/CDC fold were combined with already identified MACPF sequences from the Pfam database (version 27.0). Redundant sequences were removed after analysis of alignments generated using ClustalW. Altogether we identified 181 proteins in

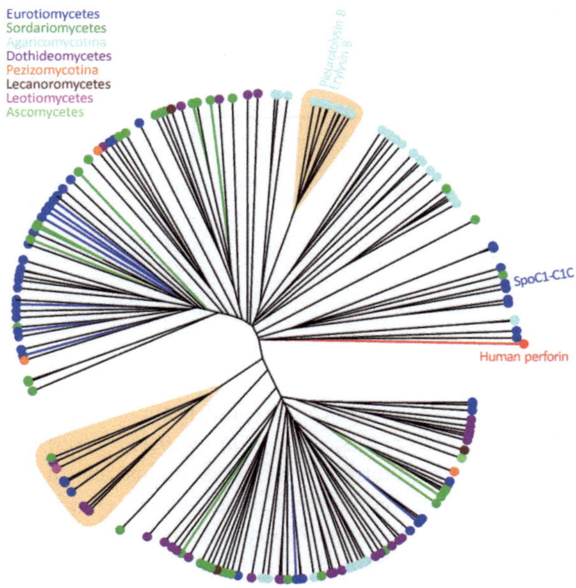

Fig. 2.4 Grouping of putative fungal MACPF proteins in relation to taxonomy, neighbour-joining tree without distance corrections was constructed after aligning sequences by using MAFFT algorithm. Taxon representation is according to JGI Mycocosm [32]. The distribution of 9 putative MACPF proteins from *Aspergillus wentii* is marked with *blue* branches and 11 putative MACPF proteins from *Fusarium graminearum* in *green*. *Orange* background highlight branches of MACPF protein-containing genes that are in neighbourhood of aegerolysin protein. The position of functionally characterised proteins is denoted by names

fungi with a filamentous growth form (Fig. 2.4). A short signature sequence Y/F-G-X_2-F/Y-X_6-G-G was identified in 151 of the putative MACPF-containing proteins (Fig. 2.3b), verifying their identity. Some MACPF coding genes have a large number of introns, 8–13, in comparison to two introns in an average fungal gene and, consequently, annotation was difficult (interestingly, note that the MACPFs of Apicomplexans are all mono-exonic except one, PPLP1/SPECT2, which has seven exons). The phylogenetic evolutionary tree was generated by using MAFFT [48] and a Neighbor-Joining approach. The sequences fall into two groups (Fig. 2.4). The first contains primarily sequences confirmed by Pfam and they are grouped together with PFN. In the second group are mostly sequences confirmed only by PHYRE2 but containing the typical MACPF/CDC signature as well.

We found the distribution of putative MACPF proteins in filamentous fungi uneven and with no clear correlation to taxonomy (Fig. 2.4). For example, in Agaricomycotina MACPF proteins were identified in 45 % of species, while in Eurotiomycetes they were present in 69 % of species (Table 2.1). The numbers of putative MACPF proteins present in a single fungal species are from one and up to nine. Some MACPF/CDC proteins within one species group together in the phylogenetic tree, while some do not, as is marked on Fig. 2.4 for 9 putative

Table 2.1 Taxonomic distribution of MACPF proteins in fungi

Taxon	Number of species in taxon	Number of species with MACPF sequence	Percent of species with MACPF sequence (%)	Number of MACPF sequences
Agaricomycotina	37	17	45	38
Pezizomycotina	4	3	75	3
Eurotiomycetes	42	29	69	60
Dothideomycetes	39	18	46	34
Lecanoromycetes	1	1	100	3
Leotiomycetes	6	2	33	2
Sordariomycetes	35	16	46	40
Ascomycetes	1	1	100	1

MACPF proteins from *Aspergillus wentii* (blue branches on Fig. 2.4) or for 11 putative MACPF proteins from *Fusarium graminearum* (green branches).

Among the strains with putative MACPF domains are those with declared pathogen lifestyles, like plant pathogens *Cochliobolus carbonum* (Northern Corn Leaf Spot), *C. heterostrophus* (Southern Corn Leaf Blight), *C. miyabeanus* (Brown Spot of rice), *C. victoriae* (Victoria Blight of oats), *Fusarium graminearum* (Wheat head blight), *F. oxysporum* (Panama disease) and *Moniliophthora perniciosa* (Witches'-broom disease), or animal pathogens like: *Arthrobotrys oligospora* (Nematode trapping), *Cordyceps militaris* (Caterpillar fungus), *Trichophyton equinum* (Horse ringworm), *T. rubrum* (Athlete's foot) and *T. tonsurans* (Scalp ringworm). Some are also saprophytic, like soil fungus *Chaetomium globosum* and *Aspergillus nidulans*, *Postia placenta* (Brown rot) or edible Oyster mushroom *Pleurotus ostreatus* (White-rot), *Pleurotus eryngii* (Boletus of the steppes) and food producing *Aspergillus oryzae* (Yellow koji).

SpoC1-C1C from *Aspergillus nidulans* is among those rare fungal MACPF proteins with assigned function, as a conidium-specific member of the SpoC1 gene cluster [33]. It has been shown for some fungal MACPF proteins, like pleurotolysin B or erylysin B, that they act in concert with additional smaller proteins, the aegerolysins [71, 82, 92]. It is interesting to note that in some cases these two components are found as neighbouring genes in the genome. The subsets of MACPF proteins with an aegerolysin component as a neighbour are denoted by an orange background on Fig. 2.4. The interactions between the components and pore-forming capacity of fungal MACPF proteins are described in detail in Chap. 14.

In conclusion, only a little is known about the function of MACPF proteins in filamentous fungi. From the taxonomic distribution one can clearly conclude that they are definitely not part of the core proteome. Fungal MACPF proteins most probably contribute to different specific processes, while some of them were found in secretomes (without a typical signal sequence recognised) [18], others being intracellular, and some of them may be involved in pathogenesis, however, most probably not all of them. Large intron number, sporadic taxonomic distribution, and different numbers of diverse MACPF proteins per fungal species may imply involvement of horizontal transfer mechanisms in specific environmental niches.

Domain Organisation of MACPF Proteins

In most cases one MACPF/CDC domain exists per protein, although some examples with two or even three MACPF/CDC domains have been identified [30]. It is not however uncommon for other protein domains to be associated with MACPF proteins. The domain architecture is quite simple in bacteria, where MACPFs are located N-terminally to other domains and as a rule there is only one additional domain present. Metazoan MACPF/CDC proteins have much more complex architectures and several other domains may be present on either side of the MACPF domain. These domains are quite often involved in molecular interactions needed to exert the biological function of a particular MACPF protein. For example many MAC proteins contain additional domains that are important for regulation of MAC assembly at the surface of the target cell (see the details in Chap. 6). Many MACPF-containing proteins are involved in membrane interactions and pore formation and, therefore, there is quite a large number of small β-sheet rich domains associated with membrane interactions. For example, the C2 domain is present at the C-terminus of PFN and is used for calcium-specific membrane attachment [74, 99]. In other proteins such domains may facilitate cell attachment through interactions with sugars, such as the MIR domain, Jacalin domain (a mannose-binding lectin domain), B-lectin domain, MABP domain, etc. Apicomplexan MACPF proteins contain 1–3 so-called APC-β domains (Apicomplexan Perforin-like proteins C-terminal β-pleated domains) [47]. These are small modules of $\sim$55 amino acids and, according to sequence comparisons at least, are unique to apicomplexan proteins (though see discussion in Chap. 4). For many of these cases the functional significance of particular domains for cell-surface interactions remains to be verified.

There are, however, many MACPF-containing proteins that do not possess additional domains (such as torso-like protein from *Drosophila melanogaster*). In such cases it needs to be determined whether these proteins are able to interact with lipid membranes and if such interactions are needed at all to exert their physiological activity. However, it is known that some MACPF proteins associate with other components in order to successfully complete formation of pores. For example, complement monomers of C9 associate with the C5b-8 complex [4, 5, 36, 62] (see Chap. 6). The MACPF protein pleurotolysin B from the fungus *Pleurotus ostreatus* needs a smaller component, pleurotolysin A for efficient membrane interactions and pore formation [71, 92] (see Chap. 14); also, an interesting combination of disulphide linked MACPF and lectin domains was reported recently [23].

Functional Properties of MACPF Proteins

Some of the described functions of MACPF-containing proteins are listed in Table 2.2. The most studied and understood role is that of MAC proteins and perforin in the immune system. These proteins help with the removal of unwanted

Table 2.2 Functions of MACPF proteins

Biological process	Function	Examples
Immunity	Membrane attack complex of vertebrates	C6, C7, C8α, C8β, C9
	Killing of unwanted cells	Perforin
	Killing of bacteria and innate immune defence in invertebrates	Mpg-1 and homologues
	Pathogen-induced immune response	CAD1
Development	Neuronal development	Astrotactin1, BRINP proteins
	Insect embryo development	Tsl
	Sea urchin embryo development	Apextrin
Pathogenesis	Tissue invasion and cell egress	Apicomplexan MACPF proteins
	Bacterial pathogenesis	Plu-MACPF
	Toxins	PsTX-60A

For references see text.

cells from the organism, either microorganisms [72], virally infected cells or cancer cells [98]. The removal is achieved through efficient formation of transmembrane pores at the cell membrane surface or within endosomes and is described in detail elsewhere in this book (Chaps. 6 and 10). In plants a MACPF protein CAD1 was shown to have a role in plant responses induced by pathogens [67].

The MACPF protein family is, however, characterised by a variety of functions that may not always be linked to pore formation or membrane association (Table 2.2). One prominent role of MACPF-containing proteins is in the development of insects [88], sea urchins [34] and mammalian neurons [1, 49, 106]. In apicomplexansMACPF proteins are implicated in parasite invasion of tissues and egress from cells [47] (see also Chap. 12 for a more detailed description of MACPF protein function in apicomplexan parasites). In bacteria it was proposed that they are important for pathogenic bacteria (e.g. *Chlamydia*), however their role in pathogenicity is not yet entirely understood [73, 77] (see also Chap. 13 for the role of MACPF protein in *Chlamydia*). MACPF-containing proteins were also suggested to have a role in protein secretion, nutrient uptake systems or to have a preventive role through molecular mimicry [104].

Involvement of MACPF proteins in the native immunity of lower eukaryotes and as a toxic arsenal for water-borne organisms is described in more detail in the following sections.

Macrophage-Expressed Gene 1 (Mpg-1)

Mpg-1 protein, termed also perforin-2, was originally identified by differential screening of a murine macrophage cDNA library. It showed macrophage and differentiation stage-specific expression and its expression was increased upon transition of murine fetal liver hematopoietic cells into macrophages [86]. Mice contain another Mpg-1 homologue, termed EPCS50, which is primarily expressed in trophoblast giant cells in the ectoplacental cone and the parietal yolk sac [39]. It was also shown that Mpg-1 is increased in prion-inoculated brain tissue of mice, however the physiological function of Mpg-1 in mice brains was not clarified and is not known [55]. Thus, very little was known about the function of Mpg-1 until recently, where two studies demonstrated its importance for the killing of intracellular bacteria. In one, it was shown that Mpg-1 has an important role in elimination of intracellular bacteria [64] because the expression of its mRNA in mouse embryonic fibroblasts and a fibroblast NIH-3T3 cell line was shown after exposure of cells to several cytokines (IFN-α, IFN-β and IFN-γ). The same study showed that Mpg-1 mRNA was induced by exposure of fibroblasts to *Escherichia coli* and *Mycobacterium smegmatis*. Increased levels of Mpg-1 mRNA were associated with increased bactericidal activity and Mpg-1 also increased the susceptibility of intracellular bacteria to lysozyme. The second recent report showed that Mpg-1 is needed for elimination of intracellular *Chlamydia trachomatis* cells in macrophages [25]. *Chlamydia* infection caused an increase in Mpg-1 mRNA and in protein levels of Mpg-1 in murine macrophages. Production of Mpg-1 mRNA could also be induced by heat-killed *Chlamydia* in HeLa epithelial cells but not with viable bacteria and this led to the realisation that *Chlamydia* can actively suppress the production of Mpg-1 mRNA in such cells. This study also showed that Chlamydiae were susceptible to killing by ectopically expressed Mpg-1. In summary, these results demonstrated that expression of the Mpg-1 gene can be induced in other cells than macrophages, i.e. fibroblasts and HeLa cells, and that it is required for elimination of intracellular bacteria.

Mpg-1 represents an important constituent of innate immune defence in evolutionarily lower organisms. For example it was discovered in the sponge *Suberites domuncula* [103], cnidarians [65], the planarian *Schmidtea mediterranea* [6], molluscs, oysters [38] and abalones [52, 63, 100]. The Mpg-1 protein is evolutionarily very conserved, for example sponge and human Mpg-1 share 46 % similar and 28 % identical residues [103]. However, the evolutionary history of Mpg-1 is not clear, as it is not found in the genomes of *Drosophila* or *Caenorhabditis*. Its presence in gastropods and bivalves indicates that it may originate from an ancient ancestral gene that was lost in insects and nematodes [38]. Recent studies on perforin evolution are in agreement with this proposition and suggest that Mpg-1 is the precursor of perforin, which arose from duplication of an ancient Mpg-1 gene [21].

Marine invertebrates are in constant contact with bacteria and many genome-encoded effectors are an important constituent of their innate immunity. Haemocytes are specialised blood cells of the mollusc's cellular immune response. They circulate in the hemolymph and are able to recognise foreign antigens through specific pattern recognition motifs. Pathways for pathogen destruction include specific transduction pathways that are similar to vertebrate pathways [44]. It was observed that Mpg-1 is kept at very low levels in haemocytes and is upregulated upon stress or infection with bacteria [52]. In different experimental settings the up-regulation of Mpg-1 was found upon challenge with bacteria [38, 52, 100, 101], ostreid herpesvirus 1 [76], parasite *Schistosoma mansoni* [45] or lipopolysaccharides in sponges and planarians [6, 103], further indicating its important and ancient role in innate immunity. Kemp et al. also showed that the protein levels of Mpg-1 were upregulated upon stress with bacterial exposure and that protein levels were upregulated in a similar manner to mRNA levels [52]. This was shown also for Mpg-1 protein from sponge [103]. Some Mpg-1 homologues have been expressed, purified to homogeneity and tested for antibacterial activity. Mpg-1 from Pacific oyster *Crassostrea gigas* was shown to possess antibacterial activity against Gram-positive and Gram-negative bacteria [38], while recombinant Mpg-1 protein from sponge exhibited antibacterial activity against Gram-negative bacteria [103].

Toxins with the MACPF Domain

MACPF proteins were described also as a part of the toxic arsenal of lower eukaryotes. Cnidarians evolved special organelles called cnidocysts, which are considered as one of the most complex cellular secretory products. It is widely believed that cnidocysts serve for storage and active delivery of toxic compounds in cnidarian venom [53, 61, 91]. Many different cytolytic protein families were identified in Cnidaria [9, 11, 28]; for example Nagai et al. have reported isolation of two toxins, PsTX-60A and PsTX-60B, from the nematocysts of the sea anemone *Phyllodiscus semoni*. These two proteins represent major venom toxins and were haemolytic (ED_{50} values 300 and 600 ng/ml using 0.8 % suspension of sheep red blood cells) and lethal to shrimps (LD_{50} value around 800 μg/kg for intraperitoneal injection) [68]. A similar protein was purified from the nematocyst venom of another sea anemone *Actineria villosa*. This toxin, termed AcTX-60A, was lethal to mice (minimal lethal dose for intraperitoneal injection less than 250 μg/kg). Its primary structure was determined from mRNA and for the first time these proteins were recognised as members of the MACPF family [70]. A subsequent report of the full length sequence of PsTX-60B reconfirmed similarity to MACPF domains [80]. Proteins similar to PsTX-60A were found by proteomic analysis also in nematocyst venom from the jellyfish *Olindias sambaquiensis* [102] and similar proteins were found in the genomes of the hydrozoan *Hydra magnipapillata*, sea anemone *Nematostella vectensis* and in coral *Acropora millepora* [65]. Furthermore, EST clones, publicly available in GenBank, of sea anemones *Aiptasia pallida* [89] and *Metridium senile* also contain PsTX-60A homologues. It was shown previously that *M. senile* contains an 80 kDa polypeptide cytolysin metridiolysin, reminiscent of CDCs based on functional and structural properties [14] and it is tempting to suggest that metridiolysin belongs to the MACPF protein family. It seems that MACPF-containing protein toxins are not only restricted to sea anemones, but are present also in other cnidarians and they clearly represent another family of cnidarian pore-forming toxins [11].

The domain organisation of cnidarian MACPF domain-containing toxins resembles that of perforin. A MACPF domain is placed at the N-terminus and it is followed by an EGF-like domain. There is, however, a notable exception that the C2 domain is missing at the C-terminus and it will be interesting to determine the molecular details of membrane attachment and which parts of the protein allow initial interaction with the membrane. The PsTX-60A protein has also a typical signal sequence and a propeptide at the very N-terminus, a structural organisation that is shared by other nematocyst toxins [10]. However, it is not clear if all cnidarian MACPF-containing proteins are toxins expressed in cnidocysts. As noted by Miller et al. it is unclear if Hy-MAC, a homologue of PsTX-60A in *Hydra*, is an orthologue, since the overall sequence identity is low [65]. Furthermore, expression of Hy-MAC was restricted to gland cells in the endoderm and not nematocytes [65]. In another study, a differentially expressed PsTX-60A homologue gene, termed *anklet*, was also shown to be expressed in the newly differentiated gland cells of the basal disk and it was suggested that this gene is involved in the formation and maintenance of the basal disk in *Hydra* by exerting a cytotoxic role [7].

Another toxin that contains the MACPF domain was reported recently in an aquatic snail [23]. Protein PcPV2 from *Pomacea canaliculata* is a perivitellin involved in egg defence against predation. PcPV2 is a neurotoxin with strong lethal effects against rodents and is used for defence purposes [23]; it is glycosylated and is composed of four 98 kDa heterodimers yielding a 400 kDa complex. The smaller 31 kDa component is similar to tachylectin-1, while the larger 67 kDa domain is similar to the MACPF domain [23]. Both domains are linked with a disulphide bond [27] and since the tachylectin-1 component of PcPV2 allows binding to cells this domain combination resembles the A-B toxin combination found in bacterial and plant lectins [23].

Distribution of CDC Proteins

CDCs (Pfam PF01289) were originally identified in a number of Gram-positive bacteria from the genera *Bacillus*, *Clostridium*, *Streptococcus* and *Listeria* where they are associated with invasion of and replication within animal hosts [29, 94]. The most well-known such CDCs include perfringolysin O from *Clostridium perfringens*, streptolysin O from *Streptococcus pyogenes*, pneumolysin from *Streptococcus pneumoniae* and listeriolysin O from *Listeria monocytogenes* [29, 94]. In each case the CDC has a distinctive role. Perfringolysin O is responsible for *Clostridia* gaining access deep into tissue to protect anaerobic respiration, leading to gangrene [40], pneumolysin is required for pneumococcal virulence via the effects it has on cell membranes that permit host invasion [97], and listeriolysin O is used to enable invasion and growth within cells [17]. There are, however, many less well-known examples and to snapshot the current state of knowledge we performed a protein sequence database search using the whole

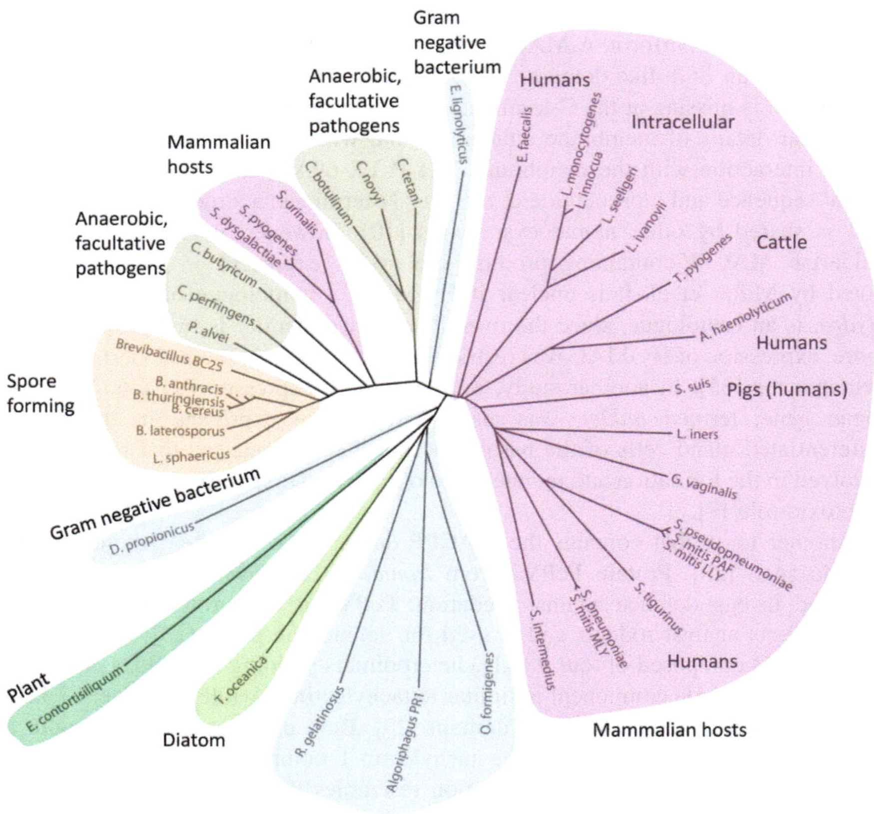

Fig. 2.5 Sequence-based phylogenetic tree of CDC proteins, *B. mycoides* and *B. weihenstephanensis* are not shown but share a branch with *B. anthracis*. *S. canis* shares a branch with *S. dysgalactiae*

perfringolysin O sequence, on the UniprotKB (http://www.uniprot.org). Of the 1,000 hits studied we found 40 unique sequences, and subsequently added the enterolysin from the Gram negative bacterium *Enterobacter lignolyticus* as identified alongside the CDC from *Desulfobulbous propionicus* by Tweten and colleagues [41], and the CDC identified in *Enterolobium contortisiliquum* (a member of the pea family, from South America) [26]. We pre-aligned the sequences in Uniprot and then used Clustal Omega [84] to perform alignments of the amino-terminal region of each protein (up to the end of the MACPF-homologous domain). We used the output alignments to construct phylogenetic trees using ClustalW2 Phylogeny [31, 56], which we then depicted using the program DRAWTREE (part of the PHYLIP package) [24] (Fig. 2.5). There are a number of interesting observations to be made on the basis of this analysis.

Firstly, in addition to the previously-identified CDCs from *E. lignolyticus* and *D. propionicus* we identified three more Gram-negative organisms which possess family members: one, *Oxalobacter formigenes* is a commensal human bacterium which relies on the degradation of oxaloacetate for most of its energetic and carbon needs and which (at 2,100 genes) has one of the smaller bacterial genomes [54]. By contrast, *Rubrivirax gelatinosus* and the *Algoriphagus* species PR1 are aquatic organisms; *R. gelatinosus* is a photosynthetic and nitrogen-fixing aquatic beta-proteobacteria [69], while *Algoriphagus* PR1 (*marchipongonensis*) is distinguished by the fact it is the prey of the choanoflagellate *Salpingoeca rosetta* [2, 3]. Choanoflagellates are the closest known relatives of animals, species of eukaryote of which some are unicellular and some multicellular colonies [95]. Even more surprisingly, we identified a form of CDC produced by the diatom *Thalassiosira oceanica* [60], which, after enterolobin from *E. contortisiliquum* is the second identified eukaryotic CDC. Genomic analysis of this species demonstrates 5 % gene acquisition from diverse taxonomic groups and so is consistent with the acquisition of a CDC by horizontal transfer from an aquatic bacterium [60]. The percent identity between the CDC of *T. oceanica* and that of *D. propionicus* is 23.9 %, that with the *Algoriphagus* PR1 26 % and that with *C. perfringens* perfringolysin 27.5 %. Its identity with the human-specific intermedilysin from *S. intermedius* is 25.2 % and to enterolobin, its neighbour in the phylogeny, is 13.9 % (see Fig. 2.6 for examples of alignment). The greater difference between the CDCs in *T. oceanica* and *E. contortisiliquum* reflects the much more developed evolutionary history of the plant versus the diatom, and the relative stasis in lifestyle and selective pressure experienced by bacterial lineages. The fact that sequences of CDCs in prokaryotic organisms which have not encountered each other for millions of years remain relatively stable at ~25 % identity suggests a strong selective pressure which is also, of course, reflected in the much greater conservation demonstrated by the MACPF/CDC fold [30]. The fact that the *T.oceanica* CDC is not more similar to those of bacteria nearby in ecological niche and within the phylogenetic tree than it is to those of bacteria much more distant phylogenetically and in ecology suggests that the diatom either acquired its CDC from the nuclear genome of its ancestor heterotrophic eukaryote, or due to an endosymbiotic event, or that any horizontal gene transfer that occurred was very ancient.

How ancient was the acquisition of a CDC by an ancestor of *T.oceanica*? A version of this phylogeny lacking just enterolobin did not present a single root, but instead a series of branches from a central common line. It is very interesting that this phylogenetic analysis based solely on sequence positions the diatom and plant CDCs adjacent to one another, and branching from a single point close to the origin of species. This leads to four of the five Gram negative CDCs alongside the diatom and plant CDCs forming a clade distinct from all other known family members, rooted at the origin of the tree. The absence of CDC proteins in many species argues against its presence in the genome of the last universal common ancestor and, therefore, the CDC gene transfer to diatoms and plants appears to have occurred from (endosymbiosis of) an ancestor of these Gram negative species

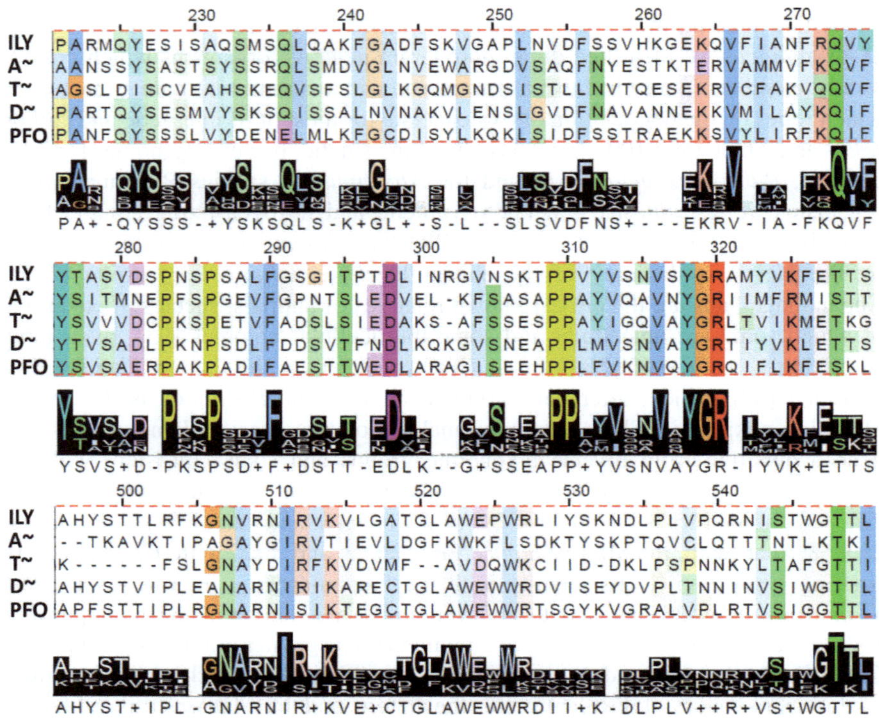

Fig. 2.6 Two regions of CDCs aligned. The first two sets of sequence comprise the core of the MACPF homology domain including the TMH helices; then final sequence comparison encompasses the carboxy-terminal domain (domain 4) undecapeptide. *ILY* intermedilysin from *S. intermedius*; *A*∼ the *Algoriphagus* species CDC; *T*∼ the *Thalassiosira oceanica* CDC; *D*∼ the *Desulfolobus* CDC and *PFO* perfringolysin O from *C. perfringens*

of which the evolutionarily closest example now known is *D. propionicus*. Another possibility is that a CDC was present in the last universal common ancestor as some sort of a housekeeping gene and was subsequently lost if not needed in particular species.

Diatoms evolved via a two-stage endosymbiotic process [12]. First, a cyanobacterium was captured by a eukaryotic heterotroph and this generated the origin of plants and red and green algae. Second, a red alga was captured by a different eukaryotic heterotroph and this gave rise to the stramenophile lineage that includes diatoms (emerging about 250 million years ago), macroalgae and plant parasites. Since our plant and diatom examples of CDCs derive from a single root this suggests that the CDC in diatoms (and in *E. contortisiliquum*) derives from the first endosymbiosis and thereby from an original cyanobacterium which shared a common (non-photosynthetic) ancestor with *D. propionicus*. In their paper on the *D. propionicus* and *E. lignolyticus* CDCs, Tweten and colleagues [41] discuss the possibility that a CDC may have helped to fend off endocytosis of primitive bacteria. It seems plausible that diatoms and plants result from a successful endosymbiotic event involving a prokaryote possessing a CDC.

Overall, the pore-forming (MACPF) domain phylogeny for CDCs shows a structure strongly consistent with the lifestyles adopted by the producing organisms. Thus, one clade encompasses the majority of the *Streptococcal* species (those of the *mitis* grouping) [19, 66, 105] alongside others infecting humans such as *Gardnerella vaginalis* and *Lactobacillus iners*. Here, *S. mitis* stands out as a species of bacterium expressing three different CDC proteins: lectinolysin, mitilysin and the platelet aggregation factor. Its clade shares a common root to the centre of the phylogeny with other bacteria with similar or the same hosts (*Arcanobacterium haemolyticum*, first isolated from Pacific islanders, and *T. pyogenes*) [16, 90] and separate branches containing the intracellular *Listeria* bacteria and the originally commensal bacterium *Enterococcus faecalis* now found also as a drug resistant pathogen [58]. The position of the *mitis* grouping of *Streptococci* at a location distant from the pyogenic group (*S. pyogenes*, *S. canis*, *S. urinalis* and *S. dysgalactiae*) [20] is in line with their relative distance based on a published analysis of seven housekeeping genes that confirmed the phylogenetic positioning of the *dysgalactiae* species, among others [46].

Interposed between the two groups of *Streptococci* lie an isolated Gram negative example (from *E. lignolyticus*) and anaerobic bacteria (*Clostridium novyi*, *C. botulinum* and *C. tetani*) which can also cause human disease [40]. This common grouping matches the shared anaerobic ecological niche of *E. lignolyticus* and the Clostridial species. The *Clostridium* producers of CDC proteins also include *P. alvei*, *C. perfringens* and *C. butyricum* which are similarly anaerobic and associated with plants, humus and rotting material and again are potential human pathogens. These share an ecological niche with the spore-forming bacilli infecting livestock (*B. antracis*) and insects (*B. thuringiensis*, *L. sphaericus*, *B. latersporus* and *Brevibacillus*) [15, 22].

Evolutionary Implications

While CDCs are primarily cytolytic proteins, not all MACPF proteins have lytic activity and other biological functions are possible. This echoes the situation with eukaryotic proteins of the Bcl-2 family, which are similar to bacterial colicins. Although some Bcl-2 proteins are able to interact with membranes and form pores they do not all exhibit lytic activity and function at different stages of apoptosis—some to promote it, others to inhibit [8]. Various proteins that contain the MACPF domain are clearly evolutionary extremely conserved. For example Mpg-1 is present in sponge where it has a clear role in the immune response [103]. Similarly apextrins are widespread in cnidaria, echinoderms and apicomplexans [65]. Interestingly, apextrins seem to be specifically expressed in the ectoderm during metamorphosis in two distant evolutionary groups, cnidaria and echinoderms [65]. It is, however, extremely difficult to reconstruct the evolutionary history of the

MACPF/CDC superfamily with the current set of available genomic information and protein structures (see for example the distribution of MACPFs in fungi in section "MACPF Proteins in Fungi"). Until recently there were no examples of the two protein families that could be retrieved by sequence searches. However, now one can find MACPFs and CDCs with 20 % identity. For example the CDC of the diatom *Thalassiosira oceanica* is 22 % identical in its MACPF/CDC domain to the MACPF of *T. pseudonana*. The former was identified as a CDC by searching with the perfringolysin O sequence, the latter as a MACPF by searching with a *Plasmodium yoelii* perforin-like protein.

When reconstructing a CDC phylogeny we have identified not only known prokaryotic examples but also examples of eukaryotic proteins which, in our view, are likely to have arisen due to an endosymbiotic event. Mirroring the finding of eukaryotic CDCs, a preliminary phylogenetic study of 82 examples of MACPFs from unique genera highlights several bacterial examples of MACPFs as well as examples in coelacanths, a sea squirt, fungi (*Arthrobotrys oligospora* and *Fusarium oxysporum*) and even the diatom *T. pseudonana*. Indeed, the bacterium *Enterococcus faecalis* appears to have a MACPF-lineage protein 19.9 % identical to its CDC (data not shown). This level of similarity alongside a pairwise alignment of the CDC from *T. oceanica* and the MACPF from *T. pseudonana* that suggests 22 % sequence identity indicates the possibility of constructing a single sequence-based phylogeny for the MACPF/CDC superfamily but this is outside the scope of this chapter. The analysis we have been able to perform does, however, show that there was indeed one original protein domain now found in at least two different sequence families, and that it was present at a point in extreme antiquity before the lineages giving rise to current bacteria, diatoms, plants, animals and fungi diverged.

Although MACPF/CDC proteins may belong to at least two different sequence families, they belong to one structural class, which has been conserved throughout evolution. Understanding why the structure has been so deeply conserved will require further work but to a first approximation it is reasonable to conclude that the capacity it has to refold from an aqueous monomeric conformation into an oligomeric membrane-inserted conformation lies at the heart of the matter. The CDCs and MACPFs may not be considered as "just" toxins or even exclusively toxins: in many cases of pathogenic bacteria, for example, they are not the distinctive protein at the heart of the mechanism of disease (e.g. not in anthrax, not in botulism). Perhaps we should, after all, think of the MACPF/CDCs as housekeeping genes, genes that conferred the capacity to remodel membranes in fundamentally helpful ways, for example in establishing colony formation and as seen today in developmental processes, and that only subsequently did they get co-opted as agents of attack and defence in bacteria, apicomplexans and humans, among other cases [30].

Elsewhere we have reported a structural phylogenetic analysis [30] and it is worth highlighting the fact that however useful such an analysis is in terms of tying together shared fold and function, it does not provide a measure of evolutionary distance in time. In that case, because there is no simple concept of a "mutational clock" in play (as there is in the sequence data we have studied here, albeit one running faster or slower in different organisms with different nucleotide base biases, accuracies of genome replication and environments) it is very hard to know what the meaning of a tree branch length is. The sequence analysis we describe here gives a more informative measure of the evolutionary process; alongside each other the sequential and structural phyogenies provide a compelling case for understanding the unity of the CDC and MACPF families. Further structural work and novel genomic information will aid in understanding further the evolutionary history of this interesting protein superfamily.

Acknowledgments We would like to thank to Sabina Belc for genome mining and The Slovenian Research Agency for financial support.

References

1. Adams NC, Tomoda T, Cooper M, Dietz G, Hatten ME (2002) Mice that lack astrotactin have slowed neuronal migration. Development 129:965–972
2. Alegado RA, Ferriera S, Nusbaum C, Young SK, Zeng Q, Imamovic A, Fairclough SR, King N (2011) Complete genome sequence of *Algoriphagus* sp. PR1, bacterial prey of a colony-forming choanoflagellate. J Bacteriol 193:1485–1486
3. Alegado RA, Grabenstatter JD, Zuzow R, Morris A, Huang SY, Summons RE, King N (2013) *Algoriphagus machipongonensis* sp. nov., co-isolated with a colonial choanoflagellate. Int J Syst Evol Microbiol 63:163–168
4. Aleshin AE, Discipio RG, Stec B, Liddington RC (2012) Crystal structure of C5b-6 suggests a structural basis for priming the assembly of the membrane attack complex (MAC). J Biol Chem 287:19642–19652
5. Aleshin AE, Schraufstatter IU, Stec B, Bankston LA, Liddington RC, DiScipio RG (2012) Structure of complement C6 suggests a mechanism for initiation and unidirectional, sequential assembly of membrane attack complex (MAC). J Biol Chem 287:10210–10222
6. Altincicek B, Vilcinskas A (2008) Comparative analysis of septic injury-inducible genes in phylogenetically distant model organisms of regeneration and stem cell research, the planarian *Schmidtea mediterranea* and the cnidarian *Hydra vulgaris*. Front Zool 5:6
7. Amimoto Y, Kodama R, Kobayakawa Y (2006) Foot formation in *Hydra*: a novel gene, anklet, is involved in basal disk formation. Mechanisms Develop 123:352–361
8. Anderluh G, Lakey JH (2008) Disparate proteins use similar architectures to damage membranes. Trends Biochem Sci 33:482–490
9. Anderluh G, Maček P (2002) Cytolytic peptide and protein toxins from sea anemones (Anthozoa: Actiniaria). Toxicon 40:111–124
10. Anderluh G, Podlesek Z, Maček P (2000) A common motif in proparts of Cnidarian toxins and nematocyst collagens and its putative role. Biochim Biophys Acta 1476:372–376
11. Anderluh G, Sepčić K, Turk T, Maček P (2011) Cytolytic proteins from cnidarians—an overview. Acta Chim Slovenica 58:724–729
12. Armbrust EV (2009) The life of diatoms in the world's oceans. Nature 459:185–192
13. Arnaud MB, Cerqueira GC, Inglis DO, Skrzypek MS, Binkley J, Chibucos MC, Crabtree J, Howarth C, Orvis J, Shah P, Wymore F, Binkley G, Miyasato SR, Simison M, Sherlock G, Wortman JR (2012) The *Aspergillus* genome database (AspGD): recent developments in comprehensive multispecies curation, comparative genomics and community resources. Nucleic Acids Res 40:D653–D659

14. Bernheimer AW, Avigad LS, Kim K (1979) Comparison of metridiolysin from the sea anemone with thiol-activated cytolysins from bacteria. Toxicon 17:69–75
15. Berry C (2012) The bacterium, *Lysinibacillus sphaericus*, as an insect pathogen. J Invertebr Pathol 109:1–10
16. Billington SJ, Jost BH, Cuevas WA, Bright KR, Songer JG (1997) The *Arcanobacterium* (Actinomyces) *pyogenes* hemolysin, pyolysin, is a novel member of the thiol-activated cytolysin family. J Bacteriol 179:6100–6106
17. Birmingham CL, Canadien V, Kaniuk NA, Steinberg BE, Higgins DE, Brumell JH (2008) Listeriolysin O allows *Listeria monocytogenes* replication in macrophage vacuoles. Nature 451:350–354
18. Braaksma M, Martens-Uzunova ES, Punt PJ, Schaap PJ (2010) An inventory of the *Aspergillus niger* secretome by combining in silico predictions with shotgun proteomics data. BMC Genomics 11:584
19. Choi SM, Cho BH, Choi KH, Nam TS, Kim JT, Park MS, Kim BC, Kim MK, Cho KH (2012) Meningitis caused by *Streptococcus suis*: case report and review of the literature. J Clin Neurol 8:79–82
20. Collins MD, Hutson RA, Falsen E, Nikolaitchouk N, LaClaire L, Facklam RR (2000) An unusual *Streptococcus* from human urine, *Streptococcus urinalis* sp. nov. Int J Syst Evol Microbiol 50:1173–1178
21. D'Angelo ME, Dunstone MA, Whisstock JC, Trapani JA, Bird PI (2012) Perforin evolved from a gene duplication of MPEG1, followed by a complex pattern of gene gain and loss within Euteleostomi. BMC Evol Biol 12:59
22. Djukic M, Poehlein A, Thurmer A, Daniel R (2011) Genome sequence of *Brevibacillus laterosporus* LMG 15441, a pathogen of invertebrates. J Bacteriol 193:5535–5536
23. Dreon MS, Frassa MV, Ceolin M, Ituarte S, Qiu JW, Sun J, Fernandez PE, Heras H (2013) Novel animal defenses against predation: a snail egg neurotoxin combining lectin and pore-forming chains that resembles plant defense and bacteria attack toxins. PLoS One 8:e63782
24. Felsenstein J (1997) An alternating least squares approach to inferring phylogenies from pairwise distances. Syst Biol 46:101–111
25. Fields KA, McCormack R, de Armas LR, Podack ER (2013) Perforin-2 restricts growth of *Chlamydia trachomatis* in macrophages. Infect Immun 81:3045–3054
26. Fontes W, Sousa MV, Aragao JB, Morhy L (1997) Determination of the amino acid sequence of the plant cytolysin enterolobin. Arch Biochem Biophys 347:201–207
27. Frassa MV, Ceolin M, Dreon MS, Heras H (2010) Structure and stability of the neurotoxin PV2 from the eggs of the apple snail *Pomacea canaliculata*. Biochim Biophys Acta 1804:1492–1499
28. Frazao B, Vasconcelos V, Antunes A (2012) Sea anemone (Cnidaria, Anthozoa, Actiniaria) toxins: an overview. Marine Drugs 10:1812–1851
29. Gilbert RJ (2005) Inactivation and activity of cholesterol-dependent cytolysins: what structural studies tell us. Structure 13:1097–1106
30. Gilbert RJ, Mikelj M, Dalla Serra M, Froelich CJ, Anderluh G (2013) Effects of MACPF/CDC proteins on lipid membranes. Cell Mol Life Sci 70:2083–2098
31. Goujon M, McWilliam H, Li W, Valentin F, Squizzato S, Paern J, Lopez R (2010) A new bioinformatics analysis tools framework at EMBL-EBI. Nucleic Acids Res 38:W695–W699
32. Grigoriev IV, Nordberg H, Shabalov I, Aerts A, Cantor M, Goodstein D, Kuo A, Minovitsky S, Nikitin R, Ohm RA, Otillar R, Poliakov A, Ratnere I, Riley R, Smirnova T, Rokhsar D, Dubchak I (2012) The genome portal of the department of energy joint genome institute. Nucleic Acids Res 40:D26–D32
33. Gwynne DI, Miller BL, Miller KY, Timberlake WE (1984) Structure and regulated expression of the SpoC1 gene cluster from *Aspergillus nidulans*. J Mol Biol 180:91–109
34. Haag ES, Sly BJ, Andrews ME, Raff RA (1999) Apextrin, a novel extracellular protein associated with larval ectoderm evolution in *Heliocidaris erythrogramma*. Dev Biol 211:77–87

35. Hadders MA, Beringer DX, Gros P (2007) Structure of C8alpha-MACPF reveals mechanism of membrane attack in complement immune defense. Science 317:1552–1554
36. Hadders MA, Bubeck D, Roversi P, Hakobyan S, Forneris F, Morgan BP, Pangburn MK, Llorca O, Lea SM, Gros P (2012) Assembly and regulation of the membrane attack complex based on structures of C5b6 and sC5b9. Cell Rep 1:1–8
37. Han MV, Zmasek CM (2009) phyloXML: XML for evolutionary biology and comparative genomics. BMC Bioinformatics 10:356
38. He X, Zhang Y, Yu Z (2011) An Mpeg (macrophage expressed gene) from the *Pacific oyster Crassostrea gigas*: molecular characterization and gene expression. Fish Shellfish Immunol 30:870–876
39. Hemberger M, Himmelbauer H, Ruschmann J, Zeitz C, Fundele R (2000) cDNA subtraction cloning reveals novel genes whose temporal and spatial expression indicates association with trophoblast invasion. Develop Biol 222:158–169
40. Hickey MJ, Kwan RY, Awad MM, Kennedy CL, Young LF, Hall P, Cordner LM, Lyras D, Emmins JJ, Rood JI (2008) Molecular and cellular basis of microvascular perfusion deficits induced by *Clostridium perfringens* and *Clostridium septicum*. PLoS Pathog 4:e1000045
41. Hotze EM, Le HM, Sieber JR, Bruxvoort C, McInerney MJ, Tweten RK (2013) Identification and characterization of the first cholesterol-dependent cytolysins from gram-negative bacteria. Infect Immun 81:216–225
42. Hotze EM, Tweten RK (2012) Membrane assembly of the cholesterol-dependent cytolysin pore complex. Biochim Biophys Acta 1818:1028–1038
43. Huang S, Yuan S, Guo L, Yu Y, Li J, Wu T, Liu T, Yang M, Wu K, Liu H, Ge J, Huang H, Dong M, Yu C, Chen S, Xu A (2008) Genomic analysis of the immune gene repertoire of *amphioxus* reveals extraordinary innate complexity and diversity. Genome Res 18:1112–1126
44. Humphries JE, Yoshino TP (2003) Cellular receptors and signal transduction in molluscan hemocytes: connections with the innate immune system of vertebrates. Integrat Comp Biol 43:305–312
45. Ittiprasert W, Miller A, Myers J, Nene V, El-Sayed NM, Knight M (2010) Identification of immediate response genes dominantly expressed in juvenile resistant and susceptible *Biomphalaria glabrata* snails upon exposure to *Schistosoma mansoni*. Mol Biochem Parasitol 169:27–39
46. Jensen A, Kilian M (2012) Delineation of *Streptococcus dysgalactiae*, its subspecies, and its clinical and phylogenetic relationship to *Streptococcus pyogenes*. J Clin Microbiol 50:113–126
47. Kafsack BF, Carruthers VB (2010) Apicomplexan perforin-like proteins. Commun Integr Biol 3:18–23
48. Katoh K, Standley DM (2013) MAFFT multiple sequence alignment software version 7: improvements in performance and usability. Mol Biol Evol 30:772–780
49. Kawano H, Nakatani T, Mori T, Ueno S, Fukaya M, Abe A, Kobayashi M, Toda F, Watanabe M, Matsuoka I (2004) Identification and characterization of novel developmentally regulated neural-specific proteins, BRINP family. Mol Brain Res 125:60–75
50. Kelley LA, Sternberg MJ (2009) Protein structure prediction on the Web: a case study using the Phyre server. Nat Protocols 4:363–371
51. Kemena C, Notredame C (2009) Upcoming challenges for multiple sequence alignment methods in the high-throughput era. Bioinformatics 25:2455–2465
52. Kemp IK, Coyne VE (2011) Identification and characterisation of the Mpeg1 homologue in the South African abalone, *Haliotis midae*. Fish Shellfish Immunol 31:754–764
53. Klug M, Weber J, Tardent P (1989) Hemolytic and toxic properties of *Hydra attenuata* nematocysts. Toxicon 27:325–339
54. Knight J, Deora R, Assimos DG, Holmes RP (2013) The genetic composition of *Oxalobacter formigenes* and its relationship to colonization and calcium oxalate stone disease. Urolithiasis 41:187–196

55. Kopacek J, Sakaguchi S, Shigematsu K, Nishida N, Atarashi R, Nakaoke R, Moriuchi R, Niwa M, Katamine S (2000) Upregulation of the genes encoding lysosomal hydrolases, a perforin-like protein, and peroxidases in the brains of mice affected with an experimental prion disease. J Virol 74:411–417
56. Larkin MA, Blackshields G, Brown NP, Chenna R, McGettigan PA, McWilliam H, Valentin F, Wallace IM, Wilm A, Lopez R, Thompson JD, Gibson TJ, Higgins DG (2007) Clustal W and Clustal X version 2.0. Bioinformatics 23:2947–2948
57. Law RH, Lukoyanova N, Voskoboinik I, Caradoc-Davies TT, Baran K, Dunstone MA, D'Angelo ME, Orlova EV, Coulibaly F, Verschoor S, Browne KA, Ciccone A, Kuiper MJ, Bird PI, Trapani JA, Saibil HR, Whisstock JC (2010) The structural basis for membrane binding and pore formation by lymphocyte perforin. Nature 468:447–451
58. Lawley TD, Walker AW (2013) Intestinal colonization resistance. Immunology 138:1–11
59. Lichtenheld MG, Olsen KJ, Lu P, Lowrey DM, Hameed A, Hengartner H, Podack ER (1988) Structure and function of human perforin. Nature 335:448–451
60. Lommer M, Specht M, Roy AS, Kraemer L, Andreson R, Gutowska MA, Wolf J, Bergner SV, Schilhabel MB, Klostermeier UC, Beiko RG, Rosenstiel P, Hippler M, Laroche J (2012) Genome and low-iron response of an oceanic diatom adapted to chronic iron limitation. Genome Biol 13:R66
61. Lotan A, Fishman L, Loya Y, Zlotkin E (1995) Delivery of a nematocyst toxin. Nature 375:456
62. Lovelace LL, Cooper CL, Sodetz JM, Lebioda L (2011) Structure of human C8 protein provides mechanistic insight into membrane pore formation by complement. J Biol Chem 286:17585–17592
63. Mah SA, Moy GW, Swanson WJ, Vacquier VD (2004) A perforin-like protein from a marine mollusk. Biochem Biophys Res Commun 316:468–475
64. McCormack R, de Armas LR, Shiratsuchi M, Ramos JE, Podack ER (2013) Inhibition of intracellular bacterial replication in fibroblasts is dependent on the perforin-like protein (perforin-2) encoded by macrophage-expressed gene 1. J Innate Immun 5:185–194
65. Miller DJ, Hemmrich G, Ball EE, Hayward DC, Khalturin K, Funayama N, Agata K, Bosch TC (2007) The innate immune repertoire in cnidaria–ancestral complexity and stochastic gene loss. Genome Biol 8:R59
66. Mitchell J (2011) *Streptococcus mitis*: walking the line between commensalism and pathogenesis. Mol Oral Microbiol 26:89–98
67. Morita-Yamamuro C, Tsutsui T, Sato M, Yoshioka H, Tamaoki M, Ogawa D, Matsuura H, Yoshihara T, Ikeda A, Uyeda I, Yamaguchi J (2005) The *Arabidopsis* gene CAD1 controls programmed cell death in the plant immune system and encodes a protein containing a MACPF domain. Plant Cell Physiol 46:902–912
68. Nagai H, Oshiro N, Takuwa-Kuroda K, Iwanaga S, Nozaki M, Nakajima T (2002) Novel proteinaceous toxins from the nematocyst venom of the Okinawan sea anemone *Phyllodiscus semoni* Kwietniewski. Biochem Biophys Res Commun 294:760–763
69. Nagashima S, Kamimura A, Shimizu T, Nakamura-Isaki S, Aono E, Sakamoto K, Ichikawa N, Nakazawa H, Sekine M, Yamazaki S, Fujita N, Shimada K, Hanada S, Nagashima KV (2012) Complete genome sequence of phototrophic betaproteobacterium Rubrivivax gelatinosus IL144. J Bacteriol 194:3541–3542
70. Oshiro N, Kobayashi C, Iwanaga S, Nozaki M, Namikoshi M, Spring J, Nagai H (2004) A new membrane-attack complex/perforin (MACPF) domain lethal toxin from the nematocyst venom of the Okinawan sea anemone *Actineria villosa*. Toxicon 43:225–228
71. Ota K, Leonardi A, Mikelj M, Skočaj M, Wohlschlager T, Kunzler M, Aebi M, Narat M, Krizaj I, Anderluh G, Sepčić K, Maček P (2013) Membrane cholesterol and sphingomyelin, and ostreolysin A are obligatory for pore-formation by a MACPF/CDC-like pore-forming protein, pleurotolysin B. Biochimie 95:1855–1864
72. Podack ER, Deyev V, Shiratsuchi M (2007) Pore formers of the immune system. Adv Exp Med Biol 598:325–341

73. Ponting CP (1999) Chlamydial homologues of the MACPF (MAC/perforin) domain. Curr Biol 9:R911–R913
74. Praper T, Beseničar MP, Istinič H, Podlesek Z, Metkar SS, Froelich CJ, Anderluh G (2010) Human perforin permeabilizing activity, but not binding to lipid membranes, is affected by pH. Mol Immunol 47:2492–2504
75. Punta M, Coggill PC, Eberhardt RY, Mistry J, Tate J, Boursnell C, Pang N, Forslund K, Ceric G, Clements J, Heger A, Holm L, Sonnhammer EL, Eddy SR, Bateman A, Finn RD (2012) The Pfam protein families database. Nucleic Acids Res 40:D290–D301
76. Renault T, Faury N, Barbosa-Solomieu V, Moreau K (2011) Suppression substractive hybridisation (SSH) and real time PCR reveal differential gene expression in the Pacific cupped oyster, *Crassostrea gigas*, challenged with Ostreid herpesvirus 1. Develop Comparat Immunol 35:725–735
77. Rosado CJ, Buckle AM, Law RH, Butcher RE, Kan WT, Bird CH, Ung K, Browne KA, Baran K, Bashtannyk-Puhalovich TA, Faux NG, Wong W, Porter CJ, Pike RN, Ellisdon AM, Pearce MC, Bottomley SP, Emsley J, Smith AI, Rossjohn J, Hartland EL, Voskoboinik I, Trapani JA, Bird PI, Dunstone MA, Whisstock JC (2007) A common fold mediates vertebrate defense and bacterial attack. Science 317:1548–1551
78. Rosado CJ, Kondos S, Bull TE, Kuiper MJ, Law RHP, Buckle AM, Voskoboinik I, Bird PI, Trapani JA, Whisstock JC, Dunstone MA (2008) The MACPF/CDC family of pore-forming toxins. Cell Microbiol 10:1765–1774
79. Rossjohn J, Feil SC, McKinstry WJ, Tweten RK, Parker MW (1997) Structure of a cholesterol-binding, thiol-activated cytolysin and a model of its membrane form. Cell 89:685–692
80. Satoh H, Oshiro N, Iwanaga S, Namikoshi M, Nagai H (2007) Characterization of PsTX-60B, a new membrane-attack complex/perforin (MACPF) family toxin, from the venomous sea anemone *Phyllodiscus semoni*. Toxicon 49:1208–1210
81. Schuster-Bockler B, Schultz J, Rahmann S (2004) HMM Logos for visualization of protein families. BMC Bioinform 5:7
82. Shibata T, Kudou M, Hoshi Y, Kudo A, Nanashima N, Miyairi K (2010) Isolation and characterization of a novel two-component hemolysin, erylysin A and B, from an edible mushroom, *Pleurotus eryngii*. Toxicon 56:1436–1442
83. Shinkai Y, Takio K, Okumura K (1988) Homology of perforin to the ninth component of complement (C9). Nature 334:525–527
84. Sievers F, Wilm A, Dineen D, Gibson TJ, Karplus K, Li W, Lopez R, McWilliam H, Remmert M, Soding J, Thompson JD, Higgins DG (2011) Fast, scalable generation of high-quality protein multiple sequence alignments using Clustal Omega. Mol Syst Biol 7:539
85. Slade DJ, Lovelace LL, Chruszcz M, Minor W, Lebioda L, Sodetz JM (2008) Crystal structure of the MACPF domain of human complement protein C8 alpha in complex with the C8 gamma subunit. J Mol Biol 379:331–342
86. Spilsbury K, O'Mara MA, Wu WM, Rowe PB, Symonds G, Takayama Y (1995) Isolation of a novel macrophage-specific gene by differential cDNA analysis. Blood 85:1620–1629
87. Stamatakis A, Hoover P, Rougemont J (2008) A rapid bootstrap algorithm for the RAxML Web servers. Syst Biol 57:758–771
88. Stevens LM, Frohnhofer HG, Klingler M, Nusslein-Volhard C (1990) Localized requirement for torso-like expression in follicle cells for development of terminal anlagen of the *Drosophila embryo*. Nature 346:660–663
89. Sunagawa S, Wilson EC, Thaler M, Smith ML, Caruso C, Pringle JR, Weis VM, Medina M, Schwarz JA (2009) Generation and analysis of transcriptomic resources for a model system on the rise: the sea anemone *Aiptasia pallida* and its dinoflagellate endosymbiont. BMC Genomics 10:258
90. Tan TY, Ng SY, Thomas H, Chan BK (2006) *Arcanobacterium haemolyticum* bacteraemia and soft-tissue infections: case report and review of the literature. J Infect 53:e69–e74
91. Tardent P (1995) The cnidarian cnidocyte, a hightech cellular weaponry. BioEssays 17:351–362

92. Tomita T, Noguchi K, Mimuro H, Ukaji F, Ito K, Sugawara-Tomita N, Hashimoto Y (2004) Pleurotolysin, a novel sphingomyelin-specific two-component cytolysin from the edible mushroom *Pleurotus ostreatus*, assembles into a transmembrane pore complex. J Biol Chem 279:26975–26982
93. Tschopp J, Masson D, Stanley KK (1986) Structural/functional similarity between proteins involved in complement- and cytotoxic T-lymphocyte-mediated cytolysis. Nature 322:831–834
94. Tweten RK (2005) Cholesterol-dependent cytolysins, a family of versatile pore-forming toxins. Infect Immun 73:6199–6209
95. Umen J, Heitman J (2013) Evolution of sex: mating rituals of a pre-metazoan. Curr Biol 23:R1006–R1008
96. van Beurden SJ, Bossers A, Voorbergen-Laarman MH, Haenen OL, Peters S, Abma-Henkens MH, Peeters BP, Rottier PJ, Engelsma MY (2010) Complete genome sequence and taxonomic position of anguillid herpesvirus 1. J Gen Virol 91:880–887
97. Vernatter J, Pirofski LA (2013) Current concepts in host-microbe interaction leading to pneumococcal pneumonia. Curr Opin Infect Dis 26:277–283
98. Voskoboinik I, Smyth MJ, Trapani JA (2006) Perforin-mediated target-cell death and immune homeostasis. Nat Rev Immunol 6:940–952
99. Voskoboinik I, Thia MC, Fletcher J, Ciccone A, Browne K, Smyth MJ, Trapani JA (2005) Calcium-dependent plasma membrane binding and cell lysis by perforin are mediated through its C2 domain: A critical role for aspartate residues 429, 435, 483, and 485 but not 491. J Biol Chem 280:8426–8434
100. Wang GD, Zhang KF, Zhang ZP, Zou ZH, Jia XW, Wang SH, Lin P, Wang YL (2008) Molecular cloning and responsive expression of macrophage expressed gene from small abalone Haliotis diversicolor supertexta. Fish Shellfish Immunol 24:346–359
101. Wang KJ, Ren HL, Xu DD, Cai L, Yang M (2008) Identification of the up-regulated expression genes in hemocytes of variously colored abalone (Haliotis diversicolor Reeve, 1846) challenged with bacteria. Develop Comparat Immunol 32:1326–1347
102. Weston AJ, Chung R, Dunlap WC, Morandini AC, Marques AC, Moura-da-Silva AM, Ward M, Padilla G, da Silva LF, Andreakis N, Long PF (2013) Proteomic characterisation of toxins isolated from nematocysts of the South Atlantic jellyfish *Olindias sambaquiensis*. Toxicon 71:11–17
103. Wiens M, Korzhev M, Krasko A, Thakur NL, Perovic-Ottstadt S, Breter HJ, Ushijima H, Diehl-Seifert B, Muller IM, Muller WE (2005) Innate immune defense of the sponge *Suberites domuncula* against bacteria involves a MyD88-dependent signaling pathway. Induction of a perforin-like molecule. J Biol Chem 280:27949–27959
104. Xu Q, Abdubek P, Astakhova T, Axelrod HL, Bakolitsa C, Cai X, Carlton D, Chen C, Chiu HJ, Clayton T, Das D, Deller MC, Duan L, Ellrott K, Farr CL, Feuerhelm J, Grant JC, Grzechnik A, Han GW, Jaroszewski L, Jin KK, Klock HE, Knuth MW, Kozbial P, Krishna SS, Kumar A, Lam WW, Marciano D, Miller MD, Morse AT, Nigoghossian E, Nopakun A, Okach L, Puckett C, Reyes R, Tien HJ, Trame CB, van den Bedem H, Weekes D, Wooten T, Yeh A, Zhou J, Hodgson KO, Wooley J, Elsliger MA, Deacon AM, Godzik A, Lesley SA, Wilson IA (2010) Structure of a membrane-attack complex/perforin (MACPF) family protein from the human gut symbiont *Bacteroides thetaiotaomicron*. Acta Crystallogr, Sect F: Struct Biol Cryst Commun 66:1297–1305
105. Zbinden A, Mueller NJ, Tarr PE, Eich G, Schulthess B, Bahlmann AS, Keller PM, Bloemberg GV (2012) *Streptococcus tigurinus*, a novel member of the *Streptococcus mitis* group, causes invasive infections. J Clin Microbiol 50:2969–2973
106. Zheng C, Heintz N, Hatten ME (1996) CNS gene encoding astrotactin, which supports neuronal migration along glial fibers. Science 272:417–419

Chapter 3
Evolution of the Complement System

Masaru Nonaka

Abstract The mammalian complement system constitutes a highly sophisticated body defense machinery comprising more than 30 components. Research into the evolutionary origin of the complement system has identified a primitive version composed of the central component C3 and two activation proteases Bf and MASP in cnidaria. This suggests that the complement system was established in the common ancestor of eumetazoa more than 500 million years ago. The original activation mechanism of the original complement system is believed to be close to the mammalian lectin and alternative activation pathways, and its main role seems to be opsonization and induction of inflammation. This primitive complement system has been retained by most deuterostomes without major change until the appearance of jawed vertebrates. At this stage, duplication of the C3, Bf and MASP genes as well as recruitment of membrane attack components added the classical and lytic pathways to the primitive complement system, converting it to the modern complement system. In contrast, the complement system was lost multiple times independently in the protostome lineage.

Keywords Complement · C3 · Factor B · Gene duplication · Mannan binding protein associated serine protease · Membrane attack complex · Thioester

Abbreviations

ANATO	Anaphylatoxin
Bf	Factor B
CCP	Complement control proteins
CPAMD8	PZP-like A2M domain-containing 8
CUB	C1r, C1s, uEGF and bone morphogenic protein
If	Factor I

M. Nonaka (✉)
Department of Biological Sciences, Graduate School of Science, The University of Tokyo, 7-3-1 Hongo, Bunkyo-ku, Tokyo 113-0033, Japan
e-mail: mnonaka@biol.s.u-tokyo.ac.jp

G. Anderluh and R. Gilbert (eds.), *MACPF/CDC Proteins - Agents of Defence, Attack and Invasion*, Subcellular Biochemistry 80, DOI: 10.1007/978-94-017-8881-6_3,

EGFCA	Calcium-binding EGF-like
FIMAC	Factor I membrane attack complex
iTEP	Insect TEP
LDLa	Low-density lipoprotein receptor domain class a
MAC	Membrane attack complex
MACPF	Membrane attack complex/perforin
MASP	Mannan binding protein associated serine protease
MG	Macroglobulin domain
PZP	Pregnancy zone protein
SR	Scavenger receptor Cys-rich
TCC	Terminal complement component
TED	Thioester domain
TEP	Thioester-bond containing protein
Tryp_SPc	Trypsin-like serine protease
TSP1	Thrombospondin type 1 repeats
VWA	Von Willebrand factor type A

Introduction

The mammalian complement system is composed of more than 30 components present mainly in serum and cell membranes and plays essential roles in innate immunity [1]. The proteolytic activation of the central component C3 (the third component of complement) by the C3 convertases is the pivotal step in complement activation, and most major physiological functions of the complement system are induced by the two activation fragments of C3, the larger C3b and smaller C3a. Upon activation, C3b forms a covalent bond with the surface molecules of microbes using its intrachain thioester bond, and bound C3b functions as a molecular tag for foreign particles enhancing phagocytosis by macrophages and neutrophils. C3b also forms a covalent bond with C3b or C4b (the activation fragment of the fourth component of complement equivalent to C3b) of the C3 convertases (alternative pathway C3bBb and classical pathway C4bC2a), switching their specificity into the C5 convertases. Proteolytic activation of C5 (the fifth component of complement) initiates assembly of the late components, C6 (the sixth component of complement) to C9 (the ninth component of complement), leading to the formation of membrane attack complexes (MACs) which disturb the integrity of cell membranes of microbes. The smaller C3a fragment is also known as an anaphylatoxin and induces inflammation. Most key components of the human complement system including C3 possess unique domain structures, and are classified into five protein families [2]; the C3 family (C3, C4 and C5), Bf (Factor B) family (Bf and C2), MASP (mannan binding protein associated serine protease) family (MASP-1, MASP-2, MASP-3, C1r and C1s), C6 family (C6, C7, C8A, C8B and C9), and If (Factor I) family (If only). The origin and evolution of the complement system have been studied by identifying the complement genes possessing these unique domain structures in various eumetazoan species.

Evolution of the TEP Superfamily

C3 family is included in the TEP (thioester-bond containing protein) superfamily whose members are characterized by the unique intrachain thioester bond. Seven members of the TEP superfamily are encoded in the human genome: C3, C4, C5, A2M, pregnancy zone protein (PZP) [3], *CD*109 [4] and the complement 3 and PZP-like A2M domain-containing 8 (*CPAMD*8) [5]. Phylogenetic analysis of many TEP superfamily genes from various eumetazoa indicated the presence of two dichotomous families, C3 and A2M (Fig. 3.1), and this classification was supported by the presence of the ANA (anaphylatoxin) and C345C domains in all members of the C3 family, but never in the A2M family members [6]. The C3 family comprises human C3, C4, C5 and their orthologs of various eumetazoa, whereas the A2M family comprises human A2M, PZP, CD109, CPAMD8 and their orthologs. Accumulation of sequence information concerning various eumetazoa TEPs indicated that the iTEP (insect TEP) once considered unique to insects or arthropods is in fact orthologous to CD109 [6]. A2M, PZP and CPAMD8 are close to each other, and the A2M family is further subdivided into the A2M subfamily and the CD109 subfamily. The basic domain structure of the TEP superfamily members is an eightfold repeat of the macroglobulin (MG) domain. In addition to these eight MG domains, a CUB domain (C1r, C1s, uEGF and bone morphogenetic protein) holding the TED (thioester domain) in the middle is inserted between the seventh and eighth MG domains [7]. Moreover, other specific domains are present in each of the TEP members, such as ANA and C345C (C-terminal of C3, C4 and C5) domains of C3 and the bait domain in A2M.

The evolutionary origin of the TEP superfamily remains to be clarified. However, no TEP gene is present in the published genome information for a sponge, *Amphimedon queenslandica* [8] and a choanoflagellate, *Monosiga brevicollis* [9], suggesting that this gene family arose in the eumetazoan lineage. The three types of TEP genes, C3, A2M and CD109 were identified in two cnidarian sea anemones, *Haliplanella lineate* [10] and *Nematostella vectensis* [11, 12], indicating that differentiation of the TEP genes into C3, A2M and CD109 had completed before the divergence of cnidaria and bilateria. All deuterostome species analyzed thus far such as various vertebrates, urochordate sea squirt [13], cephalochordate lancelet and echinoderm sea urchin [2], possess both the C3 and A2M family members, whereas many protostome genomes deciphered thus far such as fly [14], mosquito [15], honeybee [16], parasitoid wasp [17], aphid [18],

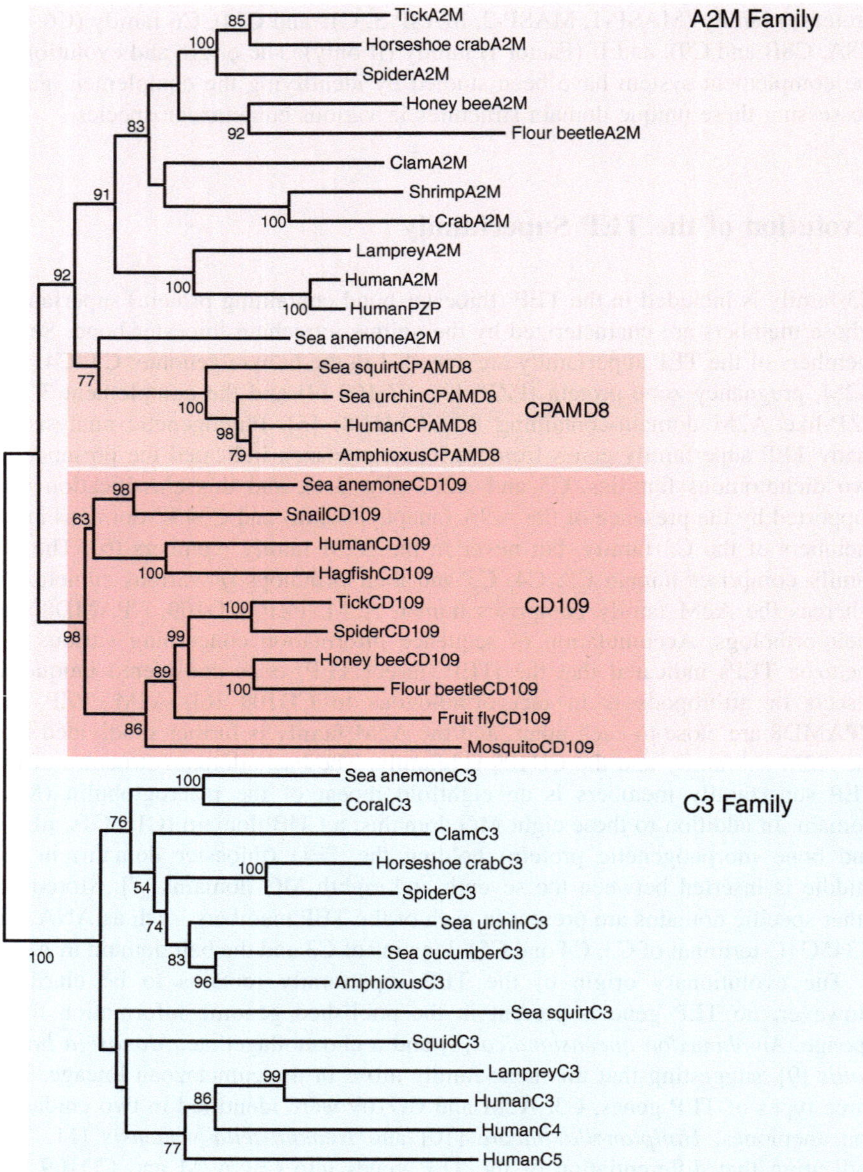
A2M Family
TickA2M
Horseshoe crabA2M
SpiderA2M
Honey beeA2M
Flour beetleA2M
ClamA2M
ShrimpA2M
CrabA2M
LampreyA2M
HumanA2M
HumanPZP
Sea anemoneA2M
Sea squirtCPAMD8
Sea urchinCPAMD8
HumanCPAMD8
AmphioxusCPAMD8
CPAMD8
Sea anemoneCD109
SnailCD109
HumanCD109
HagfishCD109
TickCD109
SpiderCD109
Honey beeCD109
Flour beetleCD109
Fruit flyCD109
MosquitoCD109
CD109
Sea anemoneC3
CoralC3
ClamC3
Horseshoe crabC3
SpiderC3
Sea urchinC3
Sea cucumberC3
AmphioxusC3
Sea squirtC3
SquidC3
LampreyC3
HumanC3
HumanC4
HumanC5
C3 Family
85
100
92
83
91
100
92
100
100
77
100
98
79
98
63
100
100
99
89
100
86
98
100
100
76
100
54
74
83
96
99
86
77
100

◀ **Fig. 3.1** Molecular phylogenetic anaylsis of TEP family genes by maximum likelihood method. The evolutionary history was inferred by using the maximum likelihood method based on the Whelan and Goldman + Freq. model. The tree is drawn to scale, with branch lengths measured in the number of substitutions per site. The analysis involved 40 amino acid sequences. All positions containing gaps and missing data were eliminated. There were a total of 873 positions in the final dataset. Bootstrap percentages with 500 replicates are given. Accession numbers of the used sequences and scientific names of animals are; human (*Homo sapiens*) C3, C4A, C5, A2M, CD109, CPAMD8, and PZP (NP_000055, P0C0L4, AAA51925, P01023, NP_598000, NP_056507 and CAA38255), squid (*Euprymna scolope*) C3 (ACF04700), sea urchin (*Strongylocentrotus purpuratus*) C3 and CPAMD8 (NP_999686 and XP_785018), sea cucumber (*Apostichopus japonicas*) C3 (ADN97000), sea squirt (*Ciona intestinalis*) C3 and CPAMD8 (NP_001027684, and XP_002124325), lamprey (*Lethenteron japonicum*) C3 and A2M (Q00685 and BAA02762), hagfish (*Eptatretus burgeri*) CD109 (BAD12264), horseshoe crab2 (*Tachypleus tridentatus*) C3 and A2M (BAH02276 and BAA19844), amphioxus (*Branchiostoma floridae*) C3 and CPAMD8 (XP_002248496 and XP_002239366), coral (*Swiftia exserta*) C3 (AAN86548), sea anemone (*Haliplanella lineate*) C3, A2M, and CD109 (AB481383, AB481385 and AB481386), fruit fly (*Drosophila melanogaster*) CD109 (*Drme*TEP1) (NP_523578), mosquito (*Anopheles gambiae*) CD109 (*Anga*TEP1) (AAG00600), clam (*Hyriopsis cumingii*) A2M (ABJ89824), clam (*Venerupis decussatus*) C3 (FJ392025), tick (*Ornithodoros moubata*) A2M (AAN10129), tick (*Ixodes scapularis*) CD109 (XP_002409560), snail (*Euphaedusa tau*) CD109 (BAE44110), spider (*Hasarius adansoni*) C3, A2M and CD109 (AB622468, AB622470 and AB622471), shrimp (*Fenneropenaeus chinensis*) A2M (ABP97431), crab (*Eriocheir sinensis*) A2M (ADD71943), honey bee (*Apis mellifera*) A2M (XP_392454) and CD109 (XP_001122599), flour beetle (*Tribolium castaneum*) A2M (EFA07508) and CD109 (XP_972838)

flour beetle [19], and *Caenorhabditis elegans* [20] possessed only the A2M family members. Among protostomes, only horseshoe crab [21, 22], spider [6], tick [23], clam [24] and squid [25] were reported to possess the C3 family, indicating that the loss of the C3 family has occurred multiple times during the protostome evolution [6]. The loss of the C3 family also occurred at least once in cnidaria, since hydra has only the A2M family gene [26]. Although the reason why the C3 family genes were lost so many times during the evolution of cnidarians and protostomes is still to be clarified.

Evolution of the Bf, MASP and If Families

The Bf, MASP and if family members are serine proteases possessing a serine protease domain at their C-termini [27]. The human genome contains two *Bf* family genes, *Bf* and *C2*, four *MASP* genes, *MASP*-1, *MASP*-2, *C1r* and *C1s*, and only one *If* family gene. The gene duplication events which played an important role in establishing a modern complement system like the mammalian one will be discussed below. In the following, the common ancestor genes of *C3*, *C4* and *C5*, *Bf* and *C2*, *MASP*-1, *MASP*-2, *C1r* and *C1s* of invertebrates, will be simply termed as *C3*, *Bf* and *MASP*, respectively. Whereas *Bf* and *MASP* are involved in the complement activation pathways, If has a regulatory function to repress complement activation by degrading activated C3 fragments in the presence of the cofactor proteins. The domain architectures of these families based on the SMART database (http://smart.embl-heidelberg.de/) are CCP (complement control proteins) × 3, VWA (von Willebrand factor type A), Tryp_SPc (trypsin-like serine protease); CUB (C1r, C1s, uEGF and bone morphogenic protein), EGFCA (calcium-binding EGF-like), CUB, CCP × 2, Tryp_SPc; and FIMAC (factor I membrane attack complex), SR (scavenger receptor Cys-rich), LDLa (low-density lipoprotein receptor domain class a) × 2, Tryp_SPc for the Bf, MASP and If families, respectively. The *Bf* family genes have been identified from most deuterostome species analyzed thus far [2], whereas only a few cnidaria [11] and protostome species belonging to arthropod [22] and mollusk [24] are reported to possess the *Bf* family gene. Actually, the *Bf* family gene is missing from the genome sequences of cnidarian hydra, *C. elegans* and several insect species. Although the only *lophotrochozoa* species reported to possess Bf thus far is a clam, the clam Bf lacks the catalytic Ser residue, suggesting that it should not activate C3 in spite of the conservation of the basic domain structure of Bf [24]. It is therefore suggested that although the common ancestor of cnidaria and bilateria had the *Bf* family gene, it was abandoned multiple times in the cnidarian and protostome lineages. Moreover, the Bf-independent complement activation pathway was well characterized in horseshoe crab, an arthropod [21], although it is suggested that horseshoe crab also has the Bf-dependent activation pathway [28]. At the present moment, information on the *Bf* gene in the protostome is very limited compared to that of the *C3* gene, and it is of interest to find out if most protostome species possessing the *C3* gene also retained the *Bf* gene or not. In contrast to the basic conservation of the domain structure throughout evolution, the primary structure of the serine protease domain of Bf shows a curious evolutionary variability. Compared to the other serine proteases, the serine protease domain of mammalian Bf and C2 has a number of structurally unique features, in particular that the bottom of the S1 pocket has a negative charge at Asp226 instead of the usual Asp189 [29]. All jawed vertebrate Bf and C2 so far analyzed have this unique structure, whereas Bf from lamprey [30], ascidian [31], lancelet [32], sea urchin [33], horseshoe crab [22] and sea anemone [11] has Asp189, not Asp226, like all other serine proteases with the trypsin-type specificity. Thus, the structural specialization of the serine protease domain of Bf/C2 seems to have occurred in the common ancestor of the jawed vertebrate, simultaneously with the appearance of the adaptive immune system. The functional consequences of this structural specialization are, if any, still to be clarified.

The *MASP* family genes have been reported from cnidarian and deuterostome except for echinoderm [2]. No *MASP* family gene has been identified in the deciphered genome sequences of protostomes. However, no comprehensive search for the *MASP* family gene has been performed in the protostome species known to possess the *C3* gene, and it is still an open question whether the MASP family gene is present in some protostome species or not. The serine protease domain of the human MASP family members are classified into two groups, the usual one termed TCN-type in which the active site serine is encoded by a TCN codon, and the other unique one termed AGY-type which is characterized by an AGY codon for the active site serine, the absence of the disulfide bond termed the histidine loop, and the absence of introns in the genomic region coding for the serine protease domain [34]. Human MASP-1 belongs to the former group, and human MASP-2, C1r and C1s belong to the latter group. The structure of the human *MASP*-1/-3 gene, which encodes both MASP-1 and MASP-3 by differential usage of the dual serine protease regions, has a usual serine protease region at its 3' end, and the second, intron-less serine protease region just upstream of the usual one [35]. Therefore, the AGY-type MASP is considered to be generated by insertion of the intron-less serine protease region into the TCN-type gene [36]. Whereas only the TCN-type *MASP* gene has been identified from sea anemone [11] and sea squirt [37], the AGY-type *MASP* is present in lancelet [38] and vertebrates. Thus, the timing of the insertion of the intron-less serine protease-encoding region is considered to be before the divergence of cephalochordates and vertebrates, although the entire evolutionary story is still to be clarified.

The If family gene has been identified in all major groups of vertebrates, cyclostomes [39], cartilaginous fish [40], teleosts [41], amphibians [42], reptiles [43], birds [44] and mammals. No If family gene has been reported from protostomes and invertebrate deuterostomes, indicating that the If gene was established in the common ancestor of vertebrates. Thus, the If-dependent regulatory mechanism of complement activation seems to be an innovation in vertebrates. Since unrestricted activation of the complement system could be harmful to the host cells and could lead to depletion of the complement components in a short time, it seems to be essential even for the invertebrate complement system to possess some regulatory mechanism.

Evolution of Terminal Complement Component Genes

The domain structures of five human terminal complement component (TCC) genes defined at the SMART site (http://smart.embl-heidelberg.de/) are; C6, TSP1 (Thrombospondin type 1 repeats)-TSP1-LDLa-MACPF (membrane attack complex/perforin)-TSP1-CCP (Domain abundant in complement control proteins)-CCP-FIMAC-FIMAC; C7, TSP1-LDLa-MACPF-TSP1-CCP-CCP-FIMAC-FIMAC; C8A, TSP1-LDLa-MACPF-TSP1; C8B, TSP1-LDLa-MACPF-TSP1; C9, TSP1-LDLa-MACPF. Thus the TSP1-LDLa-MACPF domain structure is conserved by all five TCC genes, suggesting that they originated from a common ancestor by gene duplication and following modification of the domain structure. However it is still not clear whether the common ancestor of the human TCC genes had a simple domain structure like C9 and new genes were generated by adding extra domains or it had a complex domain structure like C6 and new genes were generated by losing some domains [45]. The genes possessing exactly the same or very similar domain structures as human TCC genes have been identified from all classes of extant jawed vertebrates including cartilaginous fish [46, 47] and teleost [48]. In contrast, no such gene has been identified from lamprey. The previous functional and biochemical analysis of lamprey serum identified an opsonic complement activity but not cytolytic complement activity. Actually, a natural hemolytic activity was present in the lamprey serum, although the only molecule responsible for this hemolytic activity seems to be a 25 kDa protein without any connection to the complement system [49]. All these results indicate that the cytolytic activity of the complement system was established in the common ancestor of jawed vertebrates. Many C9-like genes have been identified from the genomes of urochordate *Ciona intestinalis* [13] and cephalochordate *Branchiostoma floridae* [50]. A typical domain structure of them is TSP1-LDLa-MACPF, indicating a close evolutionary relationship with the human TCC genes. It is highly probable that they are involved in some cytolytic process. However, it is unlikely that they are activated by the complement system of these animals. Activation of human TCC is initiated by the binding of C6 to the activated C5, C5b. The interaction between the C345c domain of C5 and the two FIMAC domains of C6 has been demonstrated to play an essential role in this binding by biochemical [51] and 3D structure [52] analyses. Since none of the C9 like molecules of urochordates and cephalochordates possesses the FIMAC domain, they seem to function as cytolytic factors independent from their complement system reported to have opsonic and preinflammatory functions. Distribution of the complement genes in various animal groups is summarized in Table 3.1.

Evolutionary Scenario of the Complement System

The tentative evolutionary process of the complement system is schematically shown in Fig. 3.2. The traceable origin of the complement system at the present moment is the primitive complement system composed of C3, Bf and MASP in the common ancestor of eumetazoa. Conservation of the amino acid sequences critical for basic functions of these components between cnidaria and humans suggests that the basic activation mechanism and physiological functions of the primitive complement system were not very different from that of the human complement system. Thus, MASP appears to be the first protease to be activated, which in turn activates Bf. Then Bf activates C3 into the C3a and C3b fragments. C3b covalently tags microbes and enhances phagocytosis, whereas C3a induces inflammation as an anaphylatoxin.

In deuterostomes, this basic complement system is retained by all members analyzed thus far; echinoderm, cephalochordate, urochordate and vertebrate. The marked development of the complement system seems to have occurred in the common ancestor of jawed vertebrates. Gene duplication and following structural and functional specialization seem to play an important role in this process. First, C4, C2 and C1r/C1s, most probably generated by gene duplication from *C3, Bf* and *MASP*, respectively, contributed to establish the classical pathway. The canonical adaptive immunity based on IgSF receptors and MHC is also believed to have been established in the common ancestor of jawed vertebrates, and the establishment of

Table 3.1 Distribution of the complement component genes in various animal groups

	Cnidaria	Protostome		Deuterostome				
		Lophotrochozoa	Ecdysozoa	Echinoderm	Cephalochordate	Urochordate	Agnatha	Gnathostome
C3	±	±	±	+	+	+	+	+
Bf	±	±	±	+	+	+	+	+
MASP	±	−	−	−	+	+	+	+
If	−	−	−	−	−	−	+	+
C9-like	−	−	−	−	+	+	−	+
C6-like	−	−	−	−	−	−	−	+

+ indicates that at least one species of that animal group possesses that complement gene. − indicates that at least one species of that animal group was shown not to possess that complement gene by genome analysis. ± indicates that that animal group contains both type species. TCC like genes with and without the FIMAC domain are termed C6-like and C9-like, respectively

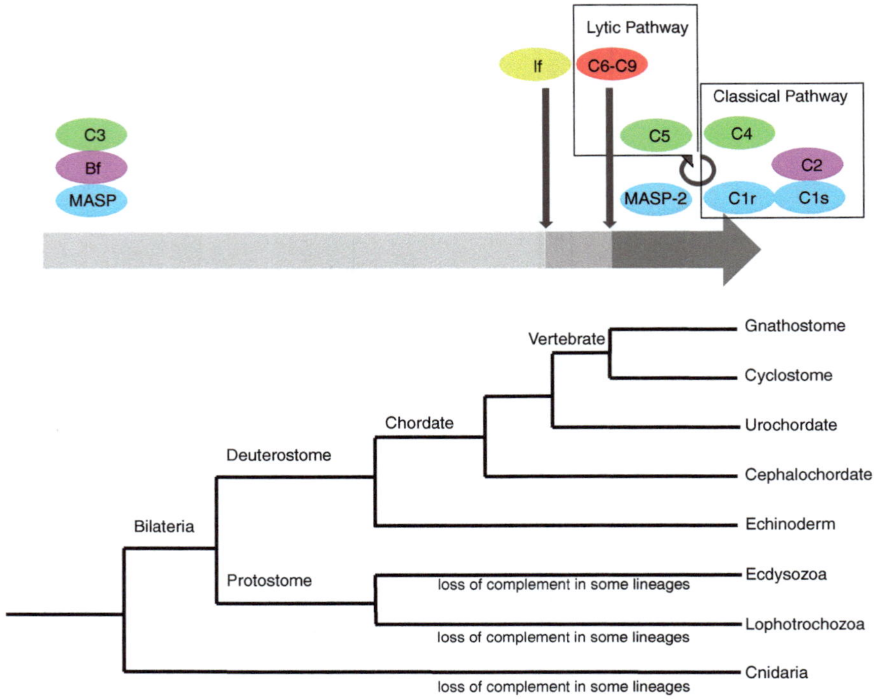

Fig. 3.2 Evolution of the complement system. The bold horizontal arrow represents evolution of the complement system. *Light gray* indicates a primitive complement system composed of C3, Bf and MASP. *Medium gray* indicates an intermediate complement system with the If-dependent regulation mechanism. The *dark gray* indicates a modern complement system found in jawed vertebrates possessing the lytic and classical pathways. Vertical arrows indicate recruitment or innovation of additional components, whereas circular arrow indicates gene duplication events

the classical pathway was epoch-making in the respect that it connected the complement system with adaptive immunity. Second, the gene duplication generating C5 from C3 on the one hand, and the recruitment of the FIMAC domain into the C9-like genes on the other hand connected the preexisting complement system and non-complement cytolytic system.

In contrast to deuterostomes which basically retained the complement system composed of C3, Bf and MASP, the loss of the complement system occurred multiple times in the protostome lineages, and several protostome genome sequences deciphered thus far do not contain any complement genes. Although the reason why the complement system was lost so many times in protostomes is still not clear, in some insects CD109, a paralogue of C3, is reported to be multiplied and play an opsonic role like C3 [53]. Thus, it is possible that some protostomes such as insects developed unique immune mechanisms, making it unnecessary to retain the complement system. Even in horseshoe crab, which retained the complement system, unique specialization is reported compared to the deuterostome complement system. Thus, factor C originally identified as an LPS-sensitive initiator of hemolymph coagulation stored within hemocytes is used as an initiating protease of the horseshoe crab complement system [21]. In this activation pathway, factor C directly activates C3 without intervention of Bf. Thus, factor C first evolved in the horseshoe crab lineage, and this alteration in the initiating mechanism of the complement system therefore occurred recently on an evolutionary time scale. Thus, as far as the complement system is concerned, the innovations of protostomes seem to be more revolutionary than those of deuterostomes.

References

1. Volanakis JE (1998) Overview of the complement system. In: Volanakis JE, Frank MM (eds) The human complement system in health and disease. Marcel Dekker, New York, pp 9–32
2. Nonaka M, Kimura A (2006) Genomic view of the evolution of the complement system. Immunogenetics 58:701–713
3. Sottrup-Jensen L, Folkersen J, Kristensen T, Tack BF (1984) Partial primary structure of human pregnancy zone protein: extensive sequence homology with human alpha 2-macroglobulin. Proc Natl Acad Sci USA 81:7353–7357
4. Lin M, Sutherland DR, Horsfall W, Totty N, Yeo E, Nayar R, Wu XF, Schuh AC (2002) Cell surface antigen CD109 is a novel member of the alpha(2) macroglobulin/C3, C4, C5 family of thioester-containing proteins. Blood 99:1683–1691
5. Li ZF, Wu XH, Engvall E (2004) Identification and characterization of CPAMD8, a novel member of the complement 3/alpha2-macroglobulin family with a C-terminal Kazal domain. Genomics 83:1083–1093
6. Sekiguchi R, Fujito NT, Nonaka M (2012) Evolution of the thioester-containing proteins (TEPs) of the arthropoda, revealed by molecular cloning of TEP genes from a spider, *Hasarius adansoni*. Dev Comp Immunol 36:483–489
7. Janssen BJ, Huizinga EG, Raaijmakers HC, Roos A, Daha MR, Nilsson-Ekdahl K, Nilsson B, Gros P (2005) Structures of complement component C3 provide insights into the function and evolution of immunity. Nature 437:505–511
8. Srivastava M, Simakov O, Chapman J et al (2010) The *Amphimedon queenslandica* genome and the evolution of animal complexity. Nature 466:720–726
9. King N, Westbrook MJ, Young SL et al (2008) The genome of the choanoflagellate *Monosiga brevicollis* and the origin of metazoans. Nature 451:783–788
10. Fujito NT, Sugimoto S, Nonaka M (2010) Evolution of thioester-containing proteins revealed by cloning and characterization of their genes from a cnidarian sea anemone, *Haliplanella lineate*. Dev Comp Immunol 34:775–784
11. Kimura A, Sakaguchi E, Nonaka M (2009) Multi-component complement system of Cnidaria: C3, Bf, and MASP genes expressed in the endodermal tissues of a sea anemone, *Nematostella vectensis*. Immunobiology 214:165–178
12. Putnam NH, Srivastava M, Hellsten U, Dirks B, Chapman J, Salamov A, Terry A, Shapiro H, Lindquist E, Kapitonov VV, Jurka J, Genikhovich G, Grigoriev IV, Lucas SM, Steele RE, Finnerty JR, Technau U, Martindale MQ, Rokhsar DS (2007) Sea anemone genome reveals ancestral eumetazoan gene repertoire and genomic organization. Science 317:86–94
13. Azumi K, De Santis R, De Tomaso A, Rigoutsos I, Yoshizaki F, Pinto MR, Marino R, Shida K, Ikeda M, Ikeda M, Arai M, Inoue Y, Shimizu T, Satoh N, Rokhsar DS, Du Pasquier L, Kasahara M, Satake M, Nonaka M (2003) Genomic analysis of immunity in a Urochordate and the emergence of the vertebrate immune system: "waiting for Godot". Immunogenetics 55:570–581

14. Adams MD, Celniker SE, Holt RA et al (2000) The genome sequence of *Drosophila melanogaster*. Science 287:2185–2195
15. Nene V, Wortman JR, Lawson D et al (2007) Genome sequence of *Aedes aegypti*, a major arbovirus vector. Science 316:1718–1723
16. Consortium HGS (2006) Insights into social insects from the genome of the honeybee *Apis mellifera*. Nature 443:931–949
17. Werren JH, Richards S, Desjardins CA et al (2010) Functional and evolutionary insights from the genomes of three parasitoid *Nasonia* species. Science 327:343–348
18. Consortium (2010) Genome sequence of the pea aphid *Acyrthosiphon pisum*. PLoS Biol 8:e1000313
19. Richards S, Gibbs RA, Weinstock GM et al (2008) The genome of the model beetle and pest *Tribolium castaneum*. Nature 452:949–955
20. Consortium TCeS (1998) Genome sequence of the nematode *C. elegans*: a platform for investigating biology. Science 282:2012–2018
21. Ariki S, Takahara S, Shibata T, Fukuoka T, Ozaki A, Endo Y, Fujita T, Koshiba T, Kawabata S (2008) Factor C acts as a lipopolysaccharide-responsive C3 convertase in horseshoe crab complement activation. J Immunol 181:7994–8001
22. Zhu Y, Thangamani S, Ho B, Ding JL (2005) The ancient origin of the complement system. EMBO J 24:382–394
23. Buresova V, Hajdusek O, Franta Z, Loosova G, Grunclova L, Levashina EA, Kopacek P (2012) Functional genomics of tick thioester-containing proteins reveal the ancient origin of the complement system. J Innate Immun 3:623–630
24. Prado-Alvarez M, Rotllant J, Gestal C, Novoa B, Figueras A (2009) Characterization of a C3 and a factor B-like in the carpet-shell clam, *Ruditapes decussatus*. Fish Shellfish Immunol 26:305–315
25. Castillo MG, Goodson MS, McFall-Ngai M (2009) Identification and molecular characterization of a complement C3 molecule in a lophotrochozoan, the Hawaiian bobtail squid *Euprymna scolopes*. Dev Comp Immunol 33:69–76
26. Chapman JA, Kirkness EF, Simakov O et al (2010) The dynamic genome of Hydra. Nature 464:592–596
27. Volanakis JE, Arlaud GJ (1998) Complement enzymes. In: Volanakis JE, Frank MM (eds) The human complement system in health and disease. Marcel Dekker, New York, pp 49–81
28. Tagawa K, Yoshihara T, Shibata T, Kitazaki K, Endo Y, Fujita T, Koshiba T, Kawabata S (2012) Microbe-specific C3b deposition in the horseshoe crab complement system in a C2/factor B-dependent or -independent manner. PLoS One 7:e36783
29. Jing H, Xu Y, Carson M, Moore D, Macon KJ, Volanakis JE, Narayana SV (2000) New structural motifs on the chymotrypsin fold and their potential roles in complement factor B. EMBO J 19:164–173
30. Nonaka M, Takahashi M, Sasaki M (1994) Molecular cloning of a lamprey homologue of the mammalian MHC class III gene, complement factor B. J Immunol 152:2263–2269
31. Yoshizaki FY, Ikawa S, Satake M, Satoh N, Nonaka M (2005) Structure and the evolutionary implication of the triplicated complement factor B genes of a *Urochordate ascidian, Ciona intestinalis*. Immunogenetics 56:930–942
32. Holland LZ, Albalat R, Azumi K et al (2008) The amphioxus genome illuminates vertebrate origins and cephalochordate biology. Genome Res 18:1100–1111
33. Smith LC, Shih CS, Dachenhausen SG (1998) Coelomocytes express SpBf, a homologue of factor B, the second component in the sea urchin complement system. J Immunol 161:6784–6793
34. Endo Y, Takahashi M, Nakao M, Saiga H, Sekine H, Matsushita M, Nonaka M, Fujita T (1998) Two lineages of mannose-binding lectin-associated serine protease (MASP) in vertebrates. J Immunol 161:4924–4930 (published erratum appears in J Immunol 164:5330)
35. Schwaeble W, Dahl MR, Thiel S, Stover C, Jensenius JC (2002) The mannan-binding lectin-associated serine proteases (MASPs) and MAp19: four components of the lectin pathway activation complex encoded by two genes. Immunobiology 205:455–466

36. Nonaka M, Miyazawa S (2002) Evolution of the initiating enzymes of the complement system. Genome Biol 3:Reviews1001
37. Ji X, Azumi K, Sasaki M, Nonaka M (1997) Ancient origin of the complement lectin pathway revealed by molecular cloning of mannan binding protein-associated serine protease from a urochordate, the Japanese ascidian, *Halocynthia roretzi*. Proc Nat Acad Sci USA 94:6340–6345
38. Endo Y, Nonaka M, Saiga H, Kakinuma Y, Matsushita A, Takahashi M, Matsushita M, Fujita T (2003) Origin of mannose-binding lectin-associated serine protease (MASP)-1 and MASP-3 involved in the lectin complement pathway traced back to the invertebrate, amphioxus. J Immunol 170:4701–4707
39. Kimura A, Ikeo K, Nonaka M (2009) Evolutionary origin of the vertebrate blood complement and coagulation systems inferred from liver EST analysis of lamprey. Dev Comp Immunol 33:77–87
40. Terado T, Nonaka MI, Nonaka M, Kimura H (2002) Conservation of the modular structure of complement factor I through vertebrate evolution. Dev Comp Immunol 26:403–413
41. Nakao M, Hisamatsu S, Nakahara M, Kato Y, Smith SL, Yano T (2003) Molecular cloning of the complement regulatory factor I isotypes from the common carp (*Cyprinus carpio*). Immunogenetics 54:801–806
42. Kunnath-Muglia LM, Chang GH, Sim RB, Day AJ, Ezekowitz RA (1993) Characterization of *Xenopus laevis* complement factor I structure–conservation of modular structure except for an unusual insert not present in human factor I. Mol Immunol 30:1249–1256
43. Alfoldi J, Di Palma F, Grabherr M et al (2011) The genome of the green anole lizard and a comparative analysis with birds and mammals. Nature 477:587–591
44. International Chicken Genome Sequencing Consortium (2004) Sequence and comparative analysis of the chicken genome provide unique perspectives on vertebrate evolution. Nature 432:695–716
45. Hobart MJ (1998) Evolution of the terminal complement genes: ancient and modern. Exp Clin Immunogenet 15:235–243
46. Aybar L, Shin DH, Smith SL (2009) Molecular characterization of the alpha subunit of complement component C8 (GcC8alpha) in the nurse shark (*Ginglymostoma cirratum*). Fish Shellfish Immunol 27:397–406
47. Kimura A, Nonaka M (2009) Molecular cloning of the terminal complement components C6 and C8beta of cartilaginous fish. Fish Shellfish Immunol 27:768–772
48. Nakao M, Uemura T, Yano T (1996) Terminal components of carp complement constituting a membrane attack complex. Mol Immunol 33:933–937
49. Nonaka M, Fujii T, Kaidoh T, Natsuume-Sakai S, Yamaguchi N, Takahashi M (1984) Purification of a lamprey complement protein homologous to the third component of the mammalian complement system. J Immunol 133:3242–3249
50. Huang S, Yuan S, Guo L, Yu Y, Li J, Wu T, Liu T, Yang M, Wu K, Liu H, Ge J, Huang H, Dong M, Yu C, Chen S, Xu A (2008) Genomic analysis of the immune gene repertoire of amphioxus reveals extraordinary innate complexity and diversity. Genome Res 18:1112–1126
51. Thai CT, Ogata RT (2004) Complement components C5 and C7: recombinant factor I modules of C7 bind to the C345C domain of C5. J Immunol 173:4547–4552
52. Aleshin AE, DiScipio RG, Stec B, Liddington RC (2012) Crystal structure of C5b-6 suggests structural basis for priming assembly of the membrane attack complex. J Biol Chem 287:19642–19652
53. Levashina EA, Moita LF, Blandin S, Vriend G, Lagueux M, Kafatos FC (2001) Conserved role of a complement-like protein in phagocytosis revealed by dsRNA knockout in cultured cells of the mosquito, *Anopheles gambiae*. Cell 104:709–718

Part II
Structures of MACPF/CDCs

Chapter 4
Structural Features of Cholesterol Dependent Cytolysins and Comparison to Other MACPF-Domain Containing Proteins

Robert Gilbert

Abstract Five different cholesterol-dependent cytolysins (CDCs) have now had their atomic structures solved. Here their structures are compared and shown to vary less in the C-terminal region than they do in their N-terminal MACPF/CDC homology region. The most variable region of the C-terminal domain is the undecapeptide, which is observed in two clusters of conformations, and comparison of this domain with the C2 domain of perforin shows that the two structures have a common ancestor. Structural studies of CDC pre-pore and pore oligomers by cryo-electron microscopy and atomic force microscopy have revealed much about their mechanism of action. Understanding the activity of CDCs has required a combination of structural, biophysical and functional assays but current models of pore formation still require development to account for variable functional pore size.

Keywords Cryo-electron microscopy · Monomeric CDC structure · Oligomeric CDC structures · Perforin · Structural phylogeny · X-ray crystallography

Abbreviations

ALO	Anthrolysin
AUC	Analytical ultracentrifugation
CDCs	Cholesterol-dependent cytolysins
ILY	Intermedilysin
LLO	Listeriolysin O
MALS	Multi-angle light scattering
PFO	Perfringolysin O
PLY	Pneumolysin
RMSD	Root mean square deviation

R. Gilbert (✉)
Division of Structural Biology, Wellcome Trust Centre for Human Genetics,
University of Oxford, Roosevelt Drive, Oxford OX3 7BN, UK
e-mail: gilbert@strubi.ox.ac.uk

G. Anderluh and R. Gilbert (eds.), *MACPF/CDC Proteins - Agents of Defence, Attack and Invasion*, Subcellular Biochemistry 80, DOI: 10.1007/978-94-017-8881-6_4,

SLO Streptolysin O
SLY Suilysin
TMH Transmembrane hairpin

Introduction

The cholesterol-dependent cytolysins (CDCs) were originally identified in the middle of the twentieth century as "thiol-activated toxins" which could be isolated directly from Gram positive bacteria such as *Streptococcus pneumoniae* (pneumolysin, PLY) and *Clostridium perfringens* (perfringolysin O also known as theta toxin, PFO). They are proteins of ~50–70 kDa molecular weight whose apparent thiol activation derives from the presentation of a free cysteine within a highly conserved region of their structure, an undecapeptide sequence towards their carboxy-terminus [1–3]. Because they are significant pathogenicity determinants for their producing organisms, much work focused on their activity and characterisation, centred on PFO, PLY, listeriolysin O (LLO, from *Listeria monocytogenes*) and most prominently streptolysin O (SLO) from *Streptococcus pyogenes*. This work mapped out some of the basic principles on which they operate—their dependency on cholesterol in particular, their high affinity for it, their assembly into large ring-shaped oligomers which appear to define pores, and some details of the kinetics of their oligomerisation.

However, the solution of the first crystal structure of PFO, published in 1997, was a landmark in the study of CDCs [4]. Up to that point structural data were limited to those obtained from techniques such as metal shadowing and hydrodynamics/bead modelling [5] or images of oligomers formed by the toxins—such as those generated for PLY [6] and PFO [7]. Other studies had used cysteine-scanning mutagenesis as a way of trying to work out which parts of the proteins inserted into membranes to form pores [8]. Such techniques would prove to be of great value, and the more fine-grained study of Tweten and colleagues identified first one [9], and then two [10] transmembrane hairpins (TMHs) as the full extent of the membrane-penetrating domain first noted by Bhakdi and colleagues [8]. In the context of the then-determined structure of perfringolysin PFO, these provided the striking insight that conversion of CDCs to a membrane-inserted form involves a transition from an alpha-helical to a beta-sheeted structure [4, 10]. The original cysteine-scanning approach lacked the crucial interpretative framework which an atomic structure provides; but techniques other than crystallography were always going to be needed if the structural basis of CDC activity was to be identified. Indeed, structural studies of PFO are an object case of ways in which a structure cannot provide all the answers: the atomic structure failed to show how the protein might act against membranes, how it bound cholesterol to target them, or even the orientation of the subunit within the oligomer which, as it turned out, was initially modelled 180° out of register [4]. These answers have now all been determined, to a greater or lesser extent, by the combination of structural, biophysical and functional data. Yet even today substantial questions remain most important of which, to this author's mind, is one of the oldest; are the incomplete (arc-shaped) oligomers observed repeatedly over many years and by a variety of techniques also involved in pore formation [11, 12]?

Monomeric Structure of CDCs

At the time of writing, five different CDC structures are available in the Protein Databank; those for PFO [4], intermedilysin (ILY) [13], anthrolysin (ALO) [14], suilysin (SLY) [15] and SLO [16]. In addition, the structure of LLO has apparently been determined [17] but is not published or released yet.

The now well-known CDC fold consists of two regions. The N-terminal ~70 % consists of the canonical MACPF/CDC fold which in the original structure was described as three domains that, clearly, are closely interleaved [4] (Fig. 4.1a). The core of the amino-terminus is dominated by a central, kinked 5-stranded antiparallel β-sheet interrupted by a total of eight α-helices. Six of these α-helices, in two clusters of three, are suspended from the ends of the β-sheet (at the tip of CDC domain 3) from where they are deployed during pore formation to generate the twin TMHs [9, 10]. The other two are longer helices which arch over the domain 1-domain 3 interface and which, it might be suggested, on pore formation provide the essential spine suspending the deployment of the TMHs into the lipid bilayer (edged with a green dashed line in Fig. 4.1a). The upper part of the other face of the the MACPF/CDC domain is capped by four further helices, one of which comprises the amino-terminus of the protein, but otherwise the MACPF domain is protected by CDC domain 2, a three-stranded antiparallel β-sheet which is packed against domain 3. Domain 2 apparently serves twin functions: linking the carboxy-terminal receptor-binding domain to the pore-forming amino-terminal region, and allowing for the insertion of the TMHs via a vertical collapse [18, 19]. The carboxy-terminal domain consists of an eight-stranded ß-sandwich against the tip of which the back of domain 2 rests in the soluble form of the structure and which contains three critical loops for interaction with cholesterol [20] in addition to the highly-conserved undecapeptide (in PFO and most Gram positive bacterial CDCs amino acids ECTGLAWEWWR) [13, 14].

Comparing the available CDC atomic structures (Fig. 4.1) it becomes clear that they display a remarkable similarity, and differ mostly with respect to the exact angle adopted by the carboxy-terminal domain with respect to the rest of the structure. It is evident both that a variety of different positions are observed (Fig. 4.1b) and that these are affected by processes such as ligand binding. Thus, the crystal structure of ILY bound to its receptor CD59 [21] shows domain 4 rotated back in line with the rest of the molecule (similar to the position seen in the structures of PFO and SLO in isolation) compared to the eccentric position it

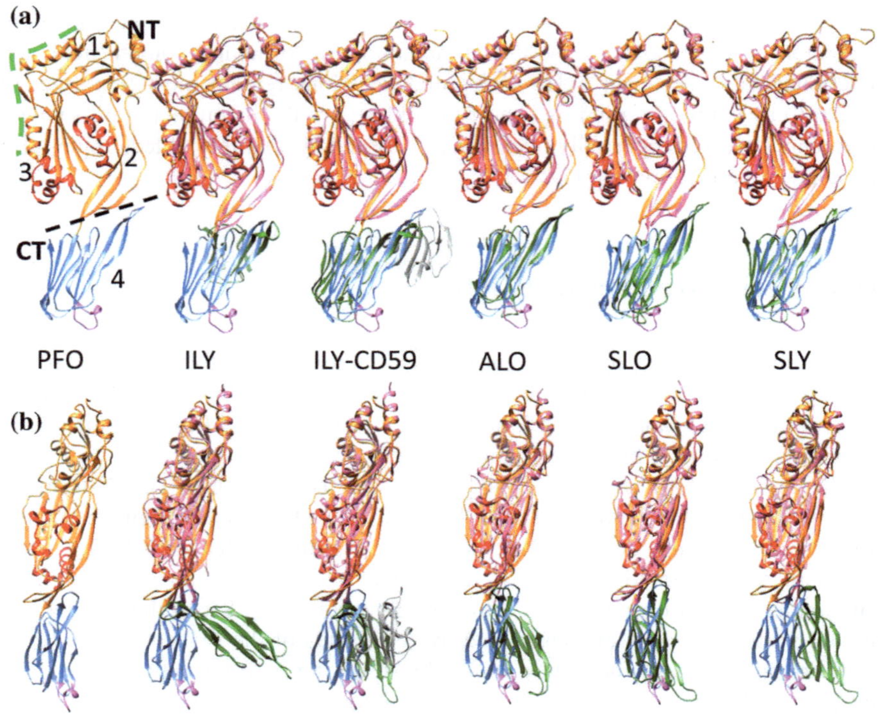

Fig. 4.1 Comparison of available CDC structures PFO [4], ILY [13], ILY with CD59 its cellular receptor [21], ALO [14], SLO [16] and SLY [15]. **a** Viewed in profile, in which the pore would be to the *left* of each subunit. PFO is shown in isolation with its amino-terminal region coloured *orange* save the six pore-forming helices which are in *red*. The carboxy-terminal domain is coloured *blue* save the undecapeptide which is coloured *pink*. A *dashed black* line marks the boundary between amino- and carboxy-terminal regions and the four domains originally described are also labelled. The *green dashed* line marks the spine dominated by two structural alpha helices which are discussed in the text. Each of the other structures is shown superimposed on PFO, and coloured with their amino-terminal (MACPF/CDC homology) region *pink* and their carboxy-terminal domain *green*. **b** The same as *panel a*, but viewed from the back of domain 2 (which forms the outer face of the oligomer) to accentuate the variable position of the carboxy-terminal domain (domain 4) with respect to the rest of the structure. The most extreme orientation is found in free ILY, brought back in line and in a rather similar position to the domain in free ALO by the binding of CD59. Altogether these structural data (and the transitions in conformation on pore formation—see below) suggest there may be little significance in the orientation of this domain with respect to the rest of the protein; instead, it appears genuinely flexible. The molecular structures in this figure and elsewhere in this chapter were depicted using UCSF Chimera software [45]

adopts in the isolated structure [13] and the intermediate orientation seen in ALO [14]. It is impossible to say, on current data, to what extent the different orientations of domain 4 are crystal artefacts; what is clear is that the domain is flexible with respect to the rest of the molecule—as seen most clearly in the alternative

ILY structures [13, 21] and the different orientations domain 4 can adopt during oligomerisation [19, 22] (see below).

Apart from its relative orientation to the rest of the molecule, the carboxy-terminal domain is very similar in structure in all the CDCs (see below) except, oddly, in what is its most conserved region; the canonical CDC undecapeptide [13, 14, 16]. This adopts a range of different conformations in the different proteins (Fig. 4.2a, b): in contrast to the furled-up conformation originally determined, in the structure of PFO, the other proteins (apart from ALO) tend to show a more unfurled conformation reminiscent of that modelled as a possible cholesterol-interacting state [23]. Two distinct conformational ensembles can be identified; the furled group comprising PFO and ALO, and the unfurled group comprising ILY (both structures), SLY and SLO. Thus, the different conformation adopted by the undecapeptide in ILY [13] compared to PFO no longer distinguishes it despite its different sequence (GATGLAWEPWR as opposed to ECTGLAWEWWR). In fact, the most similar loop conformation to that of SLO is ILY's when bound to CD59 (see below); in the unbound form of ILY Trp491 is extended from the loop but when complexed with CD59 it is folded back towards the protein backbone as in SLO.

To focus in further on the similarities and differences between the different CDCs, Fig. 4.3 shows the results of a phylogenetic analysis based on their structures alone, performed for the amino-terminal and carboxy-terminal regions in isolation. The root mean square deviation (RMSD) in backbone atoms between the four available CDC structures is 0.76 Å when one considers the amino terminal (MACPF-like) region alone (Fig. 4.3a). Comparison of this structure-based phylogeny with that based on sequence shown in Chap. 2 (Fig. 2.5) is interesting because the same phylogenetic arrangement is found in this structure-based analysis as was seen in the sequence-based one. Thus, ALO from *Bacillus anthracis*, SLO from *Streptococcus pyogenes* and PFO from *Streptococcus perfringens* occupy one side of the tree, while the CDCs from the *mitis* grouping of *Streptococci*, species *intermedius* and *suis* sit on branches away from the other streptococcal example. The same analysis focused on the carboxy-terminal domain (Fig. 4.3b) produces a very similar distribution but is also interesting because, with an RMSD of 0.29 Å, it shows that there is no evidence that this domain is any less conserved in structure than the MACPF/CDC canonical region which dominates the amino-terminal region of the proteins. Indeed, on the face of it, the carboxy-terminal domain appears more conserved than the MACPF/CDC domain—and perhaps this is unsurprising because it is responsible for the distinctive property of these proteins, their dependence on cholesterol for binding and pore formation [3, 13, 14, 20], though of course ILY binds CD59 first. Note also, for example, that the RMSD between the two ILY domain structures (free and bound to CD59) is 0.2 Å—not so different from the difference between the carboxy-terminal domains from other species. The phylogeny inferred from the carboxy-terminal domain is similar to that seen for the amino-terminal one—the differences being limited to branch length and the relative positions of SLO and ALO. These suggest some independent variation of the carboxy-terminal domain

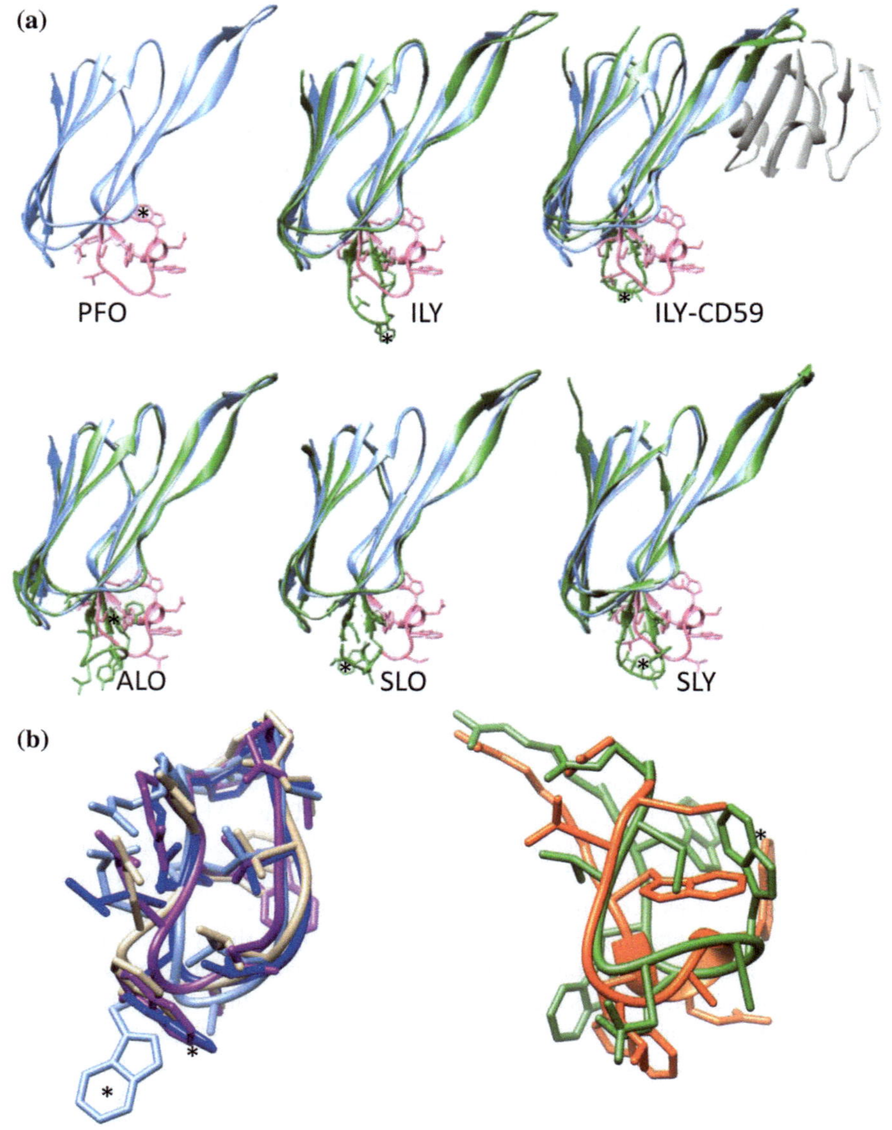

Fig. 4.2 Comparison of undecapeptides in the available CDC structures **a** Views of the whole of the carboxy-terminal domain (as in Fig. 4.1) with the sidechains of the undecapeptide residues shown. In each case the equivalent tryptophan to Trp464 in PFO is asterisked: Trp491 in ILY, Trp477 in ALO, Trp535 in SLO and Trp461 in SLY. **b** Close-ups of the undecapeptides of the CDCs, which fall into two groups. On the *left* are the conformers seen in ILY (*light blue*), ILY with CD59 bound (*dark blue*), SLO (*purple*) and SLY (*tan*). On the *right* are the conformers seen in PFO (*orange*) and ALO (*green*). In every case the equivalent tryptophan to PFO Trp464 is again asterisked

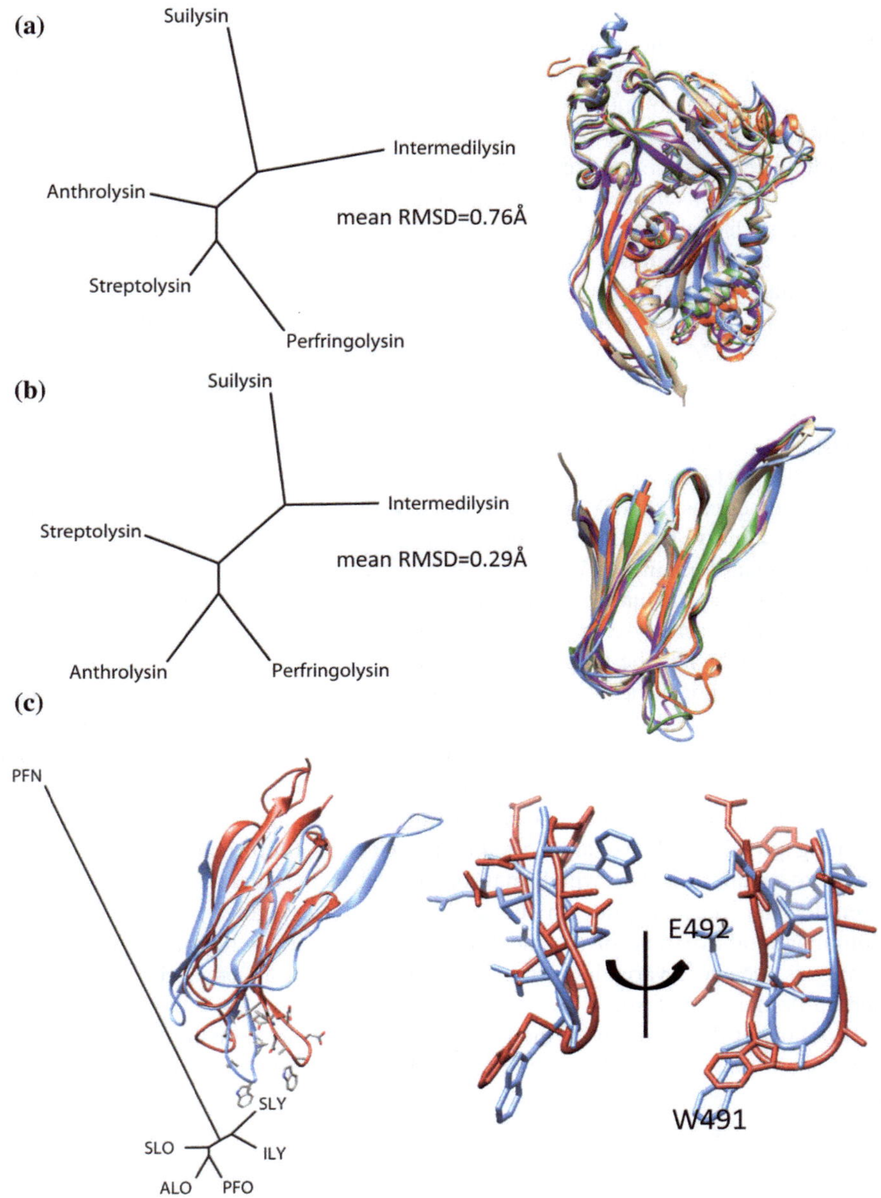

Fig. 4.3 Phylogenetic analysis of CDC proteins and comparison to the perforin C2 domain. **a** Structure-based phylogeny for the amino-terminal region constructed as previously described [12, 46] and drawn using FITCH and DRAWTREE as part of the PHYLIP package [47]. The different structures superimposed on the *right* are *coloured* as follows: PFO, *orange*; ILY, *light blue*; ALO, *green*; SLO, *purple*; SLY, tan. This phylogeny based on structure agrees with that derived from sequence data and shown in Fig. 2.5 of Chap. 2. **b** As a for the carboxy-terminal domain of each of the proteins, coloured as before. **c** A repeat of the carboxy-terminal phylogeny including the perforin C2 domain. On the *right* are shown two close-ups of the undecapeptide region of ILY only (*light blue*) superposed with an equivalent structural region from perforin. The common tryptophan (W491 in ILY) and acidic residue (at position 492 in ILY) are indicated

compared to the amino-terminal MACPF/CDC domain and that agrees with their distinctive roles in the biology of these proteins, such that they experience alternative selective pressures. Thus, phylogenies of the CDCs based on structure recapitulate their evolutionary histories and the responsibility of one module (the MAPF/CDC domain) for pore formation whilst the carboxy-terminal domain is separately responsible for membrane binding.

If the carboxy-terminal domain is so well conserved, does it display a conservation such as that previously identified for the MACPF/CDC domain [12]? For example, are the carboxy-terminal domains of the CDCs and perforin related evolutionarily? Do they have a common ancestor?

It is hard to escape the impression that they do, based on a direct comparison between the ILY and perforin carboxy-terminal domains (Figs. 4.3c and 4.4). Thus, as far as one can tell at present the combination of a MACPF/CDC domain and a carboxy-terminal ß-sheeted sandwich has been stable since the last common ancestors of perforin and PFO. Figure 4.3c shows a repeat of the phylogenetic analysis from Fig. 4.3b but including the perforin C2 domain [24]. Firstly, note that the remoteness of ILY and SLY from SLO is stably maintained in this phylogeny. Secondly, the RMSD between the perforin C2 domain and the ILY carboxy-terminal domain is 1.72 Å over 48 residues, which is a plausible agreement for orthologous structures. Thirdly, note the similarity of the undecapeptide loop between CDC carboxy-terminal domain sheets 5 and 6, which as shown in two orientations matches the loop between perforin C2 domain sheets 5 and 6 both in conformation and in the siting of two particular residues—a common tryptophan (Trp491 in ILY and Trp488 in perforin) and a glutamic acid in ILY (Glu492) which is an aspartic acid in perforin (Asp489). This similarity is quite striking and suggests that in both cases the loop conformation and the acidic and tryptophan residues have been conserved because of their role in the interaction of the pore-forming protein with its target membrane. Critical conformational changes are likely to enable this role which may be echoed in the alternative conformations seen for the undecapeptide in the available CDC crystal structures, where (as commented above) one of the most sequentially invariant regions of the protein family is the most variable region in conformation in an otherwise highly conserved domain structure.

Looking at the topology of the CDC carboxy-terminal domain and the perforin C2 domain supports the idea they are related and indicates a conformational switch by which one led to the other. As shown in Fig. 4.4, when coloured from the amino (blue) to the carboxy (red) terminus of the respective domains the strands follow each other apart from a reorientation of strands 7 and 8 which in ILY are threaded with the final strand completing the sheet and with a parallel topology for β strands 1 and 7, while in perforin they are threaded the other way round so that strand 8 is internal to the sheet and paired in an antiparallel fashion with strand 1. In a world of strand and domain swapping [25], the existence of such a topological switch at the edge of a β sandwich, on the end of a polypeptide chain, is not hard to envisage. Figure 4.4b show topology diagrams for each domain which support the claim that they derive from a common ancestor.

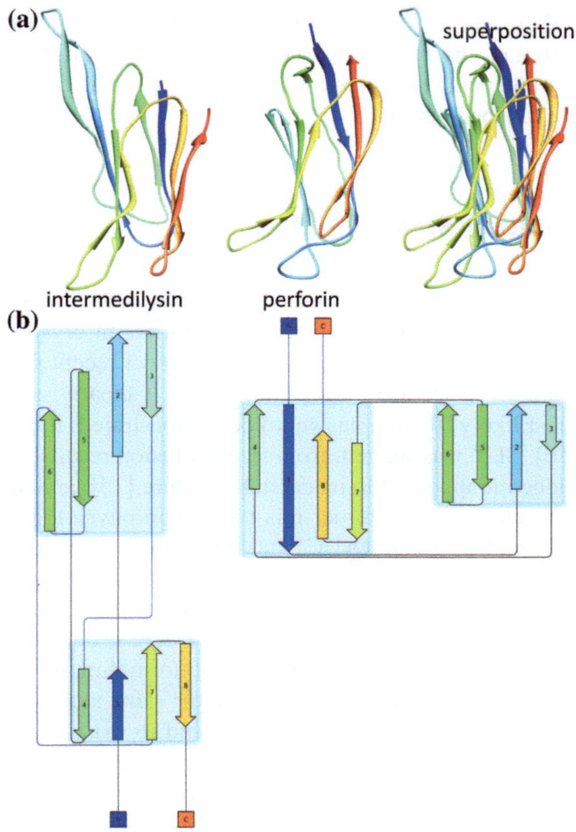

Fig. 4.4 A common topology for the CDC carboxy-terminal domain and perforin C2 domain **a** The two domains are shown, separately but aligned to each other, *coloured* along the ribbon representation of their backbones from the amino terminus (*blue*) to the carboxy terminus (*red*). The actual superposition is shown on the *right*. **b** Topology diagrams for ILY (on the *left*, as labelled) and perforin carboxy-terminal domains. These were drawn using Pro-Origami [48] (http://munk.csse.unimelb.edu.au/pro-origami/)

It therefore seems that the orthologous relationship between CDCs and perforin extends over the whole length of the molecule, encompassing both the MACPF/CDC domain and the carboxy-terminal β-sandwich. They are linked by an $\sim$17 amino acid β-strand in CDCs (one of the three stands in domain 2) and by an $\sim$29 amino acid EGF-like domain in perforin which are routed from the same region of the MACPF/CDC canonical domain. It is therefore quite likely that the CDC and perforin linking sequences also share a common evolutionary origin as well, though due to a lack of selection for its maintenance in any particular form (flexible linkers being a variable species) this is lost to analysis. But in any case, the combination of a MACPF/CDC superfamily pore-forming domain and a carboxy-terminal β-sheeted sandwich seems possibly to be as deeply conserved as the individual domains appear to be. It is worth considering that the carboxy-terminal domains of the apicomplexan perforin-like proteins—described as a novel apicomplexan perforin-like protein C-terminal ß-pleated sheet domain or APC-ß domain [26]—might also turn out to have a common evolutionary origin with the CDC carboxy-terminal domain. This newly-identified similarity between the

CDCs and perforin casts further doubt on the idea that perforin forms pores in a different orientation to the CDCs [24] but rather supports the idea that, like complement, the orientations of MACPF and CDC monomers within their pores are all the same [12].

Solution State of CDCs

Most structures of CDCs have (like perforin) turned out to be monomeric proteins, which have sometimes appears as dimers in crystals but in configurations which are clearly non-biological. Thus, the most recent crystal structure, of streptolysin, contains a head-to-head dimer which is clearly non-physiological, and thus it is most likely (though has yet to be demonstrated) that SLO is monomeric in solution [16]. SLY was judged, from size exclusion results, to be monomeric in solution as indeed it appeared in its crystal structure [15], and although ALO was observed as a crystallographic dimer this too is non-physiological, as shown using analytical ultracentrifugation (AUC) and multi-angle light scattering (MALS) [14]. Clearly, techniques such as AUC and MALS are superior to judgements based on size exclusion alone [15] or lacking experimental evidence [16], especially given the elongated conformation adopted by CDCs, but there is in truth no evidence that any of these toxins, or PLY [27–29], form dimers in solution. PFO, however, does: all three crystal forms solved have the same antiparallel dimer present within them [4, 27, 30] which has been confirmed to exist in solution using AUC and small-angle X-ray scattering [29]. There was no evidence of concentration dependence of PFO's dimeric sedimentation coefficient which indicates that the protein is mostly dimeric at all concentrations. The lowest concentration tested was approximately 5 μM (0.28 mg/ml) and thus dimerization of PFO is certainly not limited to high concentrations [29], as has recently been suggested [16]. As previously discussed, being dimeric gives PFO a highly efficient way to maintain itself in an inert state in solution before it binds to a membrane [27].

Oligomeric Structure of CDCs

The heterogeneous sizes of oligomers—both mixes of different sizes of rings and also incomplete rings (that is, arcs)—is a well-known characteristic of CDCs [5, 18, 19, 22] (Fig. 4.5a). Thus, techniques which require homogeneity—most importantly X-ray crystallography—are going to be very difficult to perform on such structures. As a result, two techniques have played a significant role in characterisation of CDC oligomeric structure, single particle cryo-electron microscopy [19, 22] and atomic force microscopy [18]. These have together provided images of both isolated CDC oligomers, in pre-pore and pore states, and of average structures with lower levels of noise and better levels of detail.

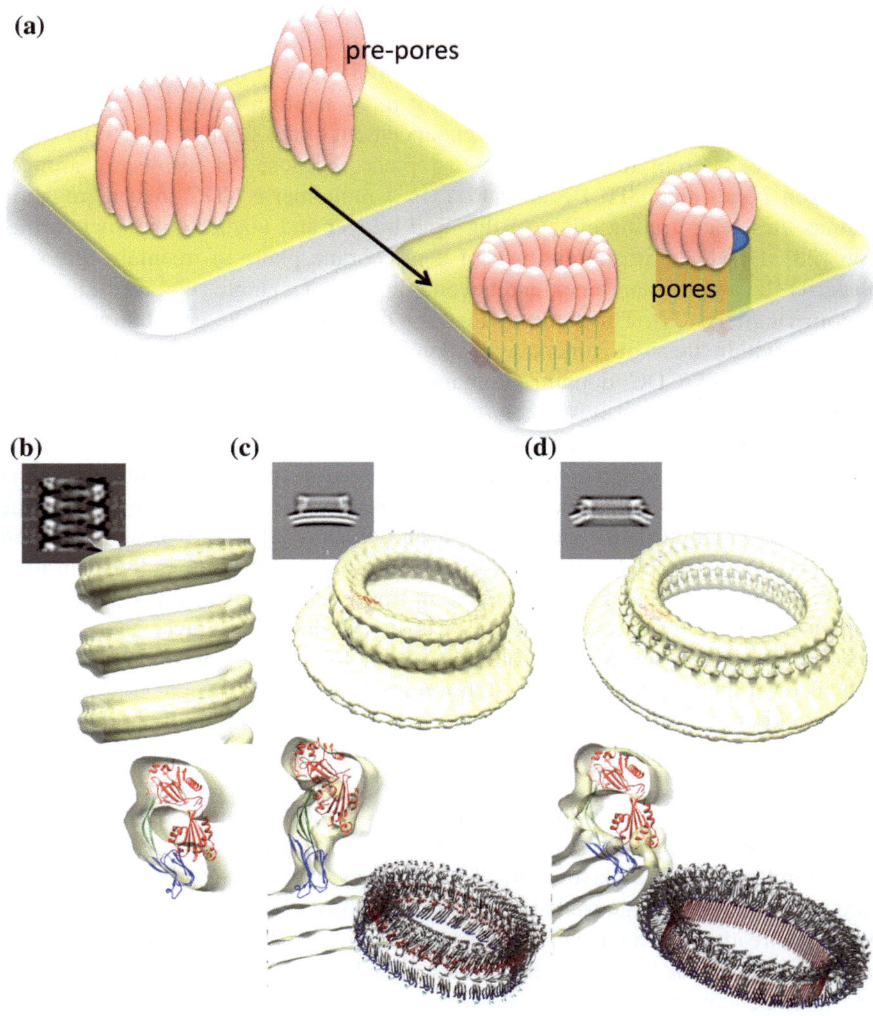

Fig. 4.5 Oligomeric structures formed by CDCs, studied using PLY **a** Pore formation involves oligomerisation on the membrane surface into *ring-shaped* oligomers of varying degrees of completeness—some are full *rings* and some *arcs*. These pre-pores convert to pores in a concerted conformational change. **b** Structure of the first CDC oligomer to have its three-dimensional structure determined, a helical form of PLY [22]. The *grayscale* inset is a projection of the structure, the main image a surface rendering of the map and below is a sectional view showing a fit with the PFO *crystal* structure [4] (domains 1 and 3 *red*, domain 2 *green* and domain 4 *blue*). c The pre-pore state (the inset again shows a projection), a fit with the PFO crystal structure and a model of the pre-pore oligomer [19]. The sectional fit is coloured as in panel b; in the oligomeric model the pore-forming TMH helices are coloured *red* and the undecapeptide *blue*. **d** The pore state, shown as in panel c [19]

The first CDC oligomeric structure was not a native state, but a helical assembly discovered by chance during the preparation of samples of PLY for small-angle neutron scattering experiments [28] (Fig. 4.5b). The heavy water (^2H_2O) into which samples are dialyzed for many neutron scattering experiments causes proteins to aggregate due to an enhancement of the same hydrophobic effect which drives protein folding in the first place, because ^2H results in stronger hydrogen bonds than ^1H. Because PLY is a monomeric CDC this aggregation mimics the concentrating effect of the lipid bilayer and leads to oligomer formation [28]. However, in the absence of a membrane plane to maintain oligomer planarity the assemblies can extend beyond the size of a single ring, resulting in an oligomeric helix [22]. This helix defined for the first time the orientation of CDC subunits within their oligomers, with domain 2 facing outwards and the pore-forming MACPF/CDC domain (domains 1 and 3) facing inwards [22]. The structure visualised was most likely close to the pore-forming state though obviously without the pore-forming TMHs being inserted into a membrane; instead they likely were disordered in structure. In the same paper a membrane-bound oligomeric state was described which, in retrospect, was the pre-pore form of the protein but which at that stage had not been defined [22] as previously discussed [27]. These pre-pore oligomers were formed at room temperature and visualisation of these two oligomeric forms revealed an important aspect of the kinetic determination of pore formation—that the viscosity of the lipid bilayer at room temperature prevented successful protein insertion to form a pore, whereas in the helical form there was nothing to stop the protein passing straight to a pore-like conformation. In turn, this means that the membrane itself has a role in imposing a prepore-to-pore transition mechanism for membrane penetration and channel formation.

In subsequent studies, in order to capture the pore structure as well as the pre-pore one, we used incubation at 37 °C to enable efficient pore formation due to a more fluid membrane bilayer [19]. The pre-pore structure (Fig. 4.5c) showed a conformation very similar to that of the protein in isolation while the pore structure was, as already indicated, very similar to that of the helical conformation and significantly refolded, with a bending over of the domain 1 region to allow the TMHs to penentrate the membrane (Fig. 4.5d). In the paper describing the pore structure [19] this was modelled as a dogleg bending of domain 2, but the reality is unknown; similarly we have modelled the transbilayer ß barrel formed out of the TMHs without a sheer to the strand orientation but, equally, there will in reality be a sheer as this is a fundamental property of such structures due to the native twist of the ß sheet [31, 32] (see also Chap. 5).

Atomic force microscopy has also, in the process of providing structural snapshots, given critical functional insights into the activity of CDCs. It was possible to image individual pre-pores on a bilayer and measure their reduction in height above the membrane following on pore formation [18]; this allowed a clear definition of pre-pore and pore states. By the introduction of a reversible disulphide bond locking the pre-pore state it was also possible, having formed pre-pores, to trigger their concerted conversion to pores [18]. The main difference between pre-pore and pore structures is the ability (if they are not yet fully circularised) to add monomers to pre-pore structures but not to the pore structures. This is shown most clearly by the fact that locking the pre-pore state produced a much greater preponderance of rings among the pre-pores rather than the incomplete arcs (Fig. 4.5a). As discussed previously [27], this structural study therefore reveals a kinetic mechanism which governs the pre-pore to pore transition, whereby conversion between the two states may occur at any point after oligomerisation has begun up to and including completion of a pre-pore ring of subunits, a complete oligomer. When it does occur is determined then by the availability of subunits to bring such a completion about.

Oligomers and Pores

As discussed elsewhere [12, 33], it is necessary to be clear on the difference between an oligomer and a pore. A "pore" is a functional description—something containing a lesion or channel across a membrane bilayer—whereas an oligomer is a kind of structure. This distinction is especially important with the CDCs because of their capacity to generate pores of various sizes, both in vitro [34, 35] and in cells [36, 37]. The chapter by Sonnen addresses some of these issues in more detail, because it is feature of the MACPF wing of the superfamily of proteins as well—as we have ourselves shown [38–40]. The basis on which this variability in pore size might be achieved seems most likely to be the incomplete oligomers—or arcs—of protein which could form pores defined by a combination of the partial protein oligomer and part of the membrane bilayer [12, 33]. In general, a lot of cold water has been poured on this idea but it is an idea whose time may now have come. Cryo-electron tomography of pneumolysin pores has provided snapshots of arcs forming pores (Andreas Sonnen and Robert Gilbert, unpublished) while in a completely different system (the Bax pro-apoptotic protein) there is already a structure determined by a similar approach for a proteolipidic pore [41, 42] confirming the insights of others [43, 44].

Conclusion

Membranes are inherently dynamic environments and thinking about how pore-forming proteins like members of the MACPF/CDC superfamily form pores needs to take this into account [33]. Thus, although structural snapshots of monomeric or oligomeric states of CDCs provide the best framework for understanding their mechanism, teasing that mechanism out in detail requires additional techniques. These techniques—such as biophysical measurements of pore formation, or functional studies, or spectroscopic studies (and see Chap. 15 later in this volume for some examples)—provide essential critiques of the structures on their own because they are able to access molecular dynamics in a way electron and atomic force microscopy, and X-ray crystallography, cannot. Insights into CDC molecular dynamics must be allowed to feedback on our understanding of what structural results mean because, after all, if a structural model for pore formation is incapable of explaining functional data then it is inadequate.

References

1. Gilbert RJ (2002) Pore-forming toxins. Cell Mol Life Sci 59:832–844
2. Menestrina G, Bashford CL, Pasternak CA (1990) Pore-forming toxins: experiments with *S. aureus* alpha-toxin, *C. perfringens* theta-toxin and *E. coli* haemolysin in lipid bilayers, liposomes and intact cells. Toxicon 28:477–491
3. Tweten RK (2005) Cholesterol-dependent cytolysins, a family of versatile pore-forming toxins. Infect Immun 73:6199–6209
4. Rossjohn J, Feil SC, McKinstry WJ, Tweten RK, Parker MW (1997) Structure of a cholesterol-binding, thiol-activated cytolysin and a model of its membrane form. Cell 89:685–692
5. Morgan PJ, Hyman SC, Byron O, Andrew PW, Mitchell TJ, Rowe AJ (1994) Modeling the bacterial protein toxin, pneumolysin, in its monomeric and oligomeric form. J Biol Chem 269:25315–25320
6. Morgan PJ, Hyman SC, Rowe AJ, Mitchell TJ, Andrew PW, Saibil HR (1995) Subunit organisation and symmetry of pore-forming, oligomeric pneumolysin. FEBS Lett 371:77–80
7. Olofsson A, Hebert H, Thelestam M (1993) The projection structure of perfringolysin O (*Clostridium perfringens* theta-toxin). FEBS Lett 319:125–127
8. Palmer M, Saweljew P, Vulicevic I, Valeva A, Kehoe M, Bhakdi S (1996) Membrane-penetrating domain of streptolysin O identified by cysteine scanning mutagenesis. J Biol Chem 271:26664–26667
9. Shepard LA, Heuck AP, Hamman BD, Rossjohn J, Parker MW, Ryan KR, Johnson AE, Tweten RK (1998) Identification of a membrane-spanning domain of the thiol-activated pore-forming toxin *Clostridium perfringens* perfringolysin O: an alpha-helical to beta-sheet transition identified by fluorescence spectroscopy. Biochemistry 37:14563–14574
10. Shatursky O, Heuck AP, Shepard LA, Rossjohn J, Parker MW, Johnson AE, Tweten RK (1999) The mechanism of membrane insertion for a cholesterol-dependent cytolysin: a novel paradigm for pore-forming toxins. Cell 99:293–299
11. Bhakdi S, Tranum-Jensen J, Sziegoleit A (1985) Mechanism of membrane damage by streptolysin-O. Infect Immun 47:52–60
12. Gilbert RJ, Mikelj M, Dalla Serra M, Froelich CJ, Anderluh G (2013) Effects of MACPF/CDC proteins on lipid membranes. Cell Mol Life Sci 70:2083–2098
13. Polekhina G, Giddings KS, Tweten RK, Parker MW (2005) Insights into the action of the superfamily of cholesterol-dependent cytolysins from studies of intermedilysin. Proc Nat Acad Sci USA 102:600–605
14. Bourdeau RW, Malito E, Chenal A, Bishop BL, Musch MW, Villereal ML, Chang EB, Mosser EM, Rest RF, Tang WJ (2009) Cellular functions and X-ray structure of anthrolysin O, a cholesterol-dependent cytolysin secreted by *Bacillus anthracis*. J Biol Chem 284:14645–14656
15. Xu L, Huang B, Du H, Zhang XC, Xu J, Li X, Rao Z (2010) Crystal structure of cytotoxin protein suilysin from *Streptococcus suis*. Protein Cell 1:96–105
16. Feil SC, Ascher DB, Kuiper MJ, Tweten RK, Parker MW (2013) Structural studies of *streptococcus pyogenes* streptolysin o provide insights into the early steps of membrane penetration. J Mol Biol (in press). doi:10.1016/j.jmb.2013.11.020

17. Koster S, Hudel M, Chakraborty T, Yildiz O (2013) Crystallization and X-ray crystallographic analysis of the cholesterol-dependent cytolysin listeriolysin O from *Listeria monocytogenes*. Acta Crystallogr Sect F Struct Biol Cryst Commun 69:1212–1215
18. Czajkowsky DM, Hotze EM, Shao Z, Tweten RK (2004) Vertical collapse of a cytolysin prepore moves its transmembrane beta-hairpins to the membrane. EMBO J 23:3206–3215
19. Tilley SJ, Orlova EV, Gilbert RJ, Andrew PW, Saibil HR (2005) Structural basis of pore formation by the bacterial toxin pneumolysin. Cell 121:247–256
20. Soltani CE, Hotze EM, Johnson AE, Tweten RK (2007) Structural elements of the cholesterol-dependent cytolysins that are responsible for their cholesterol-sensitive membrane interactions. Proc Nat Acad Sci USA 104:20226–20231
21. Johnson S, Brooks NJ, Smith RA, Lea SM, Bubeck D (2013) Structural basis for recognition of the pore-forming toxin intermedilysin by human complement receptor CD59. Cell Rep 3:1369–1377
22. Gilbert RJ, Jimenez JL, Chen S, Tickle IJ, Rossjohn J, Parker M, Andrew PW, Saibil HR (1999) Two structural transitions in membrane pore formation by pneumolysin, the pore-forming toxin of *Streptococcus pneumoniae*. Cell 97:647–655
23. Rossjohn J, Gilbert RJ, Crane D, Morgan PJ, Mitchell TJ, Rowe AJ, Andrew PW, Paton JC, Tweten RK, Parker MW (1998) The molecular mechanism of pneumolysin, a virulence factor from *Streptococcus pneumoniae*. J Mol Biol 284:449–461
24. Law RH, Lukoyanova N, Voskoboinik I, Caradoc-Davies TT, Baran K, Dunstone MA, D'Angelo ME, Orlova EV, Coulibaly F, Verschoor S, Browne KA, Ciccone A, Kuiper MJ, Bird PI, Trapani JA, Saibil HR, Whisstock JC (2010) The structural basis for membrane binding and pore formation by lymphocyte perforin. Nature 468:447–451
25. Sonnen AF, Yu C, Evans EJ, Stuart DI, Davis SJ, Gilbert RJ (2010) Domain metastability: a molecular basis for immunoglobulin deposition? J Mol Biol 399:207–213
26. Kafsack BF, Carruthers VB (2010) Apicomplexan perforin-like proteins. Commun Integr Biol 3:18–23
27. Gilbert RJ (2005) Inactivation and activity of cholesterol-dependent cytolysins: what structural studies tell us. Structure 13:1097–1106
28. Gilbert RJ, Heenan RK, Timmins PA, Gingles NA, Mitchell TJ, Rowe AJ, Rossjohn J, Parker MW, Andrew PW, Byron O (1999) Studies on the structure and mechanism of a bacterial protein toxin by analytical ultracentrifugation and small-angle neutron scattering. J Mol Biol 293:1145–1160
29. Solovyova AS, Nollmann M, Mitchell TJ, Byron O (2004) The solution structure and oligomerization behavior of two bacterial toxins: pneumolysin and perfringolysin O. Biophys J 87:540–552
30. Rossjohn J, Polekhina G, Feil SC, Morton CJ, Tweten RK, Parker MW (2007) Structures of perfringolysin O suggest a pathway for activation of cholesterol-dependent cytolysins. J Mol Biol 367:1227–1236
31. Aleshin AE, Schraufstatter IU, Stec B, Bankston LA, Liddington RC, Discipio RG (2012) Structure of complement C6 suggests a mechanism for initiation and unidirectional, sequential assembly of the membrane attack complex (MAC). J Biol Chem 287:10210–10222
32. Lovelace LL, Cooper CL, Sodetz JM, Lebioda L (2011) Structure of human C8 protein provides mechanistic insight into membrane pore formation by complement. J Biol Chem 286:17585–17592
33. Gilbert RJ (2010) Cholesterol-dependent cytolysins. Adv Exp Med Biol 677:56–66
34. Korchev YE, Bashford CL, Pederzolli C, Pasternak CA, Morgan PJ, Andrew PW, Mitchell TJ (1998) A conserved tryptophan in pneumolysin is a determinant of the characteristics of channels formed by pneumolysin in cells and planar lipid bilayers. Biochem J 329:571–577
35. Marchioretto M, Podobnik M, Dalla Serra M, Anderluh G (2013) What planar lipid membranes tell us about the pore-forming activity of cholesterol-dependent cytolysins. Biophys Chem 182:64–70

36. Birmingham CL, Canadien V, Kaniuk NA, Steinberg BE, Higgins DE, Brumell JH (2008) Listeriolysin O allows *Listeria monocytogenes* replication in macrophage vacuoles. Nature 451:350–354
37. El-Rachkidy RG, Davies NW, Andrew PW (2008) Pneumolysin generates multiple conductance pores in the membrane of nucleated cells. Biochem Biophys Res Comm 368:786–792
38. Metkar SS, Wang B, Catalan E, Anderluh G, Gilbert RJ, Pardo J, Froelich CJ (2011) Perforin rapidly induces plasma membrane phospholipid flip-flop. PLoS One 6:e24286
39. Praper T, Sonnen AF, Kladnik A, Andrighetti AO, Viero G, Morris KJ, Volpi E, Lunelli L, Dalla Serra M, Froelich CJ, Gilbert RJ, Anderluh G (2011) Perforin activity at membranes leads to invaginations and vesicle formation. Proc Nat Acad Sci USA 108:21016–21021
40. Praper T, Sonnen AF, Viero G, Kladnik A, Froelich CJ, Anderluh G, Dalla Serra M, Gilbert RJ (2011) Human perforin employs different avenues to damage membranes. J Biol Chem 286:2946–2955
41. Volkmann N, Marassi FM, Newmeyer DD, Hanein D (2013) The rheostat in the membrane: BCL-2 family proteins and apoptosis. Cell Death Differ (in press). doi:10.1038/cdd.2013.153
42. Xu XP, Zhai D, Kim E, Swift M, Reed JC, Volkmann N, Hanein D (2013) Three-dimensional structure of Bax-mediated pores in membrane bilayers. Cell Death Dis 4:e683
43. Basanez G, Nechushtan A, Drozhinin O, Chanturiya A, Choe E, Tutt S, Wood KA, Hsu Y, Zimmerberg J, Youle RJ (1999) Bax, but not Bcl-xL, decreases the lifetime of planar phospholipid bilayer membranes at subnanomolar concentrations. Proc Nat Acad Sci USA 96:5492–5497
44. Qian S, Wang W, Yang L, Huang HW (2008) Structure of transmembrane pore induced by Bax-derived peptide: evidence for lipidic pores. Proc Nat Acad Sci USA 105:17379–17383
45. Pettersen EF, Goddard TD, Huang CC, Couch GS, Greenblatt DM, Meng EC, Ferrin TE (2004) UCSF chimera—a visualization system for exploratory research and analysis. J Comput Chem 25:1605–1612
46. Riffel N, Harlos K, Iourin O, Rao Z, Kingsman A, Stuart D, Fry E (2002) Atomic resolution structure of moloney murine leukemia virus matrix protein and its relationship to other retroviral matrix proteins. Structure 10:1627–1636
47. Felsenstein J (1997) An alternating least squares approach to inferring phylogenies from pairwise distances. Syst Biol 46:101–111
48. Stivala A, Wybrow M, Wirth A, Whisstock JC, Stuckey PJ (2011) Automatic generation of protein structure cartoons with pro-origami. Bioinformatics 27:3315–3316

Chapter 5
Perfringolysin O Structure and Mechanism of Pore Formation as a Paradigm for Cholesterol-Dependent Cytolysins

Benjamin B. Johnson and Alejandro P. Heuck

Abstract Cholesterol-dependent cytolysins (CDCs) constitute a family of pore forming toxins secreted by Gram-positive bacteria. These toxins form transmembrane pores by inserting a large β-barrel into cholesterol-containing membrane bilayers. Binding of water-soluble CDCs to the membrane triggers the formation of oligomers containing 35–50 monomers. The coordinated insertion of more than seventy β-hairpins into the membrane requires multiple structural conformational changes. Perfringolysin O (PFO), secreted by *Clostridium perfringens*, has become the prototype for the CDCs. In this chapter, we will describe current knowledge on the mechanism of PFO cytolysis, with special focus on cholesterol recognition, oligomerization, and the conformational changes involved in pore formation.

Keywords β-barrel · Cholesterol · Cholesterol-dependent cytolysins · Listeriolysin O · Membrane · Perfringolysin O · Pneumolysin · Pore formation · Streptolysin O · Toxin

Abbreviations

CDCs	Cholesterol-dependent cytolysins
D1, D2, D3, and D4	Domain 1, domain 2, domain 3, and domain 4
L1, L2, and L3	Loop 1, loop 2, and loop 3
PFO	Perfringolysin O
TMH1 and TMH2	Transmembrane hairpin 1 and transmembrane hairpin 2

B. B. Johnson · A. P. Heuck (✉)
Department of Biochemistry and Molecular Biology, University of Massachusetts,
710 N. Pleasant St., Lederle GRT, Amherst, MA 01003, USA
e-mail: apheuck@biochem.umass.edu

G. Anderluh and R. Gilbert (eds.), *MACPF/CDC Proteins - Agents of Defence, Attack and Invasion*, Subcellular Biochemistry 80, DOI: 10.1007/978-94-017-8881-6_5,

Introduction

Perfringolysin O (PFO) is the prototypical example of a growing family of bacterial pore-forming toxins known as the Cholesterol Dependent Cytolysins (CDCs, [19, 25, 81]). CDCs are secreted by Gram-positive bacteria including *Bacillus, Listeria, Lysinibacillus*, *Paenibacillus*, *Brevibacillus*, *Streptococcus*, *Clostridium*, *Gardnerella*, *Arcanobacterium*, and *Lactobacillus* (see [25, 35, 64]). There are 30 members of the CDC family reported for Gram-positive bacteria and, surprisingly, two CDC-coding DNA sequences have been found in the Gram-negative *Desulfobulbus propionicus* and *Enterobacter lignolyticus*. However, in contrast with the Gram-positive bacteria that produce CDCs, the Gram-negative ones have not been shown to inhabit humans or indeed animals of any kind [29]. Despite their extremely diverse lineage, the majority of CDCs show an amino acid sequence identity greater than 39 % when compared to PFO [25]. The C-terminus (domain 4 or D4) of PFO is responsible for membrane binding and is the domain with the highest percentage of amino acid identity when sequences are compared with other CDC members.

Most CDCs possess a cleavable signal sequence which targets the toxins for secretion to the extracellular medium. The secreted water-soluble toxins diffuse until encountering their target, a cholesterol-containing mammalian cell membrane (Fig. 5.1, step I). An exception to the cholesterol requirement for targeting was found for intermedilysin which uses the human receptor CD59 for membrane targeting [17]. However this toxin still requires cholesterol to insert into the membrane and form a transmembrane pore [16]. After binding, CDC monomers diffuse across the surface of the membrane and interact reversibly with other monomers until formation of a stable dimer (Fig. 5.1, step II, [30, 54]. These initial dimers grow by the incorporation of additional monomers into a large ring shaped complex (known as the pre-pore complex (Fig. 5.1, step III, [75]. Each of these complexes contains 35–50 monomers, and upon insertion into the membrane, they form large β-barrel pores (up to 250–300 Å in diameter, Fig. 5.1, step IV, [8, 73, 74]).

In this chapter, we will discuss CDCs through the lens of one of the most studied and well understood CDCs, PFO [19, 25, 81]. We will focus on the targeting of PFO to cholesterol-containing membranes and on the multiple conformational changes the protein undergoes in order to spontaneously transition from a water-soluble monomer to a large multimeric transmembrane complex. We will also comment on the most recent findings about the PFO cytolytic mechanism.

PFO is secreted by *Clostridium perfringens* as a 52.6 kDa protein, and the crystal structure of the water-soluble monomer revealed four distinct domains (Fig. 5.2a, [68]). The overall three dimensional structure observed for PFO is conserved for all other CDCs whose high resolution structures have been solved [5, 57, 82]. Domain 1 (D1) consists of the top portion of the elongated molecule. D1 is the only domain that does not undergo large structural rearrangements during pore formation. Domain 2 (D2) adopts mostly a β-strand secondary structure that collapses vertically during pore-formation to allow the insertion of the β-hairpins that form the transmembrane β-barrel [7, 8, 63, 80]. Domain 3 (D3) contains both the

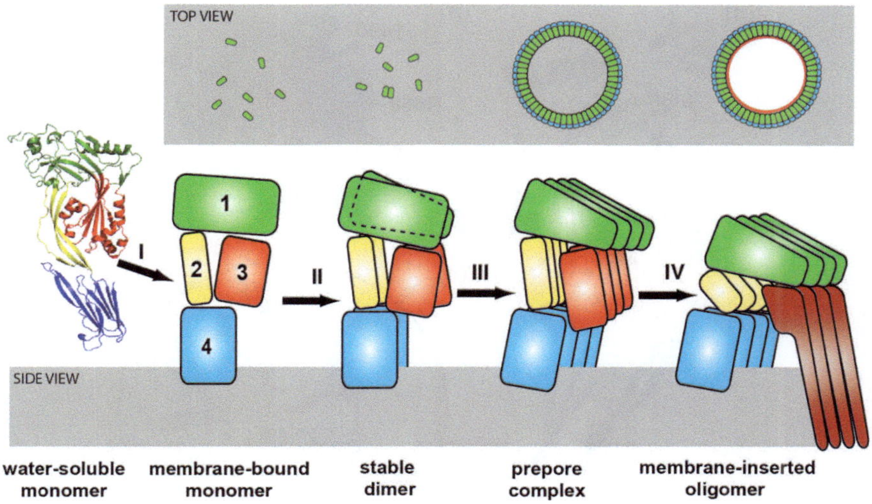

Fig. 5.1 Cartoon representation of the different steps/intermediates identified for the PFO mechanism of pore formation. A water-soluble monomer is secreted by the bacterium and binds to the target membrane via D4 (step I). Membrane-bound monomers diffuse across the membrane surface interacting transiently until they form a stable dimer (step II). The initial dimer starts growing with the addition of other monomers until completion of a circular ring or pre-pore complex (step III). In the last step, each monomer inserts two amphipathic transmembrane hairpins into the bilayer aided by the vertical collapse of D2 forming a large β-barrel pore (step IV). Domains are numbered and color coded as follows: D1 (*green*), D2 (*yellow*), D3 (*red*), and D4 (*blue*). Only a few PFO monomers are shown in the *side view* at the *bottom* to simplify the figure. On the *top* is a schematic *top view* for each step of the pore formation mechanism shown below. The membrane bilayer is depicted by a *gray rectangle*

β-sheet involved in the oligomerization of the toxin and the six short α-helixes that unfurl into two amphipathic β-hairpins to form the β-barrel [62, 73, 74]. Domain 4 (D4) consists of a β-sandwich and contains a conserved Trp rich loop as well as three other conserved loops at the distal tip (Fig. 5.2b and c). D4 is responsible for cholesterol recognition and the initial binding of the toxin to the membrane [22, 61].

Membrane Recognition and Binding

One of the unique features of the mammalian cell membrane is the presence of cholesterol. *C. perfringens* and other pathogens have exploited this property of mammalian membranes to target their CDCs without compromising the integrity of their own membranes. It has long been known that binding of PFO and other CDCs requires high levels of cholesterol in model membranes prepared with phosphatidylcholine [1, 47, 67]. Based on the requirement of high cholesterol levels, targeting of PFO to cholesterol rich domains or “lipid rafts” has been

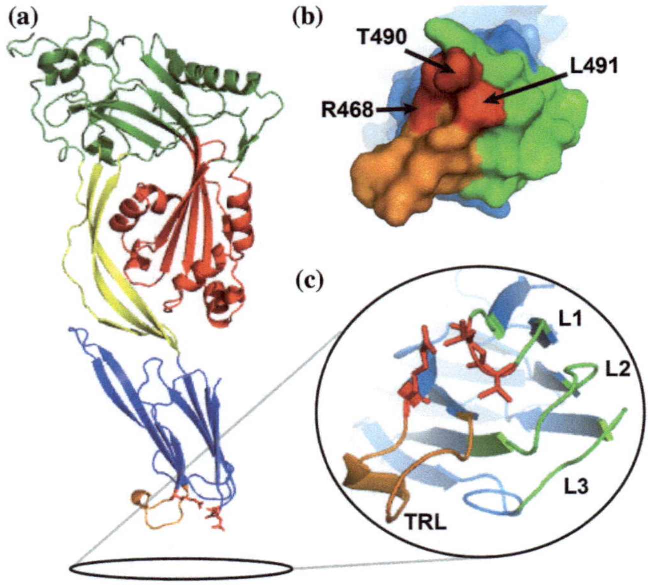

Fig. 5.2 Three dimensional structure of PFO showing the location of important elements that modulate cholesterol interaction. **a** ribbon representation of the water-soluble PFO monomer with domains colored as indicated in Fig. 5.1. Also in color are three key residues that influence cholesterol interaction T490, L491, R468 (*Red*), and the Trp rich loop (TRL, *orange*). **b** A view of the tip of D4 from the *bottom* showing the exposed surface of the Trp rich loop residues (*orange*), the three small loops (*green*), and the residues indicated in A (*red*). **c** The ribbon rendering of the same *bottom* view of D4 shown in B. PFO (1PFO) structure representation was rendered using PyMol (DeLano Scientific LLC)

suggested [49]. However, it has become clear that exposure of cholesterol at the membrane surface is a key factor to trigger PFO binding, and "lipid rafts" may not be necessary for toxin binding [15, 26, 44, 46, 52, 76]. Moreover, the localization of PFO oligomers on the membrane surface may change from the original binding site after insertion of the β-barrel [39, 45].

It has also been shown that the binding of PFO to cholesterol containing membranes is modulated by amino acids located in the loops that connect the β-strands at the bottom of D4 (Fig. 5.2c, [11, 13, 34, 44, 78, 79]), however the precise molecular mechanism of CDC-cholesterol interaction remains poorly understood.

Cholesterol Recognition

The first step in the binding of a water-soluble CDC to the membrane involves the formation of a non-specific collisional complex between a monomer and the lipid bilayer. This step is diffusional and electrostatic interactions may play an important role (e.g. introduction or elimination of negative charges alters binding, [34, 79]). While on the membrane surface, insertion of non-polar and aromatic amino acids and/or specific interactions with membrane lipids may anchor the protein to the membrane [6]. However, non-polar amino acids are rarely exposed on the surface of water-soluble proteins, and therefore conformational changes are often required to expose these residues to the hydrophobic core of the membrane bilayer. As a result, multiple conformational changes are triggered during the transition of PFO from a water-soluble monomer to a membrane-inserted oligomer.

In model membranes prepared exclusively with phosphatidylcholine >30 mol% cholesterol is required to trigger binding of PFO [22, 47], streptolysin O [67], lysteriolysin O [3], or tetanolysin [1] but the amount of cholesterol needed does vary depending on membrane phospholipid composition. The "cholesterol threshold" can be reduced by the presence of double bonds in the acyl chains of the phospholipids or by the presence of phospholipids with smaller head groups [14, 15, 46]. Therefore, it has become clear that modifications to the phospholipids that form the membrane can alter the ability of PFO to detect cholesterol at the membrane surface [44]. Despite their influence on membrane binding the presence of phospholipids is not required, since cholesterol alone (in the absence of any other lipid) is sufficient to trigger PFO oligomerization and formation of ring-like complexes ([26] and references therein). Accessibility of cholesterol at the membrane surface seems to be the key to trigger the binding of PFO to membranes [15, 44, 52, 76].

What is Cholesterol Accessibility?

It has long been recognized that cholesterol modulates important membrane properties including permeability, fluidity, thickness, and domain formation, among others. The cholesterol-dependent association of certain proteins and peptides with membranes has often been associated with the effect of cholesterol on one or more of these membrane physical properties. More recently, studies with molecules that directly interact with cholesterol, like cyclic sugar polymers (e.g. cyclodextrins, [60]), enzymes (e.g. cholesterol-oxidase, [38]), and bacterial toxins (e.g. PFO, [15, 44, 46, 76]) have shown that the accessibility of cholesterol at the membrane surface also plays a critical role in cell biology.

Cholesterol is insoluble in aqueous solutions, but it is readily soluble in phospholipid bilayers. The solubility limit of cholesterol in lipid bilayers is dictated by the nature of the phospholipids (acyl chain length and saturation, and head group size, [50]). If the concentration of cholesterol in a bilayer increases to levels above its solubility limit, cholesterol aggregates would form crystals and precipitate out into the aqueous solution [2, 40, 83].

Given its hydrophobic nature, in a lipid bilayer cholesterol orients parallel to the acyl chains of the phospholipids with the only polar group (an OH) facing the surface of the membrane, in close proximity to the phospholipid head groups (Fig. 5.3). At low concentrations, the interaction of cholesterol with other

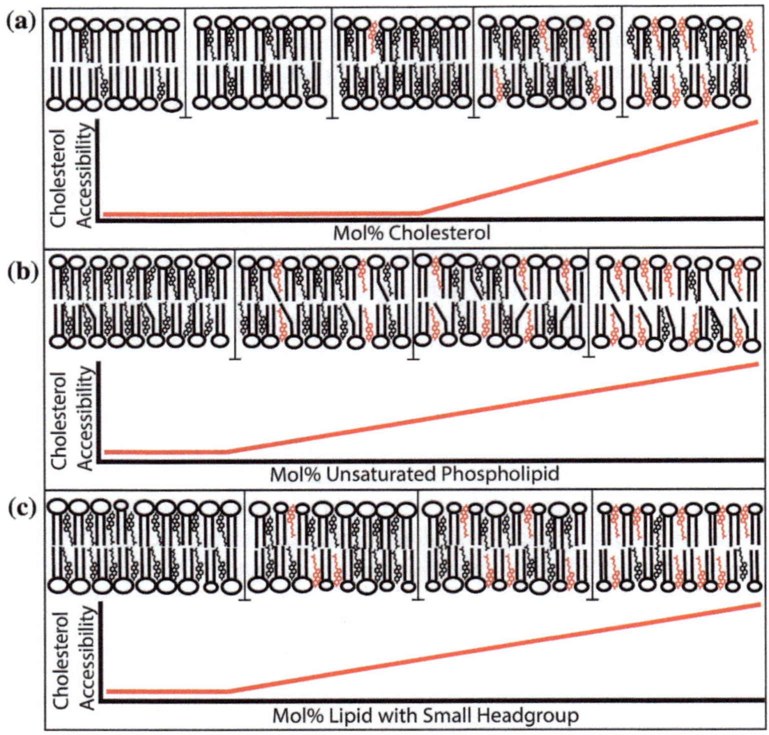

Fig. 5.3 Cholesterol accessibility changes at the membrane surface as a function of the lipid composition. **a** when interactions with other membranes components saturate the accessibility of cholesterol increases at the membrane surface. **b** At constant cholesterol concentration, an increase in the number of double bonds on the acyl chains of the phospholipids increases cholesterol accessibility. **c** At constant cholesterol concentration, an increase in the concentration of phospholipids with smaller head groups increases cholesterol accessibility. The *red lines* depict a hypothetical increase on cholesterol accessibility. The actual change on cholesterol accessibility for each schematic graph may differ from a simple linear response. Some cholesterol molecules are colored *red* to visualize the increase on accessibility but they are indistinguishable from other cholesterol molecules in the membrane

membrane components (lipids, proteins, etc.) reduces the ability of cholesterol to interact with water-soluble molecules at the membrane surface. In other words, when present in low amounts, cholesterol is not accessible to interact with molecules like PFO or cyclodextrins. As the concentration of cholesterol increases, its accessibility remains low until a saturation point is reached. The concentration of cholesterol at the saturation point will depend on the phospholipid or phospholipid mixture present in the membrane (Fig. 5.3a). At this point, a small increase in the sterol concentration causes a sharp increase in the ability of water-soluble molecules to interact with cholesterol [22, 38, 60]. Different models have been proposed to explain changes on cholesterol accessibility at the membrane surface: the cholesterol:phospholipid complex model and the umbrella model [32, 41]. Despite their thermodynamics or steric basis, the models are not mutually exclusive [37, 42]. Recent molecular dynamics simulations of simple membrane models [52] suggested that cholesterol accessibility is related to the overall cholesterol depth within the membrane bilayer and not to the appearance of a new pool of cholesterol molecules (sometimes referred as free cholesterol or active cholesterol). In favor of clarity in this chapter we will refer to the effect that cause an increase in the interaction of cholesterol with water-soluble molecules, as an increase in cholesterol accessibility at the membrane surface (Fig. 5.3).

Domain 4 and the Conserved Loops

PFO D4 consists of two four-stranded β-sheets located at the C-terminus of the protein (Fig. 5.2). There are four loops that interconnect the eight β-strands at the distal tip of the toxin, three short loops (L1, L2, and L3) and a longer Trp rich loop (also known as the conserved undecapeptide). These loops insert into the membrane upon binding and are presumably responsible for the interaction of the toxin with cholesterol [13, 61, 79]. Two of these loops (L2 and L3, Fig. 5.2c) connect β-strands from opposite β-sheets, while L1 and the Trp rich loop connect β-strands from the same β-sheet. L1 and the Trp rich loop are parallel to each other and abutted perpendicularly by L2 forming a pocket in the bottom of the protein. The loops that form the pocket are the most conserved segments in D4, and modifications to any of these loops affects the cholesterol binding properties of PFO ([13, 34, 44, 58], see below). The remaining L3 is far less conserved and distant from the pocket formed by the other three loops.

The Trp rich loop is the longest of the D4 loops, containing 11 residues (E C T G L A W E W W R). It is a signature feature of the CDCs and is highly conserved among species. The three-dimensional structure of this loop seems to be more variable [5, 57, 68, 82], but this may simply reflect its flexibility [57]. Initially, the Trp rich loop was thought to be responsible for cholesterol recognition and binding, and this idea was supported by several studies showing that modifications in it greatly decreased the pore-forming activity of the protein [4, 36, 43, 55, 58, 71, 72]. However, recent studies showed that the other loops in D4 are also responsible for cholesterol recognition [13]. The Trp rich loop has now been suggested to play a role in both the pre-pore to pore transition [79] and the coupling of monomer binding with initiation of the pre-pore assembly [11]. Dowd and colleagues recently showed that modification of a charged amino acid in the Trp rich loop (Arg468, Fig. 5.2b) resulted in complete elimination of the pore-forming activity of PFO and had a significant effect on the membrane binding of the toxin [11, 58]. The R468A PFO derivative was not able to oligomerize after membrane binding, suggesting that this modification disrupts the previously reported allosteric coupling between D4 and D3 [22]. Despite the novel functions assigned to the Trp rich loop, its role in binding cannot be neglected since many modifications to this segment have been shown to have a significant effect in toxin-membrane interaction [58].

Unlike the flexible Trp rich loop, the three-dimensional structure of the other three short loops is more conserved. The L3 is located on the far edge of D4, away from a pocket formed by the Trp rich loop, L1, and L2 (Fig. 5.2c). Modifications introduced into L3 have been shown to have either a negligible effect on cholesterol interaction, or to decrease the amount of cholesterol required for binding [13, 34]. For example, the elimination of the charge of D434 in L3 reduced the amount of cholesterol required to trigger binding [34]. These results suggest that L3 plays a limited role in cholesterol recognition, and its effect on binding may be related to nonspecific interactions with the membrane that stabilize the bound monomer at lower cholesterol levels.

Proposed Cholesterol Recognition Motif

It has been proposed that PFO contains a cholesterol recognition motif composed of only two adjacent amino acids in L1, Thr490 and Leu491 [13]. These amino acids are completely conserved throughout all reported CDCs, and modifications to them greatly affect the binding of the protein to both cell and model membranes [13]. These data suggest a prominent role for these two amino acids in cholesterol recognition, however other well conserved amino acids in that region have not been analyzed yet (e.g. H398, Y402 and A404). Moreover, no direct interaction between cholesterol and these two residues has been shown so far. The fact that both amino acids must be mutated to eliminate binding in a motif containing only two amino acids, coupled with the fact that there are many additional conserved amino acids in the vicinity, suggest that other amino acids may also play a role in cholesterol recognition and form part of the cholesterol binding site. Further studies are required in this area.

The Effect of Cholesterol Accessibility on Perfringolysin O Binding

While cholesterol accessibility is necessary for PFO binding, the analysis of PFO derivatives with modifications on D4 revealed that sterol accessibility is not sufficient to trigger stable PFO-membrane association [34]. As mentioned above, native PFO readily binds to model membranes containing 40 mol% cholesterol (and an equimolar mixture of other phospholipids, see [34]), revealing that cholesterol is accessible at the membrane surface. However, the $PFO^{C459A-L491S}$ derivative was not able to bind to the same membranes, clearly indicating that the cholesterol molecules were not sufficiently accessible to trigger toxin binding.

Binding of the $PFO^{C459A-L491S}$ derivative was recovered when the cholesterol concentration was increased to 50 mol%, suggesting that the affinity of this derivative for cholesterol is lower than that of native PFO, and more cholesterol was required at the membrane surface to trigger stable binding (note that L491 is one of the two amino acids proposed to be essential for cholesterol recognition). It is not clear how cholesterol accessibility varies with increasing amount of cholesterol in the membranes. For simplicity we have represented this variation as a linear function of cholesterol concentration (Fig. 5.3) however cholesterol accessibility may have a non-linear dependence in these systems. Further investigations are required in this area to establish the precise mechanism of PFO-cholesterol interaction as a function of cholesterol accessibility.

Mutations in Domain 4 Affect the Cholesterol Threshold Required to Trigger Binding

The effect that a particular amino acid modification has on PFO activity is often characterized by alterations to the hemolytic properties of the toxin (i.e. pore formation). The EC_{50} or effective toxin concentration required for 50 % lysis is a good indicator of these effects. Another method frequently used to characterize the effect of modifications in PFO derivatives is the percentage of hemolysis as compared with the one obtained for the native toxin under the same experimental conditions. However, it is worth noticing that the latter method is highly influenced by the toxin/red blood cell ratio used in the assay (see Fig. S2 in [34]). Similarly, when the characterization of PFO derivatives is done using model membranes, the protein/lipid ratio should be carefully taken into account.

The effect of a particular amino acid modification is dependent on how much cholesterol is accessible at the membrane surface ([34], and see below). As mentioned above, the $PFO^{C459A-L491S}$ derivative showed negligible binding to liposomes containing 40 mol% cholesterol, but the binding of $PFO^{C459A-L491S}$ was indistinguishable from that of native PFO when membranes containing 50 mol% cholesterol were used. The deleterious effect of many mutations to the toxin can be overcome by an increase in the cholesterol content in the membrane [34, 44]. Interestingly, while most modifications to the D4 loops do not affect the sharp sigmoidal shape of the binding isotherm, the amount of cholesterol required for 50 % binding (or "cholesterol threshold") may change significantly for different PFO derivatives. Therefore, when comparing PFO derivatives using model membranes it is more accurate to quantify the effect of a particular modification as the relative change in the "cholesterol threshold" compared to one obtained for native PFO using the same batch of membranes [34].

We have shown recently that modifications to the binding domain of PFO were able to increase or decrease the "cholesterol threshold" of a PFO derivative [34]. These derivatives were successfully used to detect changes in the cholesterol content of cells and model membranes. While PFO has long been put forth as a probe for cholesterol-rich membranes, the advent of new PFO derivatives with varied "cholesterol thresholds" adds a layer of selectivity to the cholesterol sensing measurements.

Oligomerization on the Membranes Surface

Upon binding to a cholesterol containing membrane, PFO diffuses across the surface of the lipid bilayer and oligomerizes into a large ring shaped complex (Fig. 5.1). This complex contains 35–50 individual PFO monomers ($\sim$250–300 Å inner diameter) and it is referred to as the pre-pore complex [8, 51, 75]. Transition of the pre-pore complex to the final membrane-inserted complex occurs by the insertion of numerous β-hairpins (two per monomer) that perforate the membrane forming a large transmembrane β-barrel [73]. The conformation of the individual PFO monomers in the pre-pore complex is not vastly changed from that of their water-soluble form. There are subtle structural changes triggered by membrane binding and oligomerization of the protein that allow for proper alignment of the monomers and the overall geometry of the pore [62]. Formation of complete rings at the membrane surface seems to be regulated by the relatively slow formation of an initial CDC dimer [30, 54].

Nucleation of the Pre-pore Complex

Oligomerization of the CDCs is triggered by membrane binding and interaction with cholesterol (or exceptionally by interaction with a protein receptor for intermedilysin). Cholesterol binding is sufficient to trigger the conformational changes that unblock the hidden oligomerization interface in the water-soluble monomer [26, 62]. Blockage of the oligomerization interface in the monomer prevents premature oligomerization of the toxin in solution. This regulatory mechanism can be overridden if the monomers are present at high concentration in solution (e.g. for pneumolysin, [20, 77]), but oligomerization is rare at physiological concentrations (i.e. nM range or lower).

The most significant of the conformational changes that follows membrane binding involves the exposure of the core β-sheet that comprises a large part of D3. A short β-strand (β5) separates from the core β-sheet in D3 and exposes β4 for its interaction with the always-exposed β1 strand of another PFO molecule, promoting oligomerization [31, 62]. This conformational change is thought to be facilitated by a pair of Gly residues, Gly324 and Gly325, located in the loop between β4 and β5. These Gly residues are highly conserved, and act as a hinge between the two β-strands [62]. In addition to the separation of β5 from β4, it has been suggested that there is a disruption of the D2 and D3 interface. This disruption is thought to be caused by the rotation of D4 which breaks the weak interactions between D2 and D3. These conformational changes cause the rotation of D3 away from D1 and ultimately the unfurling of the transmembrane hairpins [30, 69].

Hotze et al. [30] have recently suggested that the initial interaction between two membrane-bound PFO monomers is weak and transient and rarely of sufficient length to allow for the transition to a stable dimer with $\beta 1$ and $\beta 4$ strands properly aligned. However, if the transition occurs, addition of further PFO monomers to the complex becomes favorable and oligomerization ensues. Therefore, formation of a stable initial dimer constitutes the rate limiting step in oligomerization that diminishes the formation of uncompleted rings on the membrane surface (Fig. 5.1, step II, [30]). While it has been originally proposed that the separation of $\beta 5$ from $\beta 4$ happens upon membrane binding [62], it is still unclear whether these structural changes are caused by toxin binding or as a consequence of monomer-monomer oligomerization.

Alignment of Core β-Sheets

Addition of monomers to the growing oligomer requires the proper alignment of the core β-strands of the newly added PFO monomer with a β strand at the edge of the oligomer. Formation of hydrogen bonds between adjacent β-strands is energetically favorable but non-specific in nature. If the alignment is not correct, proper growing of the oligomer would not be possible. Thus, it is critical to regulate the alignment of neighbor β-strands to prevent the formation of truncated pre-pore complexes. It has been suggested that the correct alignment of adjacent β-strands among individual PFO monomers is dictated by π-stacking interactions between aromatic residues located in $\beta 1$ (Tyr181) and $\beta 4$ (Phe318) [62]. Modifications on either of these residues have proven to be extremely deleterious to the ability of PFO to form pores [34, 62]. Interestingly, despite being a critical interaction, it appears that only Tyr181 is completely conserved among the CDCs. A few CDC family members do not contain an aromatic residue in the corresponding location of Phe318 in PFO, suggesting that proper alignment of adjacent β-strands may follow another regulatory mechanism for these members (i.e. lectinolysin, intermedilysin, vaginolysin, pneumolysin, mitilysin, pseudoneumolysin, and the two newly identified members, see [25, 35, 64]).

Mechanism of Pore Formation

The last step in the cytolytic mechanism of PFO is the formation of the transmembrane pore. The pre-pore complex transitions into a membrane-inserted complex forming a large transmembrane β-barrel (Fig. 5.1, step IV). This transition involves the unfurling of six short α-helixes located in D3 down to two

Fig. 5.4 A schematic view that depicts the position and orientation of the transmembrane hairpins (TMH1 and TMH2) of PFO in the membrane-inserted complex as determined by Sato et al. [70]. The tilted membrane and the rectangle representing the rest of the PFO molecule are depicted in *gray* and *blue*, respectively. The amino acids that compose the D3 β–sheet and the transmembrane hairpins are depicted by their single letter code and color-coded according to conservation in the CDC family. Amino acids conserved in more than 90 % of the 28 CDC members are shown in *red*, in *orange* if conservation was higher than 70 % but lower than 90 %, and in *black* if not highly conserved. Highlighted in *green* are the aromatic amino acids that are thought to be involved in π-stacking interaction that stabilize PFO pre-pore confirmation and help to align individual PFO monomers for pore formation (see text for details)

amphipathic β-hairpins, and the collapse of D2 to bring down the β-hairpins so they can span the hydrophobic core of the membrane. Large secondary and tertiary structural changes are required to coordinate the insertion of more than 140 individual β-strands and the removal of thousands of lipid molecules to form a β-barrel pore. The use of two β-hairpins per monomer to create a transmembrane β-barrel was first described for PFO [27, 73], and it is likely that this mechanism is also employed by other important pore-forming proteins like the Membrane Attack Complex/Perforin (MACPF) proteins [12, 21, 66].

A key step in the pore formation mechanism of the CDCs is the unfurling of six short α-helixes in D3 to form two extended amphipathic β-hairpins [26, 70, 73]. These conformational changes are necessary to minimize the exposure of hydrophobic residues in the water-soluble form of the PFO monomer [24, 73]. After insertion, the hydrophobic side of the amphipathic hairpin faces the non-polar lipid core, and the hydrophilic side faces the aqueous pore (Fig. 5.4, [73, 74]). The exact molecular mechanism for the pre-pore to pore conversion remains unknown, but thermal energy plays a key factor since at low temperatures (e.g. 4 °C) the PFO oligomer remains locked at the pre-pore complex state [28, 75].

Sato et al. [70] have recently shown that in the pre-pore complex the β-strands that form the transmembrane pore are flexible and mobile. These transmembrane β-hairpins are located high above the membrane in the pre-pore complex [63, 80] and are able to extend and test hydrogen bonding arrangements, but they do not fully form a β-barrel structure [28, 70]. This partially unfolded state of the β-hairpins is thought to represent an intermediate step in pre-pore to membrane-inserted complex transition for PFO [70]. The partial alignment of the β-hairpins in the pre-pore complex may constitute a kinetic barrier that deters the insertion of incomplete rings favoring the formation of complete pre-pore complexes.

The unfurling of the two α-helical bundles into two β-hairpins is favored by the formation of multiple hydrogen bonds, both between hairpins within a PFO monomer and between hairpins on adjacent monomers (Fig. 5.4). Crosslinking experiments revealed that the β-hairpins in the inserted β-barrel adopt a $\sim 20°$ angle to the plane of the membrane, and the adjacent inter-monomer strands align themselves with a shift of two amino acids (Fig. 5.4, [70]). As mentioned above, PFO oligomerization is aided by the proper alignment of β-strands from adjacent monomers via π-stacking interaction between the completely conserved Tyr181 and the highly conserved Phe318. Inspection of the extended hairpins in the β-barrel conformation (Fig. 5.4) revealed another potential π-stacking interaction that may act to stabilize the hairpins in their extended conformation. These are the completely conserved Phe211 in the transmembrane hairpin 1 (TMH1) and highly conserved Phe294 (present in all but vaginolysin, lectinolysin, and intermedilysin of the 30 members) in the transmembrane hairpin 2 (TMH2). Interestingly, the F211C modification decreased the hemolytic activity of PFO [74] and the PFO derivative containing the F294C modification could not be stably produced [73].

The vertical collapse of D2 to bring D3 closer to the membrane surface is another important step in pore formation [7, 63]. In the pre-pore complex, PFO is positioned perpendicular to the membrane leaving D3 about 40 Å above the membrane surface [63, 80]. In this position, the β-strands that form the pore would barely reach the membrane surface and could not penetrate the membrane. The required vertical collapse of D2 would drop D3 to the membrane surface and allow the β-hairpins to punch through the membrane and form a β-barrel. Unfortunately, little is known about the mechanism of the transmembrane β-barrel insertion.

Formation of a pre-pore complex and formation of hydrogen-bonds between adjacent β-strands helps the toxin to overcome the energetic barrier of inserting non-hydrogen bonded β-hairpins [25]. The insertion of incomplete rings may also occur, especially when free monomers are no longer available to complete the circular complex. Trapped metastable arc-like structures may form a pore by themselves, but the formation of a lipid edge at one side of the pore is not energetically favored, and the arcs would have a tendency to associate with other arcs or any proximal complete rings [18, 53, 59].

One of the most intriguing aspects of the CDCs cytolytic mechanism is what happens to the lipids that are displaced to form the pore. The insertion of the β-barrel requires the displacement of more than 1,000 lipid molecules from the membrane [27]. It is not clear how such a large amount of molecules are removed from the center of the pre-pore complex, but the hydrophilic nature of the inner portion of PFO the β-barrel could aid in this process.

Conclusions and Future Perspectives

Despite the lack of high resolution structures for intermediates or for the final membrane-inserted complex, the pore formation mechanism of the CDCs is becoming one of the better understood mechanisms among pore-forming toxins. The elucidation of the three-dimensional structure of the water-soluble PFO monomer [68] in combination with the use of site-directed mutagenesis and fluorescence spectroscopy techniques played a critical role in these advances [23, 33].

More recently a lot of attention has been focused on the study of PFO-cholesterol interaction. These studies revealed that not only changes in cholesterol levels may increase or decrease the ability of the toxin to bind to membranes [22, 47], but also changes in the phospholipid composition of the lipid bilayer [14, 15, 44, 46, 47, 76].

Accessibility of cholesterol at the membrane surface seems to be a key factor to trigger PFO binding. However, simply having accessible sterol molecules is not enough to stabilize PFO monomers at the membrane surface. Different "grades" of cholesterol accessibility are required to trigger binding of PFO derivatives with modifications at the conserved loops of D4 [34, 44]. More studies are required to elucidate the molecular details of how cholesterol accessibility modulates PFO binding.

A detailed understanding of the cytolytic mechanism combined with the ability to modify the toxin and produce novel PFO derivatives, have now opened the door for the development of tools to study cell biology. Non-lytic PFO derivatives have been developed to study cholesterol accessibility at the surface of cell membranes [34, 48]. Other lytic PFO derivatives have been used to specifically permeabilize the plasma membrane of cells to study the biochemistry of intact organelles in their native environment (e.g. mitochondria [10]). Moreover, the striking on-and-off binding properties of PFO have been exploited to study the role of cholesterol in cell physiology and the intracellular traffic of cholesterol [9, 56]. It is clear that the understanding of the molecular mechanism of these fascinating proteins secreted by pathogens has had, and will have a great impact on the studies of biochemistry and physiology in whole cells [10], as well as on the studies of cholesterol-dependent mechanisms in cell biology [9, 56, 65].

Acknowledgments Work in the author's laboratory was supported by Grant Number GM 097414 from the National Institute of Health (A.P.H). B.B.J. was partially supported by the National Science Foundation, Integrative Graduate Education and Research Traineeship (IGERT), Institute for Cellular Engineering (DGE-0654128).

References

1. Alving CR, Habig WH, Urban KA, Hardegree MC (1979) Cholesterol-dependent tetanolysin damage to liposomes. Biochim Biophys Acta 551:224–228
2. Bach D, Wachtel E (2003) Phospholipid/cholesterol model membranes: formation of cholesterol crystallites. Biochim Biophys Acta 1610:187–197
3. Bavdek A, Gekara NO, Priselac D, Gutierrez Aguirre I, Darji A, Chakraborty T, Macek P, Lakey JH, Weiss S, Anderluh G (2007) Sterol and pH interdependence in the binding, oligomerization, and pore formation of listeriolysin O. Biochemistry 46:4425–4437
4. Billington SJ, Songer JG, Jost BH (2002) The variant undecapeptide sequence of the *Arcanobacterium pyogenes* haemolysin, pyolysin, is required for full cytolytic activity. Microbiology 148:3947–3954
5. Bourdeau RW, Malito E, Chenal A, Bishop BL, Musch MW, Villereal ML, Chang EB, Mosser EM, Rest RF, Tang W-J (2009) Cellular functions and x-ray structure of anthrolysin O, a cholesterol-dependent cytolysin secreted by *Bacillus anthracis*. J Biol Chem 284:14645–14656
6. Cho W, Stahelin RV (2005) Membrane-protein interactions in cell signaling and membrane trafficking. Annu Rev Biophys Biomol Struct 34:119–151
7. Czajkowsky DM, Hotze EM, Shao Z, Tweten RK (2004) Vertical collapse of a cytolysin prepore moves its transmembrane beta-hairpins to the membrane. EMBO J 23:3206–3215
8. Dang TX, Hotze EM, Rouiller I, Tweten RK, Wilson-Kubalek EM (2005) Prepore to pore transition of a cholesterol-dependent cytolysin visualized by electron microscopy. J Struct Biol 150:100–108
9. Das A, Goldstein JL, Anderson DD, Brown MS, Radhakrishnan A (2013) Use of mutant 125I-Perfringolysin O to probe transport and organization of cholesterol in membranes of animal cells. Proc Natl Acad Sci USA 110:10580–10585
10. Divakaruni AS, Wiley SE, Rogers GW, Andreyev AY, Petrosyan S, Loviscach M, Wall EA, Yadava N, Heuck AP, Ferrick DA, Henry RR, McDonald WG, Colca JR, Simon MI, Ciaraldi TP, Murphy AN (2013) Thiazolidinediones are acute, specific inhibitors of the mitochondrial pyruvate carrier. Proc Natl Acad Sci USA 110:5422–5427
11. Dowd KJ, Tweten RK (2012) The cholesterol-dependent Cytolysin Signature Motif: a critical element in the Allosteric pathway that couples membrane binding to pore assembly. PLoS Pathog 8:e1002787
12. Dunstone MA, Tweten RK (2012) Packing a punch: the mechanism of pore formation by cholesterol dependent cytolysins and membrane attack complex/perforin-like proteins. Curr Opin Struct Biol 22:342–349
13. Farrand AJ, LaChapelle S, Hotze EM, Johnson AE, Tweten RK (2010) Only two amino acids are essential for cytolytic toxin recognition of cholesterol at the membrane surface. Proc Natl Acad Sci USA 107:4341–4346
14. Flanagan JJ, Heuck AP, Johnson AE (2002) Cholesterol-Phospholipid interactions play an important role in Perfringolysin O binding to membrane. FASEB J 16:A929
15. Flanagan JJ, Tweten RK, Johnson AE, Heuck AP (2009) Cholesterol exposure at the membrane surface is necessary and sufficient to trigger perfringolysin O binding. Biochemistry 48:3977–3987
16. Giddings KS, Johnson AE, Tweten RK (2003) Redefining cholesterol's role in the mechanism of the cholesterol-dependent cytolysins. Proc Natl Acad Sci USA 100:11315–11320
17. Giddings KS, Zhao J, Sims PJ, Tweten RK (2004) Human CD59 is a receptor for the cholesterol-dependent cytolysin intermedilysin. Nat Struct Mol Biol 11:1173–1178
18. Gilbert RJ (2005) Inactivation and activity of cholesterol-dependent cytolysins: what structural studies tell us. Structure 13:1097–1106
19. Gilbert RJ (2010) Cholesterol-dependent cytolysins. Adv Exp Med Biol 677:56–66

20. Gilbert RJC, Rossjohn J, Parker MW, Tweten RK, Morgan PJ, Mitchell TJ, Errington N, Rowe AJ, Andrew PW, Byron O (1998) Self-interaction of pneumolysin, the pore-forming protein toxin of *Streptococcus pneumoniae*. J Mol Biol 284:1223–1237
21. Hadders MA, Beringer DX, Gros P (2007) Structure of C8 a-MACPF reveals mechanism of membrane attack in complement immune defense. Science 317:1552–1554
22. Heuck AP, Hotze EM, Tweten RK, Johnson AE (2000) Mechanism of membrane insertion of a multimeric β-barrel protein: perfringolysin O creates a pore using ordered and coupled conformational changes. Mol Cell 6:1233–1242
23. Heuck AP, Johnson AE (2002) Pore-forming protein structure analysis in membranes using multiple independent fluorescence techniques. Cell Biochem Biophys 36:89–101
24. Heuck AP, Johnson AE (2005) Membrane recognition and pore formation by bacterial pore-forming Toxins. In: Tamm LK (ed) Protein-lipid interactions. From membrane domains to cellular networks. Wiley-VCH, Weinheim, pp 163–186
25. Heuck AP, Moe PC, Johnson BB (2010) The cholesterol-dependent cytolysins family of Gram-positive bacterial toxins. In: Harris JR (ed) Cholesterol binding proteins and cholesterol transport, Subcellular biochemistry, vol 51. Springer, The Netherlands pp 551–577
26. Heuck AP, Savva CG, Holzenburg A, Johnson AE (2007) Conformational changes that effect Oligomerization and initiate pore formation are triggered throughout Perfringolysin O upon binding to cholesterol. J Biol Chem 282:22629–22637
27. Heuck AP, Tweten RK, Johnson AE (2001) Beta-barrel pore-forming toxins: intriguing dimorphic proteins. Biochemistry 40:9065–9073
28. Heuck AP, Tweten RK, Johnson AE (2003) Assembly and topography of the prepore complex in cholesterol-dependent cytolysins. J Biol Chem 278:31218–31225
29. Hotze EM, Le HM, Sieber JR, Bruxvoort C, McInerney MJ, Tweten RK (2013) Identification and characterization of the first cholesterol-dependent cytolysins from Gram-negative bacteria. Infect Immun 81:216–225
30. Hotze EM, Wilson-Kubalek E, Farrand AJ, Bentsen L, Parker MW, Johnson AE, Tweten RK (2012) Monomer-Monomer interactions propagate structural transitions necessary for pore formation by the cholesterol-dependent cytolysins. J Biol Chem 287:24534–24543
31. Hotze EM, Wilson-Kubalek EM, Rossjohn J, Parker MW, Johnson AE, Tweten RK (2001) Arresting pore formation of a cholesterol-dependent cytolysin by disulfide trapping synchronizes the insertion of the transmembrane beta-sheet from a prepore intermediate. J Biol Chem 276:8261–8268
32. Huang J, Feigenson GW (1999) A microscopic interaction model of maximum solubility of cholesterol in lipid bilayers. Biophys J 76:2142–2157
33. Johnson AE (2005) Fluorescence approaches for determining protein conformations, interactions and mechanisms at membranes. Traffic 6:1078–1092
34. Johnson BB, Moe PC, Wang D, Rossi K, Trigatti BL, Heuck AP (2012) Modifications in Perfringolysin O domain 4 alter the cholesterol concentration threshold required for binding. Biochemistry 51:3373–3382
35. Jost BH, Lucas E, Billington S, Ratner A, McGee D (2011) Arcanolysin is a cholesterol-dependent cytolysin of the human pathogen *Arcanobacterium haemolyticum*. BMC Microbiol 11:239
36. Korchev YE, Bashford CL, Pederzolli C, Pasternak CA, Morgan PJ, Andrew PW, Mitchell TJ (1998) A conserved tryptophan in pneumolysin is a determinant of the characteristics of channels formed pneumolysin in cells and planar lipid bilayers. Biochem J 329:571–577
37. Lange Y, Steck TL (2008) Cholesterol homeostasis and the escape tendency (activity) of plasma membrane cholesterol. Prog Lipid Res 47:319–332
38. Lange Y, Ye J, Steck TL (2005) Activation of membrane cholesterol by displacement from phospholipids. J Biol Chem 280:36126–36131
39. Lin Q, London E (2013) Altering hydrophobic sequence lengths shows that hydrophobic mismatch controls affinity for ordered lipid domains (rafts) in the multitransmembrane strand protein Perfringolysin O. J Biol Chem 288:1340–1352

40. Mason PR, Tulenko TN, Jacob RF (2003) Direct evidence for cholesterol crystalline domains in biological membranes: role in human pathobiology. Biochim Biophys Acta 1610:198–207
41. McConnell HM, Radhakrishnan A (2003) Condensed complexes of cholesterol and phospholipids. Biochim Biophys Acta 1610:159–173
42. Mesmin B, Maxfield FR (2009) Intracellular sterol dynamics. Biochim Biophys Acta 1791:636–645
43. Michel E, Reich KA, Favier R, Berche P, Cossart P (1990) Attenuated mutants of the intracellular bacterium *Listeria monocytogenes* obtained by single amino acid substitutions in listeriolysin O. Mol Microbiol 4:2167–2178
44. Moe PC, Heuck AP (2010) Phospholipid hydrolysis caused by *Clostridium perfringens* α-toxin facilitates the targeting of perfringolysin O to membrane bilayers. Biochemistry 49:9498–9507
45. Nelson LD, Chiantia S, London E (2010) Perfringolysin O association with ordered lipid domains: implications for transmembrane protein raft affinity. Biophys J 99:3255–3263
46. Nelson LD, Johnson AE, London E (2008) How interaction of Perfringolysin O with membranes is controlled by sterol structure, lipid structure, and physiological low pH: insights into the origin of Perfringolysin O-lipid raft interaction. J Biol Chem 283:4632–4642
47. Ohno-Iwashita Y, Iwamoto M, Ando S, Iwashita S (1992) Effect of lipidic factors on membrane cholesterol topology—mode of binding of θ-toxin to cholesterol in liposomes. Biochim Biophys Acta 1109:81–90
48. Ohno-Iwashita Y, Shimada Y, Hayashi M, Iwamoto M, Iwashita S, Inomata M (2010) Cholesterol-binding toxins and anti-cholesterol antibodies as structural probes for cholesterol localization In: Harris JR (ed) Cholesterol binding and cholesterol transport proteins. Subcellular Biochemistry. vol 51, Springer, The Netherlands pp 597–621
49. Ohno-Iwashita Y, Shimada Y, Waheed A, Hayashi M, Inomata M, Nakamura M, Maruya M, Iwashita M (2004) Perfringolysin O, a cholesterol-binding cytolysin, as a probe for lipid rafts. Anaerobe 10:125–134
50. Ohvo-Rekilä H, Ramstedt B, Leppimäki P, Slotte JP (2002) Cholesterol interactions with phospholipids in membranes. Prog Lipid Res 41:66–97
51. Olofsson A, Hebert H, Thelestam M (1993) The projection structure of Perfringolysin O (*Clostridium perfringens* θ-toxin). FEBS Lett 319:125–127
52. Olsen BN, Bielska AA, Lee T, Daily MD, Covey DF, Schlesinger PH, Baker NA, Ory DS (2013) The structural basis of cholesterol accessibility in membranes. Biophys J 105: 1838–1847
53. Palmer M, Harris R, Freytag C, Kehoe M, Tranum-Jensen J, Bhakdi S (1998) Assembly mechanism of the oligomeric streptolysin O pore: the early membrane lesion is lined by a free edge of the lipid membrane and is extended gradually during oligomerization. EMBO J 17:1598–1605
54. Palmer M, Valeva A, Kehoe M, Bhakdi S (1995) Kinetics of Streptolysin O self-assembly. Eur J Biochem 231:388–395
55. Pinkney M, Beachey E, Kehoe M (1989) The thiol-activated toxin streptolysin O does not require a thiol group for cytolytic activity. Infect Immun 57:2553–2558
56. Pocognoni CA, De Blas GA, Heuck AP, Belmonte SA, Mayorga LS (2013) Perfringolysin O as a useful tool to study human sperm physiology. Fertil Steril 99(99–106):e102
57. Polekhina G, Feil SC, Tang J, Rossjohn J, Giddings KS, Tweten RK, Parker MW (2006) Comparative three-dimensional structure of cholesterol-dependent cytolysins. In: Alouf JE, Popoff MR (eds) The comprehensive sourcebook of bacterial protein toxins, 3rd edn. Academic Press, Oxford, England, pp 659–670
58. Polekhina G, Giddings KS, Tweten RK, Parker MW (2005) Insights into the action of the superfamily of cholesterol-dependent cytolysins from studies of intermedilysin. Proc Natl Acad Sci USA 102:600–605
59. Praper T, Sonnen A, Viero G, Kladnik A, Froelich CJ, Anderluh G, Dalla Serra M, Gilbert RJC (2010) Human perforin employs different avenues to damage membranes. J Biol Chem 286:2946–2955

60. Radhakrishnan A, McConnell HM (2000) Chemical activity of cholesterol in membranes. Biochemistry 39:8119–8124
61. Ramachandran R, Heuck AP, Tweten RK, Johnson AE (2002) Structural insights into the membrane-anchoring mechanism of a cholesterol-dependent cytolysin. Nat Struct Mol Biol 9:823–827
62. Ramachandran R, Tweten RK, Johnson AE (2004) Membrane-dependent conformational changes initiate cholesterol-dependent cytolysin oligomerization and intersubunit beta-strand alignment. Nat Struct Mol Biol 11:697–705
63. Ramachandran R, Tweten RK, Johnson AE (2005) The domains of a cholesterol-dependent cytolysin undergo a major FRET-detected rearrangement during pore formation. Proc Natl Acad Sci USA 102:7139–7144
64. Rampersaud R, Planet PJ, Randis TM, Kulkarni R, Aguilar JL, Lehrer RI, Ratner AJ (2011) Inerolysin, a cholesterol-dependent cytolysin produced by *Lactobacillus iners*. J Bacteriol 193:1034–1041
65. Reid PC, Sakashita N, Sugii S, Ohno-Iwashita Y, Shimada Y, Hickey WF, Chang T-Y (2004) A novel cholesterol stain reveals early neuronal cholesterol accumulation in the Niemann-Pick type C1 mouse brain. J Lipid Res 45:582–591
66. Rosado CJ, Buckle AM, Law RHP, Butcher RE, Kan W-T, Bird CH, Ung K, Browne KA, Baran K, Bashtannyk-Puhalovich TA, Faux NG, Wong W, Porter CJ, Pike RN, Ellisdon AM, Pearce MC, Bottomley SP, Emsley J, Smith AI, Rossjohn J, Hartland EL, Voskoboinik I, Trapani JA, Bird PI, Dunstone MA, Whisstock JC (2007) A common fold mediates vertebrate defense and bacterial attack. Science 317:1548–1551
67. Rosenqvist E, Michaelsen TE, Vistnes AI (1980) Effect of streptolysin O and digitonin on egg lecithin/cholesterol vesicles. Biochim Biophys Acta 600:91–102
68. Rossjohn J, Feil SC, McKinstry WJ, Tweten RK, Parker MW (1997) Structure of a cholesterol-binding, thiol-activated cytolysin and a model of its membrane form. Cell 89:685–692
69. Rossjohn J, Polekhina G, Feil SC, Morton CJ, Tweten RK, Parker MW (2007) Structures of Perfringolysin O suggest a pathway for activation of cholesterol-dependent cytolysins. J Mol Biol 367:1227–1236
70. Sato TK, Tweten RK, Johnson AE (2013) Disulfide-bond scanning reveals assembly state and beta-strand tilt angle of the PFO beta-barrel. Nat Chem Biol 9:383–389
71. Saunders FK, Mitchell TJ, Walker JA, Andrew PW, Boulnois GJ (1989) Pneumolysin, the thiol-activated toxin of *Streptococcus pneumoniae*, does not require a thiol group for in vitro activity. Infect Immun 57:2547–2552
72. Sekino-Suzuki N, Nakamura M, Mitsui K-I, Ohno-Iwashita Y (1996) Contribution of individual tryptophan residues to the structure and activity of θ-toxin (perfringolysin O), a cholesterol-binding cytolysin. Eur J Biochem 241:941–947
73. Shatursky O, Heuck AP, Shepard LA, Rossjohn J, Parker MW, Johnson AE, Tweten RK (1999) The mechanism of membrane insertion for a cholesterol-dependent cytolysin: a novel paradigm for pore-forming toxins. Cell 99:293–299
74. Shepard LA, Heuck AP, Hamman BD, Rossjohn J, Parker MW, Ryan KR, Johnson AE, Tweten RK (1998) Identification of a membrane-spanning domain of the thiol-activated pore-forming toxin *Clostridium perfringens* perfringolysin O: an α-helical to β-sheet transition identified by fluorescence spectroscopy. Biochemistry 37:14563–14574
75. Shepard LA, Shatursky O, Johnson AE, Tweten RK (2000) The mechanism of pore assembly for a cholesterol-dependent cytolysin: formation of a large prepore complex precedes the insertion of the transmembrane beta-hairpins. Biochemistry 39:10284–10293
76. Sokolov A, Radhakrishnan A (2010) Accessibility of cholesterol in endoplasmic reticulum membranes and activation of SREBP-2 switch abruptly at a common cholesterol threshold. J Biol Chem 285:29480–29490
77. Solovyova AS, Nollmann M, Mitchell TJ, Byron O (2004) The solution structure and oligomerization behavior of two bacterial toxins: pneumolysin and perfringolysin O. Biophys J 87:540–552

78. Soltani CE, Hotze EM, Johnson AE, Tweten RK (2007) Specific protein-membrane contacts are required for prepore and pore assembly by a cholesterol-dependent cytolysin. J Biol Chem 282:15709–15716
79. Soltani CE, Hotze EM, Johnson AE, Tweten RK (2007) Structural elements of the cholesterol-dependent cytolysins that are responsible for their cholesterol-sensitive membrane interactions. Proc Natl Acad Sci USA 104:20226–20231
80. Tilley SJ, Orlova EV, Gilbert RJ, Andrew PW, Saibil HR (2005) Structural basis of pore formation by the bacterial toxin pneumolysin. Cell 121:247–256
81. Tweten RK (2005) Cholesterol-dependent cytolysins, a family of versatile pore-forming toxins. Infect Immun 73:6199–6209
82. Xu L, Huang B, Du H, Zhang X, Xu J, Li X, Rao Z (2010) Crystal structure of cytotoxin protein suilysin from *Streptococcus suis*. Protein Cell 1:96–105
83. Ziblat R, Leiserowitz L, Addadi L (2010) Crystalline domain structure and cholesterol crystal nucleation in single hydrated DPPC:cholesterol:POPC Bilayers. J Am Chem Soc 132:9920–9927

Chapter 6
Structural Biology of the Membrane Attack Complex

Andreas F.-P. Sonnen and Philipp Henneke

Abstract The complement system is an intricate network of serum proteins that mediates humoral innate immunity through an amplification cascade that ultimately leads to recruitment of inflammatory cells or opsonisation or killing of pathogens. One effector arm of this network is the terminal pathway of complement, which leads to the formation of the membrane attack complex (MAC) composed of complement components C5b, C6, C7, C8 and C9. Upon formation of C5 convertases via the classical or alternative pathways of complement activation, C5b is generated from C5 by proteolytic cleavage, nucleating a series of association and polymerisation reactions of the MAC-constituting complement components that culminate in pore formation of pathogenic membranes. Recent structures of MAC components and homologous proteins significantly increased our understanding of oligomerisation, membrane association and integration, shedding light onto the molecular mechanism of this important branch of the innate immune system.

Keywords Complememt system · Lipid membrane pore · Membrane attack complex · Pore-forming proteins · Structural biology

Abbreviations

Aa	Amino acid
CCP	Complement control protein repeat
CDCs	Cholesterol-dependent cytolysins
CH1 and CH2	Helical cluster 1 and helical cluster 2
CVF	Cobra venom factor
d1, d3 and d4	Domain 1, domain 3 and domain 4
EGF	Epidermal growth factor-like repeat

A. F.-P. Sonnen (✉) · P. Henneke
Center for Chronic Immunodeficiency, University Medical Center Freiburg, Breisacher Straße 117, 79106 Freiburg, Germany
e-mail: andreas.sonnen@uniklinik-freiburg.de

G. Anderluh and R. Gilbert (eds.), *MACPF/CDC Proteins - Agents of Defence, Attack and Invasion*, Subcellular Biochemistry 80, DOI: 10.1007/978-94-017-8881-6_6,

FIM	Factor I/membrane attack complex 6/7 module
LR	Low density lipoprotein receptor class A repeat
MAC	Membrane attack complex
MACPF	Membrane attack complex/perforin
MBLs	Manose binding lectins
MG	Macroglobulin
PFPs	Pore-forming proteins
poly-C9	polymeric-C9
TMH1 and TMH2	Transmembrane hairpin 1 and transmembrane hairpin 2
TS	Thrombospondin type 1 repeat

Introduction

Both prokaryotic and eukaryotic cells rely on a lipid membrane as a defining barrier against detrimental solute or water diffusion out of or into the cell. Without the plasma membrane as selectively permeable insulator against the outside world, cells cannot survive. Accordingly, membrane pore formation is a very efficient means of attacking cells. Pore-forming proteins (PFPs) are secreted by all kingdoms of life—often as toxins—to form aqueous pores in membranes of opponents and competitors [3]. Bacteria, for instance, express pore-forming toxins, most likely to destroy host epithelial cells to gain access to underlying tissues or to invade cells [39, 48, 147]. Although pore formation is a localised event often targeting specific membranes and tissues, it can contribute to the manifestation of systemic diseases [160]. The mammalian immune system on the other hand, uses the complement system membrane attack complex (MAC) or perforin from cytotoxic T-cells as potent pore forming systems against bacteria or diseased cells [105, 180].

PFPs are secreted as soluble molecules into the extracellular space. Subsequently, they usually bind to a lipidic or proteinaceous receptor on the target membrane, which is followed by 2-dimensional circular polymerisation on the bilayer surface. In order to integrate stably into the membrane, PFPs must undergo a dramatic conformational change to expose amphiphilic segments and drive them into the membrane. Often this conformational change is accompanied by an intricate trans-folding of secondary structural elements, e.g. the folding of α-helices in the soluble, monomeric state into β-sheets in the membrane inserted form [52, 150, 151, 170]. The necessary energy to breach the integrity of the lipid bilayer by forming an aqueous, solute-conductive pore is solely stored in the fold of the PFP and/or supplied by the association reactions preceding membrane integration. As the mechanism of pore formation is independent of ATP, PFP monomers can be thought of as *loaded springs* ready to unfold and integrate into the membrane.

Pore formation and cell lysis by the innate immune system is mediated by the terminal complement pathway, which leads to assembly of the MAC on membranes of bacterial pathogens [105]. The ability of complement to lyse cells was examined early on and complement lesions were first visualised by Dourmashkin and co-workers on sheep erythrocytes treated with Forssman antibody and guinea pig serum [21]. Agents of the terminal pathway are the complement components C5b, C6, C7, C8 and C9; multidomain glycoproteins that circulate in the bloodstream. Upon generation of a C5 convertase via either the classical or the alternative pathway of complement activation, C5 is proteolytically cleaved to C5b and C5a. C5b nucleates a series of association reactions: C5b binds to C6, exposing a binding site for C7, the resulting C5b-7 trimeric heterocomplex can already associate with the membrane and binding of C8, a complex composed of C8β and C8$\alpha\gamma$, firmly anchors the complex. C8 then binds C9, which homopolymerises in circular fashion ultimately leading to a membrane pore. C9 alone, in the form of polymeric-C9 (poly-C9), can form pores in model lipid membranes, and the formation of the C5b-8 initiation complex is essentially the generation of a membrane receptor for C9 polymerisation. Rather than binding to a receptor inherent to the membrane, the complement system affords its own membrane receptor, which becomes part of the pore structure itself. The final MAC is composed of C5b-8 and multiple copies of C9, all arranged side-on, in circular fashion.

All terminal pathway components except C5 contain the membrane attack complex/perforin (MACPF) domain, which constitutes the actual pore-forming domain. It refolds two helical clusters (CH1 and CH2) in the soluble state to two transmembrane hairpins (TMH1 and TMH2), in order to integrate into the membrane. The remaining domains of the MAC components are accessory in nature and regulate polymerisation, initial membrane attachment and the actual pore-forming step (for the domain arrangement of MAC components see Fig. 6.1). The MACPF domain is widespread and defines a family of homologous proteins diverse in function [40, 54, 74, 144]. The SMART non-redundant database currently lists 377 MACPF domains in 376 proteins [84]. However, pore formation is not a defining moment of the family underlining the functional flexibility of the domain. Perforin, an agent of adaptive immunity, is also part of the family, indicating a common pore-forming ancestor *prior* to the emergence of the two branches of the immune system. Initial structures of MACPF proteins [55, 143] found a structural homology to cholesterol dependent cytolysins (CDCs), such as pneumolysin from *Streptococcus pneumoniae* or listeriolysin from *Listeria monocytogenes*, leading to the definition of the MACPF/CDC superfamily [6, 53, 74, 91, 144] and possibly dating the common ancestor even further back in evolution (see Chaps. 2 and 4 for discussion of the evolutionary significance of the MACPF/CDC superfamily of proteins).

Structural biology of individual MAC components and of their complexes started around 1970 with the first electron microscopy (EM) images obtained of serum-purified proteins [130]. This provided insights into the ultrastructure of pore formation, also in context of the target membrane. Although structures of accessory domains of MAC components were available [22, 110], it was not until 2007

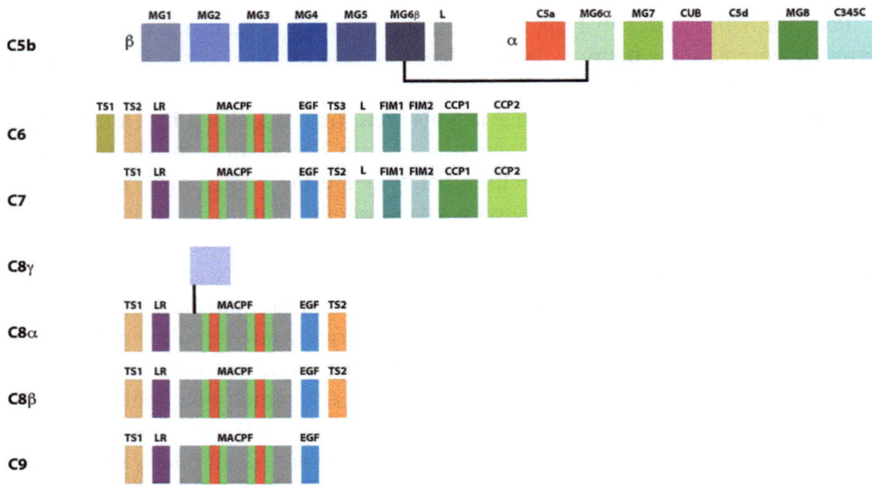

Fig. 6.1 Domain arrangement of the MAC components. The colouring scheme is maintained throughout the following figures. Approximate domain sizes correlate with box width

that detailed structural information at atomic resolution of the MACPF domain was achieved. The crystal structures of Plu-MACPF from *Photorhabdus luminescens* [143] and the MACPF domain of C8α [55] were the first of a series of eye-opening papers finally shedding light onto the structures of this important branch of the immune system. Together with cryo-EM and crystal structures of the structurally homologous CDC toxins and perforin [52, 82, 145, 170], a mechanism for pore-formation could be proposed, which is still under intense investigation and debate. The following pages will detail these recent advances in the structural biology of the MAC and aim to set them in context to functional data, proposing a molecular mechanism for membrane engagement and integration.

Assembly Pathways Leading to Formation of the MAC on Membranes

Assembly of the MAC is initiated by the formation of C5 convertases (see Fig. 6.2 for an overview and for an excellent review of the structural biology of the convertases and their assembly cascades see [45]). C5 convertases are complexes of complement components with serine protease activity [154]. They can be generated by the classical [141] or alternative [140] pathways of complement activation and cleave serum C5 at position Arg74-Leu75 to C5b and C5a [30, 106]. In the classical pathway direct or antibody-mediated binding of C1q to microbes initiates a complex cascade of autoactivation and component recruitment, resulting

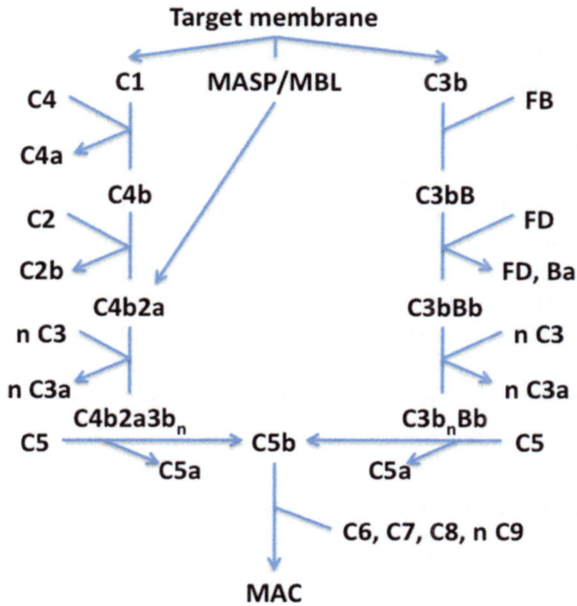

Fig. 6.2 Complement activation pathways leading to MAC formation. Classical pathway on the left and alternative pathway on the right.For a description see text

in the deposition of C4b2a on the microbial membrane. The C4b2a complex, is a serine protease, functioning as C3 convertase. Binding of multimeric manose binding lectins (MBLs) and associated proteases (MASPs) to carbohydrate moieties on the surface of microbes, can also lead to formation of the C4b2a complex (lectin pathway). C4b2a then cleaves multiple copies of C3 to the anaphylatoxin C3a and C3b, thus decorating the microbe with C3b. In the alternative pathway, C3b on the target cell membrane gains proteolytic activity by recruitment of factor B (FB) and its cleavage to Bb by factor D (FD). The resulting C3 convertase C3bBb proteolytically activates serum C3, leading to the opsonisation with multiple copies of C3b, as in the classical pathway. The membrane bound C3 convertases C4b2a and C3bBb can bind several copies of C3b, shifting their substrate spectrum from C3 to C5. The formed C5 convertases engage and cleave C5 to C5a, which diffuses as inflammatory stimulant away from the microbial surface, and C5b, which remains close to the membrane and initiates the terminal complement cascade.

The formation of C5b is essentially an interfacial amplification process, which keeps the important players close to the target. C5b alone is metastable and must bind to C6 within minutes to form the stable C5b-6 [36, 37], otherwise, the terminal assembly process is halted and cannot commence once C5b has decayed to a C6 binding-incompetent conformation. The bimolecular complex then engages C7, which exposes a metastable, hydrophobic membrane-binding site [135, 136], presumably by unfolding of the MACPF domain (see below), thus anchoring the trimolecular complex to the membrane without macro-pore formation [10]. The subsequent binding of C8 makes the initiation complex C9-binding competent. C8 is a dimer composed of C8β and the disulfide linked C8$\alpha\gamma$ complex. Whereas C8β engages C5b7, C9 binds to the C8$\alpha\gamma$ side to initiate autopolymerisation of further C9 monomers [23, 163]. The C5b9 complex, composed of seven independently expressed molecular entities with multiple copies of C9, forms the actual macromolecular pore of innate immunity geared towards invading bacteria.

Under certain disease-related conditions C5 can also be activated in the fluid phase, here also by other routes than the C5 convertases, which might contribute to disease progression in complement consuming disorders [65, 94, 95]. Similarly, release of activated C5b or C5b6 into solution can result in assembly and lysis of unsensitized neighbour cells [77]. The assembly of the MAC in solution is regulated by binding of clusterin and vitronectin, which arrest the assembly at the stage of C5b9 by forming pore-forming incompetent sC5b9 [123], whose tissue deposition is implicated in a variety of diseases such as glomerulonephritis [9].

Individual Complement Components of the MAC

In the following sections individual MAC components will be discussed in terms of their structure, but also in the context of the assembled complex.

Complement Component C5

Human complement component 5 (C5) is expressed as a single-chain, glycoprotein precursor of 1,676 amino acids (196 kDa) mainly in hepatocytes [172]. It is processed intracellularly by removal of a propetide (Met1-Gly18) and a linker peptide (Arg674-Arg677) that separates the N-terminal β-chain (Gln19-Leu673) from the C-terminal α-chain (Thr678-Cys1676). In the mature, secreted form that circulates the bloodstream, the α- and β-chains are held together by a disulfide bond (Cys567-Cys810) and non-covalent interactions that manifest themselves in the formation of a tightly folded macroglobulin domain (see below). Upon activation of the complement system and formation of the C5 convertases (serine protease complexes: C4bC2aC3b and/or $(C3b)_2Bb$), C5 is cleaved between Arg751 and Leu752 to generate the C5a anaphylatoxin (Tre678-Arg751) from the N-terminus of the α-chain and the remaining C5b. While the proinflammatory C5a anaphylatoxin functions mainly in chemotaxis and activation of neutrophils and monocytes [49, 50], the generation of C5b initiates the formation of the membrane attack complex in a sequence of complex associations [120, 129].

As a member of the α_2-macroglobulin superfamily, C5 is a homologue of C3 [66] and C4 [71], but lacks the internal Cys-Gln thioester of C3 and C4 [46]. The C5 β-chain is composed of six macroglobulin domains (see Fig. 6.3; MG, MG1 Glu20-Asp122, MG2 Asn123-Leu224, MG3 Pro225-Ser350, MG4 Pro351-Ser458, MG5 Ser459-Lys566, MG6α Cys567-Asp606, MG6β Glu771-Lys821),

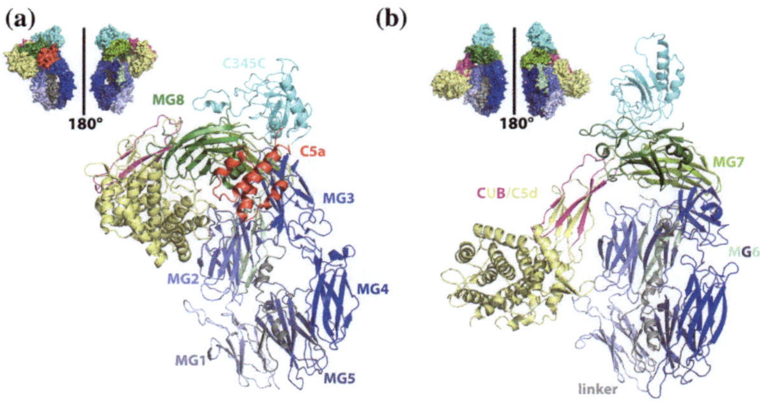

Fig. 6.3 C5 crystal structures. Crystal structure of C5 [47] (**a**) and of C5b from the complex crystal structure with C6 [56] (**b**), exemplifying the structural changes upon activation of C5 by C5 convertases and C6 binding

where MG6 is composite domain with contributions from the α- and β-chain, followed by a linker region after MG6β. The N-terminal domain of the α-chain is the anaphylatoxin, followed by MG6α, MG7 (Asp822-Pro931), a CUB domain/ C5d composite (Glu932-Glu1372), MG8 (Glu1373-Ile1516) and C345C (Lys1517-Cys1676). The CUB domain was named after the first three members identified to carry the domain: complement subcomponents C1r/C1s, embryonic sea urchin protein Uegf and bone-morphogenic protein Bmp1 [20]. The first structural data on C5 was obtained by electron microscopy [31, 37] and solution scattering techniques [114], revealing a multilobal, "heart-shaped" ultrastructure with the longest dimension 152 ± 17 Å, a width of 150 ± 15 Å, and a thickness of 93 ± 10 Å. Recent studies have now shed light on the detailed molecular architecture of C5 and its activation mechanism. Fredslund and coworkers [47] solved the crystal structure of C5 (see Fig. 6.3), indicating that C5 is composed of two structural units: the β-sandwich MG1-6 superhelix (the "β-ring") and the CUB-C5d-MG8 superdomain. As structures of homologous C3 [66] and C4 [71] as well as solution scattering data indicate, the C345C domain rides flexibly on top of the main C5 body. Interestingly, the thioester bond of C3 and C4 is not required for the typical positioning of C5d to MG8. The C5 anaphylatoxin domain is a 4-helix bundle [187] that sits at the angle bisector between MG8 and MG3.

Complex structures of C5 with cobra-venom factor (CVF [80]) and with SSL7 from *Staphylococcus aureus* [81] point towards a possible interaction of C3b and C4b—the non-catalytic components of the C5 convertases and homologues of CVF—with C5 in the C5/C5 convertase complex. CVF docks onto MG5 and MG7 of C5 resulting in a side-on, almost parallel arrangement [80]. Waves of C5 convertase formation, amplification and inhibition run across the surface of the target cell, at least in the alternative pathway, depositing cytolytic C5b-9 on the membrane [140]. As high affinity C5 convertases are bound to the target

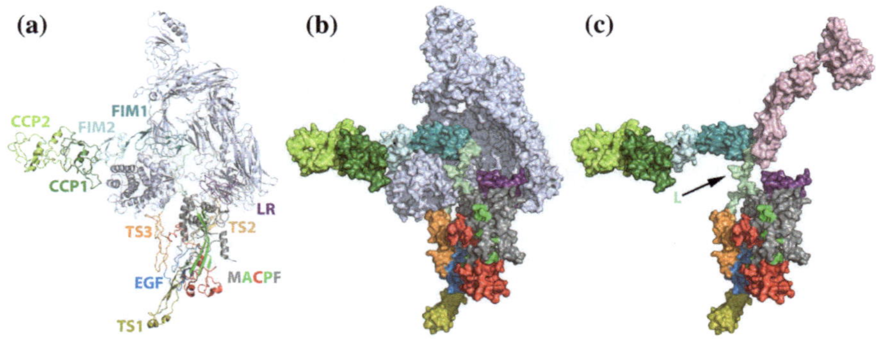

Fig. 6.4 Crystal structures of C5b-6 and C6. Complex cyrstal structure of C5b-6 (**a**) [56], C5b (*lightblue*) and C6 coloured as in Fig. 6.1. Surface representation of C5b-6 (**b**), coloured as in (a). Superposition, based on superposition of the MACPF domain, of C6 crystal structures (**c**). The region from TS1 to the linker domain (L) is shown for C6 from C5b-6, domains from CCP1 to FIM2 of the sole C6 structure [1, 2] are coloured in *lightpink*

membrane surface initially [141], C5 binding side-on to the convertase results in almost perpendicular alignment of C5b with MG1, MG4 and MG5 pointing towards the bilayer. Although C5b does not stably bind to membranes itself, it is in this way already aligned in the approximate orientation it attains in the MAC pore [56], possibly also facilitating the binding of C6.

The structure of C5b, from a complex structure with C6 [1, 56] revealed that akin to the C3 to C3b conversion, the stable arrangement of the β-chain provides the anchor for an extensive opening of the C5d-CUB-MG8 interface, which results in the movement of C5d away from the MG8 domain, albeit only approximately 40 Å and without direct contact to MG1 as in C3b (see Figs 6.3 and 6.4) [66]. The crevice created by this opening accepts the CCP1 domain of C6 resulting in a considerable movement of the CCP1-CCP2-FIM1-FIM2 arm of C6 around the TS3-CCP1 linker, which itself forms extensive contacts with C5d.

The position of C5b in the MAC has been investigated by electron microscopy. C5b harbours no MACPF domain and was suggested to be part of an elongated structure protruding away from the membrane and the tubular C9 cylinder [175, 179]. Indeed, C5b can be stripped from the assembled MAC by chaotrophic agents [118]. The 3-dimensional cryo-EM analysis of sC5b-9 confirms this location in the MAC and suggests that C5b does not contact the hydrophobic part of the membrane [56]. Although labelling with lipid-restricted probes [32], proteolytic treatment [16] and a model proposed on the basis of a crystal structure and solution scattering of C5 suggested otherwise [47], the C5bα and C5bβ domains are located approximately equidistant to the membrane, as the C5d domain moves downward, when C5 is activated to C5b. The C345C domain points away from the membrane surface. C5b does not sterically restrict side-on alignment of the final C9 monomer for ring closure, which accordingly must interact with C6 and potentially C5d of C5b for MAC ring closure.

Complement Component C6

While C5 belongs to the C3/C4/C5 family, C6 together with C7/C8/C9 forms the terminal pathway components of complement, all with a similar domain makeup (see Fig. 6.4). As the longest of the terminal pathway components the 934 amino acids of C6, including a leader peptide and several glycosylation sites, harbour the central and defining MACPF domain (Arg182-Phe518) and nine regulatory domains, namely three thrombospondin type 1 repeats (TS1 Cys22-Cys78, TS2 Asn81-Cys133, TS3 Val564-Cys611), a low density lipoprotein receptor class A repeat (LR Cys140-Ala175), an epidermal growth factor like repeat (EGF Pro519-Ser554), two complement control protein repeats (CCP1 Gly643-Ile705, CCP2 Val708-Cys761) and two Factor I/membrane attack complex 6/7 modules (FIM1 Cys773-Gln838, FIM2 Glu860-Ala934). C6 sequesters C5b resulting in a stable C5b-6 complex that is then engaged by C7. The CCP1 and FIM domains are implicated in the regulation of this process; only C6 and C7 of the terminal pathway components contain them.

A structural homology of C6 with the other members of the terminal pathway has been suggested early on [25]. The ultrastructure of C6 was first determined by DiScipio and Hugli [35], who already then noted that most of the α-helical portion of the protein resides in a domain that is devoid of cysteins and shares considerable homology with C9 and perforin, now known as the MACPF domain. The ultrastructure determined by negative stain EM visualized a two-domain structure with a more-globular region and an extend-tail region, both arranged like a letter C. It was speculated that the globular region would contain the C9 and perforin homology domain, which later proved to be true. Although Haefliger and coworkers showed that in fully hydrated, vitreous cryo-specimen the two-domain architecture with globular components could not be visualised, it could in their negative-stain images [58]. In both studies a similarity with the ultrastructure of C7 [32] was clearly evident.

The crystal structure of native C6 revealed a "seahorse" like arrangement, akin to the images obtained by EM, in which the nine domains of C6 fold into two superdomains that are connected by a flexible (and poorly resolved) linker [2]. TS1, TS2, LR, MACPF, EGF and TS3 form a globular, but flat main body superdomain, from which a long arm composed of CCP1, CCP2, FIM1 and FIM2 protrudes on the opposite side of TS1, almost like an antenna (see Fig. 6.4). By comparison with other MACPF domain structures (C8α, C8β and perforin) and EM images, it was deduced that the MACPF domain of C6 in solution possibly adopts a closed conformation, which is regulated by the movement of a Y-shaped module composed of TS1, TS2, TS3 and by EGF. Furthermore, the LR domains of C6 and C7 were proposed to pack tightly against each other due to shape and charge complementarities, placing the TS domains to the outside of a growing MAC and resulting in a domain orientation that is similar to the one found in CDC pores, but opposite to that found for perforin [52, 82, 170]. The FIM1/2 organisation in the crystal structure of C6 is poorly resolved but markedly different from the homology model derived from a C7 FIM1/2 NMR structure [116, 117]. In the C5b-6 crystal structures [1, 56] the FIM1/2 domains are resolved as a single folded domain, clearly different from the C7 FIM1/2 solution structure [117], most likely owing to an α-helical insertion motif in C6 and a charge difference on one side. Modelling of the TMHs formed by CH1 (Asp257-Lys291) and CH2 (Ser383-Gln434) of C6, suggests that they might insert into the membrane but not span it entirely, which is consistent with radiolabelling experiments [161] and the absence of pore formation at the C5b-6 stage of MAC complex assembly [153].

Helical, paracrystalline forms of the C5b-6 complex had already been observed in 1980 [120], but only recently have good diffracting three-dimensional crystals been obtained [1, 56]. From these data complex interactions of C6 with C5b and conformational changes of the unliganded states can be rationalised (see Fig. 6.4). The C-terminal region of C6 forms the major interface with C5b, with the total interface burying approximately 3,100 Å^2 of solvent-accessible area. Limited proteolysis and binding studies had previously identified this region, i.e. CCP1-FIM2, as the C5b-interacting site of C6 [31, 58]. In the C5b-6 crystal structures the interaction is mediated by three distinct contacts: (1) wedging of the CCP1 domain into the crevice created by opening of the C5b CUB-MG8 superdomain, (2) wrapping of the evolutionarily highly conserved C6 linker around the C5d domain, (3) interaction of the N-terminal C6 body with a conserved patch of C5b formed by MG1, MG4, and C5d. The TS3-CCP1 linker harbours a hydrophobic patch (^{592}PheSerIleMet^{595}) distinct from the corresponding region of C7, which is proposed to be the main discriminating factor promoting C5b binding to C6 rather than C7 [1]. Importantly, C6 undergoes a number of subtle changes that presumably prime it for engagement with C7. As modelled earlier [2], binding of C5b indeed induces the movement of the Y-shaped regulatory module and EGF. An untwisting and partial opening of the MACPF β-sheet, i.e. the first step in releasing the membrane binding TMHs, similar to the configuration seen in C8α (see below), was also observed. Although a direct intracomplex association of C6 FIM1/2 with C5b C345C was not seen, a domain-swapping interaction with a crystallographic neighbour complex, hints towards a possible engagement of the two domains in the same complex, as was proposed by functional interaction studies [36, 167].

Complement Component C7

Complement component C7 is an 843 amino acids-long glycoprotein component of the terminal complement pathway. C7 is homologous to the other members of the terminal pathway [61] and the biochemical properties are similar to those of C6 [121], as is its domain architecture, with only one N-terminal TS domain missing after the leader peptide. The domains with their approximate amino acid boundaries are: leader peptide (Met1-Ala22), TS1 (Asn27-Cys79), LR (Cys85-Arg124), MACPF (Cys128-Asp453), EGF (Pro454-Val489), TS2 (Val498-ys545), CCP1 (Phe569-Val632), CCP2 (Val635-Val688), FIM1 (Cys702-Cys763) and FIM2

(Lys771-Gln843). Negative stain EM analysis of C7 in solution, imaged an elongated molecule with approximate dimensions of 150 × 60 × 40 Å, that consisted, akin to C6, of a globular domain with a long arm attached [32]. In the absence of a full length C7 atomic structure, structural modelling using the C5b-6 crystal structure [1], suggested the interdomain linker between TS2 and CCP1 to be the main discriminating factor that allows C5b to bind C6 rather than C7. In radiolabelling experiments C7 resists stripping by trypsination in the presence of erythrocyte membranes [60], thus indicating that C7 inserts, at least partially, into the lipid membrane. Consistently, C7 binding of C5b-6 initiates initial membrane anchoring in the MAC assembly pathway via formation of a metastable, hydrophobic "binding site" [135, 136, 169] originating presumably from a concerted refolding in CH1 (S203-K246) and CH2 (S321-N368) of C7 (and C6) into partially hydrophobic TMHs that anchor the complex to the outer lipid leaflet, but do not form a pore at this stage [153]. This transition of the C7 CH helices to a β-sheet structure has been observed indirectly early on [135] by an increase in β-sheet content upon C5b-7 complex formation, but also when C7 oligomerised to a hemolytically inactive dimer in the presence of desoxycholate [136]. This dimer could be transferred to lipid membranes without pore formation.

Solution state C7 is able to transiently bind non-activated C5 [72], presumably via direct association of FIM1-2 with C345C of C5 [167], albeit in competition to but in preference over FIM1-2 of C6. This association is suggested to contribute to MAC formation [168] and a solution structure of C7 FIM1-2 was recently solved [116, 117].

Complement Component C8

Among the terminal pathway components C8 is special in that it is a heterotrimeric glycoprotein complex of genetically distinct but tightly associated protomers. C8α is a 62 kDa protein composed of a TS1 (after the leader peptide, Thr38-Cys90), a LR (Gln93-Ala135), the family defining MACPF (Glu138-Asn495), an EGF (Ala496-Gln532), and another TS domain (Ala536-Cys584). C8α is linked via a disulfide bond to C8γ (Cys194 of C8α to Cys60 of C8γ), which belongs to the lipocalin family of homologous β-barrel proteins. C8β, as C8α, is a member of the MACPF family and has an identical domain arrangement (TS1 Asp64-Cys116, LR Val120-Lys160, MACPF His164-Val500, EGF Ser501-Tyr538, TS2 Pro543-Ser591). C8$\alpha\gamma$ is believed to associate intracellularly with independently expressed C8β [108], but both subunits can also be secreted to associate in the bloodstream [149]. C8 binds to a newly established site on C7, once the C5b-7 complex is formed. C8 then mediates the circular polymerisation of C9 as a membrane-anchored nucleation site.

Electron microscopy analysis of full length C8 revealed a globular body with a globular protrusion [24]. Recently several crystal structures of C8, either full length or individual domains, have been obtained: C8γ [110], C8γ with C8α peptide [89],

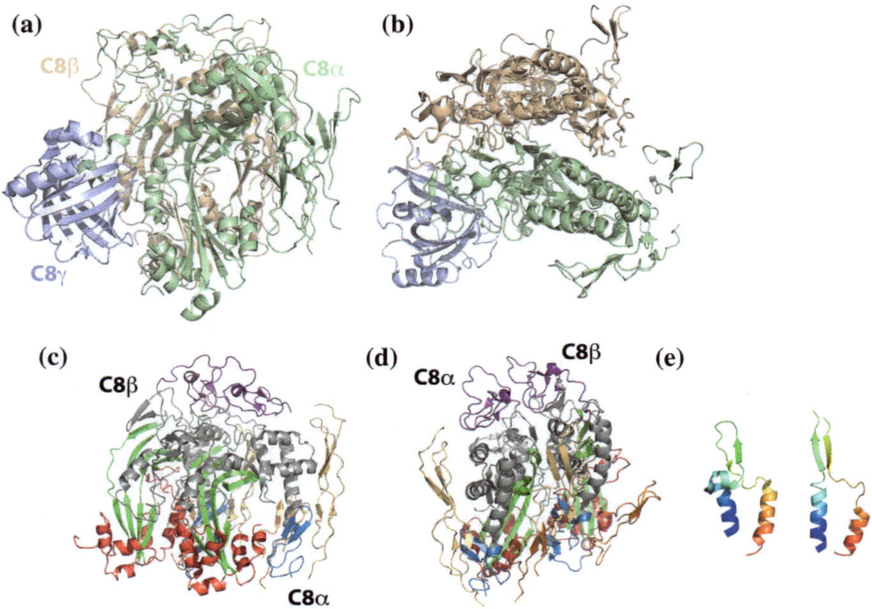

Fig. 6.5 Crystal structure of full length C8 [90], *side-view* (**a**) and *top-view* (**b**), showing the rotational angle of the subunits. Structure coloured according to domains with C8γ removed (**c**, side-view onto the C9 binding site; **d**, view along TS1 from the putative exterior rim of the MAC). The MACPF and LR interfaces can be appreciated in (**d**). Different arrangements of CH1 in the crystal structures of C8α-MACPF (*left* [55]) and C8$\alpha\gamma$ (*right* [158]) (**e**)

C8γ with laureate [26], C8α MACPF [55], C8$\alpha\gamma$ [158] and full length C8 [90]. In full length C8, the four-stranded, anti-parallel β-sheet of the MACPF domains of C8α and C8β are aligned almost co-planar along the long axis, with a rotational angle of 22° (see Fig. 6.5). C8γ is held towards the convex side of the twisted β-sheet by a cysteine bond to a unique insertion region (indel, Arg196-Lys204) of C8α and hydrophobic interactions between the indel and to the lipocalin ligand-binding site of C8γ [158]. The total interface responsible for the strong interaction of C8α with C8β is constituted by approximately 30 residues from each protomer and is reasonably hydrophilic in nature. C8α and C8β each contain four accessory domains, two towards the N-terminus and two towards the C-terminus of the MACPF domain. TS1 and TS2 flank the concave side of the twisted β-sheet, while the EGF domain, as in C6, closes it off towards the tip of the sheet (see Fig. 6.5). TS1 of C8α and C8β sits at the same position as TS2 of C6 [2] and the location of TS2 resembles that of C6 TS3, respectively. In analogy to the proposed Y-shaped regulatory element of C6, TS1 and TS2 form a V-shaped structure possibly fulfilling a similar function, i.e. regulating the unfolding of the MACPF domain. The LR domains of C8α and C8γ, which sit at the top of the MACPF domain, interact with each other. A similar interaction was proposed to contribute to the alignment of C6 and C7 in the C5b-7 complex [2]. While the overall structure of C8γ is well defined (root mean square deviation maximally 1.1 Å for pairwise structure alignment of all published structures), a comparison of the location of C8γ in the full length complex [90] with that of the C8αγ crystal structure [158] and the cryo-EM reconstruction of sC5b-9 [56] suggests a flexible attachment to C8α and a possible involvement in regulation of MAC assembly. While in the full length C8 structure it partly occludes the C9 binding site on the free face of the C8α MACPF domain, it presumably swings out by almost 90° upon C5b-8 complex formation thereby aligning in a straight line with C8α and freeing the C9 binding site. All crystal structures of C8γ, whether derived from human serum or *E. coli* expressed, show an empty hydrophobic binding pocket in the β-barrel, which in other lipocalins is usually occupied by small hydrophobic ligands. Although lipids can enter the binding pocket (co-crystal structure with laureate [26]), C8γ seems not to bind to small hydrophobic ligands *in vivo* [90], possibly also because the indel of C8α inserts into the upper compartment of the hydrophobic cavity thus precluding access to the lower one. C8γ is not essential for haemolytic activity but enhances it [112]. Based on the full length crystal structure of C8 and a molecular MAC model generated, Lovelace and coworkers proposed that this enhancement of haemolytic activity could originate from a more effective recruitment of C9 due to a direct contact of C8γ and C9 in the initial C5b-9 complex. However, the cryo-EM reconstruction of sC5b-9 indicates that the effect of C8γ on haemolytic activity could also originate from an inhibition of non-functional C8/C9 association in solution, as haemolytic activity assays were performed by complementation of factor-depleted sera.

Interestingly, C5b-8 at high concentration can partially disrupt erythrocyte membranes [186], indicating that the organisation of the TMHs generated upon C8 binding can induce bilayer leakage (see below). The MACPF domain of C8α, was proposed to exist in a partially activated state that is primed for the release of the CH helices as TMHs into the membrane, once the C5b-7 complex has been engaged [90]. The CHs of C8α are more hydrophobic than those of C8β, consistent with the notion that C8α is the main anchor of the C5b-8 complex in the membrane [161]. The β-hairpin of CH1 that connects the two helices of this segment is considerably hydrophobic, suggesting that it might function as a refolding nucleus that facilitates the formation of β-sheets from the two adjoining α-helices upon membrane integration. The inherent flexibility and this un/refolding propensity of CH1 is illustrated by alternate conformations in the crystal structures of C8α-MACPF [55], C8αγ [158] and full-length C8 (see Fig. 6.5 [90]).

Complement Component C9

Complement component 9 is a 61 kDa glycoprotein that acts as the final component in the assembly of the MAC. First described as a constituent of guinea pig serum [87], it was purified from human serum by Müller-Eberhard and colleagues in 1969 [57]. The domain structure of C9 was described on the basis of its sequence [33] and resembles that of C8α and C8β, but without the second TS (TS1 Asp41-Cys93, LR Gly101-Aps129, MACPF Arg153-Ser505, EGF Val506-Lys542). C9 is the component that functionally and structurally resembles perforin the most [86, 152, 177]. Similar to perforin, it is able to form pores on its own by polymerisation into poly-C9, which can be induced by heat or metal ion treatment, exposure to liposomes of high membrane curvature or treatment with low concentrations of proteolytic enzymes [29, 126, 178]. N-linked glycosylation at position N205 and N549 is required to attain full haemolytic activity [75]. Poly-C9 has been imaged extensively in solution and assembled on lipid membranes by EM and even electron tomography [33, 124–128]. The ultrastructure of poly-C9 closely resembles that of the fully assembled MAC, lacking the C5b-8 protrusion. A first three-dimensional structure, derived from cryo-EM micrographs of ZnCl-treated and heat-polymerised C9, revealed a hollow, tubular polymer composed of 13 monomers with an inner diameter of approximately 113 Å, an outer diameter of 254 Å and a cylinder height of 160 Å [18], which agreed well with a molecular weight determination of poly-C9 using scanning transmission EM [176]. Radial spokes of individual monomers protruded away from the radius of the cylinder at an angle of 23°, which is strikingly similar to the 22° proposed for the MAC on the basis of a recent C8 crystal structure [90] and a cryo-EM reconstruction of sC5b-9 [56]. An earlier two-dimensional study had proposed a monomer multiplicity of 12–14 with a similar architecture [34]. Monomeric C9 was imaged as a globular, ellipsoid with dimensions of 50 × 70 Å and a crevice along the long axis, akin possibly to an enclosure seen in the crystal structure of C8α, here formed by the TS1, EGF and the upper part (CDC d1-like, see below) of the MACPF domain on one side and the lower portion (d3-like) on the other. Although a solution scattering analysis overestimated the maximum dimension of C9, a V-like structure with an angle of around 10° was proposed, possibly reflecting the crevice formation [159]. Unfolding of the creviced ellipsoid accompanied by extension to more than twice its length was suggested to be required for membrane insertion [18, 34, 128]. Indeed, this indirect evidence for an open domain arrangement of C9 and its propensity to polymerise in solution suggest that C9 circulates the bloodstream in a partially activated conformation ready to refold once it sees C5b-8, its membrane receptor. Several antibodies raised against different C9 peptides could access epitopes on membrane-inserted C9 that were occluded in the solution state, hinting towards a dramatic refolding step [78, 102]. Aleshin and coworkers proposed a model in which the TS and EGF domains regulate the initial assembly stages of the C5b-9 complex [1]. It had already been shown earlier that partial removal of the TS1 module induced inactivation of C9 by detrimental self-association in solution [166], indicating that this domain indeed plays a role in regulation of C9 polymerisation.

There has been considerable debate over whether full circular polymerisation of C9 in the context of the MAC is a necessary requirement for membrane pore formation [28]. The MAC is heterogeneous in structure, with arciform as well as full ring oligomers apparently able to form a pore ([175], see below). A limited amount of C9 relative to C8 in the serum, resulting in a relative shortage of C9 for full ring polymerisation, was put forward as a main cause for pore heterogeneity [13]. Ratios of 4-8 C9 molecules per C8 were found to be sufficient in lysis experiments [13, 73, 155, 164]. Indeed, studies suggested that one C9 per C5b-8 is sufficient to generate haemolytically active, stable transmembrane pores [14]. Esser and co-workers proposed that polymerisation of C9 was required for lysis, but that there was no requirement for full circular polymerisation [28]. Despite the outer membrane of *E. coli* becoming permeable upon decoration with C5b-9 complexes carrying one C9 per C8, a loss of viability was not induced under these conditions, suggesting that C9 polymerisation onto the C5b-8 membrane receptor may be required to kill *E. coli* [12, 69]. Detailed sizing experiments using erythrocytes as well as *in vitro* LUV systems with different lipids, indicated that the size of the channel formed by C5b-9, was dependent on the ratio of C8:C9, with the channel size reaching that of poly-C9 with increasing amounts of C9 [92, 186]. Indeed EM images of MAC complexes assembled at different C9:C8 ratios on liposomes could image a clear size difference in side views between MACs at high and low C9:C8 ratios. Low ratios of 3:1 generated smaller pores with liposomes apparently narrowing towards the pore [175].

The MACPF Domain of Complement

The MACPF domain of complement components C6, C7, C8α, C8β and C9 is the actual pore-forming domain that inserts into the membrane. The MACPF domain is distributed widely throughout the domains of life, even existing in viruses [54]. Not all members are pore-forming and they have widely different functions, such as parasitic invasion (Plasmodium PLPs [5]) or egress (*Toxoplasma gondi* PLP1 [70]), anterior-posterior structure development in Drosophila (torso-like [162]) and neuronal migration during brain development (astrotactin [41]). Crystal structures of MACPF domains from human complement components C6 (three individually structures, PDBids 3T50, 4EOS, 4A5W), C8α (three individually structures, PDBids 2QQH, 2RD7, 3OJY) and C8β (PDBid 3OJY) are available, and comprise together with the domain of perforin from *Mus musculus* (PDBid 3NSJ) and the bacterial domains from *Photorhabdus luminescens* (PDBid 2QP2) and *Bacteriodes thetaiotamicron* (PDBid 3KK7) the current MACPF structural dataset [1, 2, 55, 56, 82, 90, 143, 158, 183]. Initial crystal structures of human C8α MACPF [55] and Plu-MACPF from bacterial *Photorhabdus luminescens* [143], showed a striking structural homology to domains 1 (d1) and 3 (d3) of bacterial CDCs, such that the MACPF domain family may now be extended to a MACPF/CDC superfamily (see Fig. 6.6).

MACPF domains share a limited sequence similarity of approximately 20% across the family, all containing the signature motif (Y/W)-G-(T/S)-H-(F/Y)-X_6-GG (X can be any amino acid, [131]). Indeed, amino acid sequence conservation between CDCs and MACPF members, now distils down to a glycine located at the

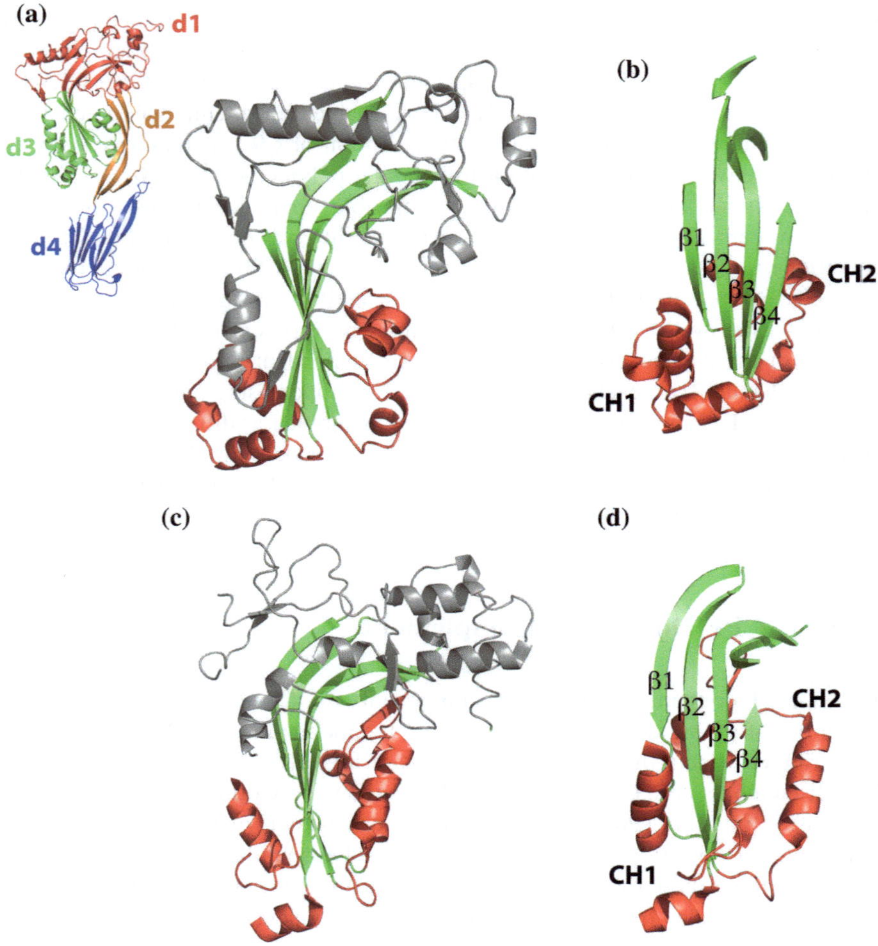

Fig. 6.6 Crystal structures of the MACPF/CDC domains of perfringolysin (**a**, **b** [145]) and C8α (**c**, **d** [55]). The inset in (**a**) shows the canonical CDC domains of perfringolysin (d1–d4). The strands of the curved central β-sheet and the adjoining CH segments are shown in (**b**) and (**d**)

hinge region of the central twisted-sheet [2]. The conservation of structure rather than sequence apparent in the evolution of the MACPF/CDC superfamily, is a striking example of a similar paradigm recently described for the evolution of viruses [7]. As in the MAC components, most MACPF/CDC superfamily domains are part of multidomain proteins, in which accessory domains may for instance function in membrane targeting and control of pore formation.

The MACPF domain is around 320 amino acids long, flat, almost rectangular in shape and harbours a central curved and twisted β-sheet, containing four anti-parallel β-strands (β1-β4, see Fig. 6.6). The portion above the kink corresponds to

CDC domain d1 and the lower portion to CDC domain d3. Not all strands of the curved four-stranded β-sheet continuously progress as β-folds through the bend. Strand β4 of all MACPF domains contains a helical insertion about midway through the strand immediately below the bend of the sheet. In the structures of C6 this insert is approximately 36 amino acids long, in C8α 30, in C8β 31, and in perforin 32 amino acids. All CDC domain structures solved to date (perfringolysin O PDBid 1PFO, 1M3 J, 1M3I; anthrolysin O PDBid 3CQF; intermedilysin PDBid 1S3R, 4BIK; streptolysin O PDBid 4HSC, suilysin PDBid 3HVN) also contain an insertion at this position, but in addition to the α-helical elements also a fifth β-strand (β5) immediately adjacent to β4, which extends the bottom of the four-stranded β-sheet to a five-stranded one. In CDCs β5 rotates away to free β4 for side-on oligomerisation upon membrane attachment [137]. The flexibility of the insert, which in CDCs functions as a gatekeeper to oligomerisation, is mediated by two glycines at the end of β4. The location of these glycines is conserved across the MACPF/CDC superfamily, which suggests a common function in maintaining the flexibility of this region. The absence of β5 in all MACPF structures available, indicates that the oligomerisation mechanism is potentially regulated differently. Mutagenisis studies using perforin showed that a salt-bridge between Arg213 and Glu343 in the region homologous to CDC domain D1 is necessary for efficient polymerisation [8], and a computational analysis also indicated this for oligomerisation of MAC components [54].

The flexibility of the β-sheet is maintained by the two conserved glycines of the MACPF signature motif. In all MACPF crystal structures, they are located in the middle of β3 at a hinge-like region immediately below the bend of the β-sheet. Straightening of the sheet is implicated in pore formation [2], which might explain the strict evolutionary requirement for glycines at this position. Aleshin and co-workers analysed the relative conformations around the hinge, finding that the degree of openness of the β-sheet correlated with the activation state of the MAC components along the assembly path, e.g. C6 in complex with C5b is less compact than free C6 in solution and MACPF domains of C8 are even more open, circulating in a stable but activated form in the blood [1, 2].

In the MACPF/CDC superfamily the α-helical clusters between β1 and β2 (CH1) and between β3 and β4 (CH2), insert as transmembrane β-hairpins (TMH1 and TMH2) into the lipid bilayer during pore formation. This intricate trans-folding of the α-helical CH1 and CH2 in the soluble form to membrane-inserted β-hairpins was inferred from the mechanism proposed for CDCs. Here, cysteine scanning mutagenesis of perfringolysin O together with fluorescence spectroscopy in the presence of liposomes suggested this dramatic refolding [150]. Cryo-EM and atomic force microscopy could evidence a concomitant vertical collapse of the prepore complex and straightening of the kinked β-sheet [27, 170]. CH1 and CH2 regions of MAC members are more than twice as long (usually more than 50 amino acids) as those of CDCs (approximately 25 amino acids), suggesting that a vertical collapse is not associated with membrane integration. The prepore and pore states of perforin were imaged by cryo-EM to attain the same height above the surface of the membrane [132]. Accordingly, the central section of each CH presumably integrates as TMH into the membrane, while the adjoining amino acids connect to $\beta 1$–$\beta 4$ above the lipid bilayer.

CH1 and CH2 of the MACPF domains of C6, C7, C8α, C8β and C9 show a marked difference in hydrophilicity and length (approximate amino acid (aa) length, C6: CH1 53 aa, CH2 53 aa; C7: CH1 44 aa, CH2 47 aa; C8α: CH1 54 aa, CH2 57 aa; C8β: CH1 53 aa, CH2 57 aa; C9: CH1 70 aa, CH2 57 aa). The central sections of CH1 and CH2, i.e. predicted TMH1 and TMH2, of C8α and C8β are hydrophobic in nature, explaining the firm anchorage of the initiation complex in the membrane once C8 is bound. The TMHs of C8α are more hydrophobic than those of C8β, consistent with the notion that C8α is the main anchor of C5b-8 in the membrane [161]. Interestingly, the central section of CH1 of C8α contains a β-hairpin region that can exist in two alternative conformations. In the structure of lone C8α MACPF [55] it is aligned parallel to $\beta 4$ due to a bend in the helix preceding the hairpin and in the C8$\alpha\gamma$ structure it engages $\beta 4$ in antiparallel fashion without helix bending [158]. The hairpin might function as a refolding nucleus that facilitates TMH formation. The alternative folding of C8α CH1 exemplifies the inherent physico-chemical adaptability and refolding potential of the helical inserts, which is necessary for unfolding and pore formation.

The CHs of C6 and C7 are less hydrophobic than those of C8 and C9 and in the case of C7 additionally also approximately 10 amino acids shorter, suggesting that their TMHs might not penetrate the membrane entirely [153], potentially only interacting with the outer leaflet of the bilayer. The predicted TMHs of C9 display a clear, alternating β-strand amphiphilicity, possibly explaining why addition of C9 establishes formation of stable, aqueous transmembrane pores. A 17 amino acid insert in C9 TMH1 may result in a lip formed at the cytosolic face of the membrane. It is tempting to speculate that the diverse functions of the other MACPF family members in part also originate, from a difference in the polarity pattern and length of TMH1 and TMH2.

Mechanism of MAC Pore Assembly

Membrane pore formation by pore-forming proteins is a dynamic process. MACPF/CDC superfamily proteins are modelled to form large β-barrel pores, with two membrane spanning β-hairpins contributed by each protomer. Earlier studies using photoaffinity labels suggested, wrongly, that each monomer contributes two amphipathic helices separated by a turn to the membrane pore [113]. The detailed membrane interaction and pore formation mechanism of CDCs has been elucidated almost to atomic detail by various complementary structural and functional techniques (for reviews see [51, 53, 62]). Briefly, CDCs, except the CD59-dependent intermedilysin (see below), bind with the cholesterol-recognition motif of domain 4 to cholesterol in the target membrane [38], anchor to the membrane with the conserved unadecapeptide, oligomerise to a prepore atop of the bilayer, partially unfold their CH regions generating a dynamic folding intermediate and

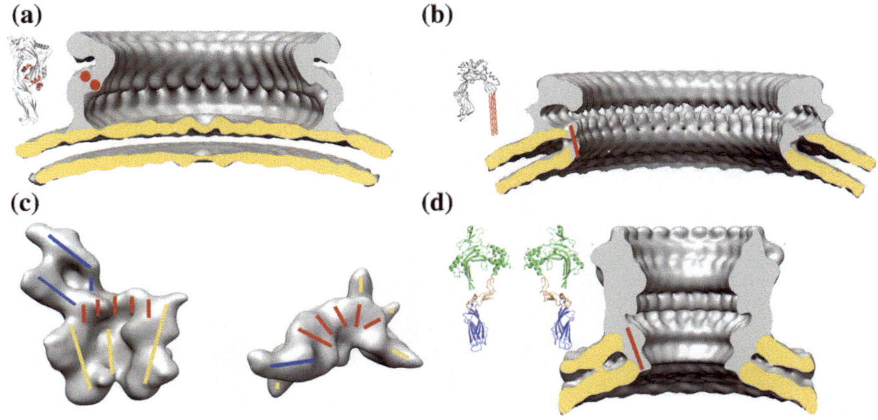

Fig. 6.7 Cryo-EM structures of MACPF/CDC proteins. Sections through the cryo-EM structure of the prepore (**a**) and pore (**b**) structures of pneumolysin [170], with the membrane region in *yellow* and the protein atop in *grey*. Location of the CH segments in the soluble form (**a**) and of the TMHs in the membrane-inserted pore (**b**) are indicated in *red*. Structures to the *left* depict the configuration of the individual monomer in the prepore and pore states, based on fits to the EM map [170]. CHs and TMHs in *red*. Cryo-EM structures of sC5b-9 [56], *side view* (**c**, on *left*) and view from *top* (**c**, on *right*). Approximate locations of C5b and of CCP1-FIM2 of C6 are indicated by *blue* lines. *Red* lines give approximate locations of MACPF domains of C6–C9, location of S-proteins as *yellow* lines. In the *top* view the arc-like arrangement of the MACPF domains can be appreciated. Cryo-EM reconstruction of perforin (**d**, [82]). Crystal structures of perforin (*solution state*) on the *left* with the C2 domains (*blue*), EGF and linker (*orange*) and MACPF domain (*green*), CH segments have been removed for clarity. The monomer orientation on the *left* is that proposed by Law et al. [82], on the *right* that by Gilbert et al. [54] (see text)

finally undergo a vertical collapse driving the TMH regions into the membrane to form a giant β-barrel-like pore with β-strands inclining at an angle from the membrane vertical (see Fig. 6.7 [148]). The MAC mechanism is different in that it requires generation of a receptor—C5b-7—on the target membrane (for a cartoon of the assembly see Fig. 6.8). Moreover, the receptor becomes an integral part of the pore itself and its assembly process is inherent to the mechanism of pore formation. Cleavage of C5 by C5 convertases results in opening of the C5d-CUB-MG8 interface, moving C5d towards the membrane and opening a crevice that accepts CCP1 of C6 resulting in the formation of C5b-6. At this stage the complex is not firmly anchored to the membrane; in the crystal structure of C5b-6 the CH segments of the MACPF domain are not transfolded to TMHs, suggesting that the transition occurs at a later stage of assembly [1, 56]. The bimolecular complex then engages C7, also stabilised by interactions of C7 FIM1-2 with C345C of C5b [167, 168] and presumably by side-on alignment of the C6/C7 MACPF domains mediated by complementary-charged/hydrophilic MACPF surfaces, akin to the salt-bridge interactions identified to be necessary for perforin oligomerisation [1, 8, 54]. LR domain packing contributes to the C6/C7 binding interface, and also to that of other protomer components along the ring. Subsequently, formation of

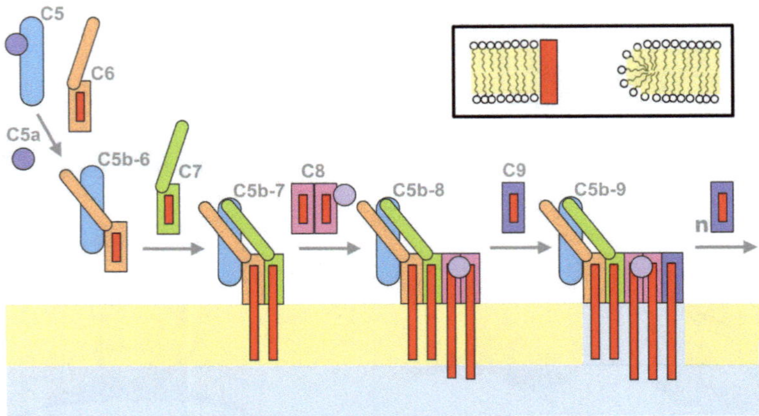

Fig. 6.8 Cartoon showing a possible assembly pathway of the MAC (see text). *Red* boxes depict *both* CH segments in the soluble and *both* TMHs/adjoining regions in the membrane inserted state. C5b-8 might be pore-forming (here not shown), C5b-9 is pore forming, addition of n C9 leads to polymerisation in circular fashion. The inset shows a section though a putative toroidal pore, where one edge of the pore is lined by lipids (*right*) that curve round and the other by protein segments (TMHs, *red*). For clarity only one lipid acyl chain is drawn

TMHs, at least from C7, anchors the complex to the membrane but without pore formation. Aleshin and co-workers suggested that the TMHs of C6 and C7 are amphiphilic enough to integrate into the membrane, but not long enough to span it entirely [2]. Note that the dissociation rate of C5b-7 from lipid membranes is similar to that of membrane proteins such as cytochrome b5 and cytochrome b5 reductase [43, 83]. When C5b-6 and C7 combine in the absence of a lipid bilayer, the resultant complexes form rosettes, which is indicative of CH unfolding/TMH formation and shielding of hydrophobic amino acids from the aqueous environment [135]. Indeed, C5b-6 at high concentration (above 1.5 mg/ml) can form helical paracrystals [120] that resemble the helical oligomers of pneumolysin [52], in which the CDC attains a domain arrangement similar to that of the pore, presumably with the TMHs oriented towards the centre of the helix; which exemplifies the potential of the MACPF domain of C5b-6 to refold and expose hydrophobic segments. These data suggest that the C5b-7 might already be integrated into the lipid bilayer via partial TMH formation. In the next step, C8β mediates the binding of C8 to a C5b-7 complex, in which the MACPF domains have already refolded. As MAC components have not been imaged to vertically collapse upon membrane pore formation, ample complementary interaction surfaces (see above) are available for C8β alongside C7 in membrane-integrated C5b-7, owing to the fact that they both attain the same height above the bilayer surface. C5b-8 is able to form small pores in artificial lipid membranes and a variety of target membranes [96, 138, 165, 186], suggesting that the eight TMHs of the four MACPF domains of the complex are already, at least in part, aligned side-on, similar to the configuration in the β-barrel pore. The relocation of C8γ upon C5b-8 formation frees a binding interface for C9 on C8α, nucleating addition of C9 and macromolecular pore formation [56]. Hadders and colleagues suggested that C8 membrane integration locally destabilises the lipid bilayer, facilitating the integration of C9 [55]. A mechanism that possibly reflects the ability of C5b-8 to form small, dynamic pores and lyse cells. The unfolding of MACPF domains following iterative addition of MAC components was suggested to be regulated by the movement of a module composed of TS and EGF domains, located towards the outside of the growing ring [2].

Three-dimensional cryo-EM reconstructions of pneumolysin, perforin, sC5b9 and the crystal structure of full-length C8, provide clues about the MACPF orientation in the MAC [52, 56, 82, 90, 170]. The cryo-EM reconstruction of the perforin pore together with atomic fitting of the crystal structure, suggested that perforin was oriented with the central four-stranded β-sheet and the CH segments facing towards the outer perimeter of the pore (see Fig. 6.7). This is in marked difference to the orientations of the β-sheet in the pneumolysin prepore and pore (see Fig. 6.7), of the MACPF domains of C6, C7, C8α, C8β and C9 in the soluble, but functional sC5b9 arc and of C8α and C8β in the full-length C8 crystal structure, in which the CH segments face the concave *interior* of the oligomer, placing the d2/TS domains towards the convex outside. As noted by Gilbert et al. [54], Hadders et al. [56] and Aleshin et al. [2] the outside-in orientation of perforin is astonishing and seems unlikely. However, as compellingly analysed and suggested earlier [54], the important salt-bridges that mediate oligomerisation of perforin [8], may be maintained in a *CDC-like* orientation, if one considers rotation of the C2 domain towards the centre or the pore.

A prepore oligomer forms atop of an intact membrane before CDCs can integrate into and disrupt the lipid bilayer. As elucidated above, considerable evidence points towards gradual pore enlargement as a mechanism for MAC assembly, at least in the initial stages: C5b-7 and C5b-8 insert into the membrane prior to macro-pore formation, under limiting C9 concentration partial arc-like pores are formed [139, 175], at low C9/C5b-8 ratios the pore size is proportional to the C9 content [92], addition of one C9 monomer can lead to pore formation without the need to assemble to typical tubular C9 [14], lipid probes indicate that full circular C9 assembly is not required for pore formation [4], C5b-9 channel sizes are considerably heterogeneous [10, 155, 157], and in the cryo-EM structure of sC5b-9 the CH segments are unfolded and detergent removal of protecting S-proteins can restore pore formation [11, 123]. If one accepts that gradual addition of C9 monomers to a membrane inserted C5b-8 result in pore formation, the immediate questions is: how can the mechanism be reconciled structurally in the context of a lipid membrane? Indeed Bhakdi [14], Podack [119], Tschopp [175], Malinski [92], Mayer [98] and co-workers have provided a possible explanation almost 30 years ago in the initial structural studies of MAC assembly: an incompletely-circularised, arc-like MAC oligomer can form a membrane pore in which the side not lined by protein is stabilised by a toroidal arrangement of lipids (see Fig. 6.8, inset). How the edges of the TMH β-sheet might be stabilised by the lipid head groups of the toroid or water molecules in the channel awaits rigorous experimental testing, but Gilbert et al. [54] have put forward a possible explanation. Gradual addition of C9 might also be achieved by constant opening and closing of a circular, membrane-embedded oligomer, to let in the *next* C9. However, such a scenario is unlikely, as the curvature of the MAC oligomers imaged so far is constant, even at limiting C9 concentrations.

Although the concept of MAC pore formation is now widely accepted, it should be noted that in the beginning of the mechanistic investigations into MAC cytotoxicity, considerable debate existed whether the MAC functioned via formation of an aqueous pore or via induction of leaky membrane patches in the immediate vicinity of the MAC oligomer [44]. In a critical review analysing seemingly contradictory results, Sucharit Bhakdi and Jørgen Tranum-Jensen stated that: … *the pore concept* [of complement lysis] *has never contended that the pores must inevitably be represented by circularised protein structures* […] *that pores must be visualised as rings in the electron microscope was never stated by any of the proponents of the pore concept* [15].

Perforin [97, 107, 119, 184, 185] and CDCs [17, 103, 111], were also imaged as incompletely-circularised oligomers on membranes. It was proposed that they also form arc-shaped pores with a toroidal lipid arrangement contributing to the pore [17, 51, 53, 54, 97, 107, 111]. However, the model of consecutive monomer addition seems at least in the case of CDCs unlikely, as the collapse of oligomeric prepores to pores is an inherent feature of the CDC mechanism [27, 170]. A mixed mechanism might apply here, where metastable, arc-shaped prepores can insert in an all-or-none event into membranes under limiting concentrations of monomer in solution (for extensive reviews see [51, 53, 54]).

Inhibition of MAC Assembly

Inhibition of MAC pore formation prevents lysis of host cells, especially in situations of complement activation. Vitronectin [171], clusterin [67] and apolipoproteins A-I and A-II [59] bind as fluid phase inhibitors to MAC components. The S-proteins vitronectin and clusterin engage C5b-7, forming sC5b-7, once the metastable, hydrophobic membrane binding site is formed [122, 134, 135]. The sC5b-7 complex allows binding of C8 and C9 but not polymerisation of C9 and assembly to a functional pore [125]. Detergent removal of the S-proteins from the sC5b-9 complex restores membrane binding and pore formation [11, 123]. Negative stain EM analysis together with biotin/streptavidin gold-labelling studies could image sC5b-9 as a wedge-shaped structure with vitronectin bound to the thicker side of the wedge [133]. The recent cryo-EM three-dimensional reconstruction of sC5b-9 showed that the complex was composed of a protrusion-carrying arc consisting of C5b-C6-C7-C8β-C8$\alpha\gamma$-C9 bound to a butterfly-like structure of S-proteins (see Fig. 6.7 [56]). Fitting of crystal structures into the cryo-EM map indicated that the 22° rotation of the C8α and C8β domains in the structure of C8 was maintained by all five MACPF domains of the arc [90]. The analysis suggested that inhibition of pore formation was mediated by binding of S-proteins to the unfolded CH segments below the arc and capping of the ends. Although earlier studies showed that S-proteins bound to newly formed hydrophobic regions (at least of C7) and that sC5b-9 carried neoepitopes resembling those of functional MAC [102], it was not at all clear that *all* components of the arc would exist with unfolded CH segments primed for membrane integration.

Interfacial inhibition of MAC assembly on homologous membranes is mediated by CD59, a GPI-anchored surface protein that binds to C8 and C9 thus preventing cell lysis [100, 101, 109]. CD59 is also the membrane receptor for intermedilysin (ILY), the CDC toxin from the human commensal *Streptococcus intermedius* that has been implicated in abscess formation in the brain, the liver and the lung [173]. Strikingly, CD59 prevents pore formation by the MAC, but enables it when the structurally homologous ILY is present. Free C8 and C9 in solution could only be bound by CD59 when they were partially unfolded [109] and it was suggested that CD59 binds to C8α in the C5b-8 complex and/or to C9 in the C5b-9 complex once they have undergone structural rearrangements [100, 142]. The binding site was mapped to a 6-residue sequence of C9 located in CH2 spanning residues 365–371 of the predicted TMH2 [63]. Similarly, C8α epitopes reside within residues 334–385 [88] most likely around amino acids 350–355 of TMH2 [63]. Binding of C8 to CD59 could be inhibited by incubation with peptides of the C9 epitope, suggesting that C8α and C9 interact with overlapping regions on CD59 [63]. Indeed, ILY and C8α/C9 share an overlapping binding site on CD59 [19, 64, 76, 181]. Hence, a recent co-crystal structure of ILY with CD59 also provided important insights into regulation of MAC assembly [68]. ILY bound with an extended β-hairpin of (d4) to two interfaces of CD59. The two interfaces of CD59 can be engaged simultaneously by two ILY monomers. Although MAC components C8α and C9 lack a domain that structurally resembles d4, mapping of epitopes to the predicted TMH2 and the requirement of partial unfolding, suggest that CD59 intercepts C8α and C9 *en route* to membrane integration when TMH2 has already partially formed. As suggested by Johnson et al. [68], CD59 might be located in between CD8α and C9 of the C5b-9 complex, simultaneously binding to their partially-formed TMH2, thus preventing further addition of C9 and efficient pore formation. The conformation of TMH2 that is recognised by CD59, might be similar to that postulated for a low temperature-trapped prepore intermediate of the MACPF structural homologue perfringolysin [148].

Conclusion and Outlook

Recent years have seen a wealth of new structural data providing snapshots of the MAC pore assembly mechanism. The structural homology of MACPF proteins to CDCs, despite their limited sequence similarity, is astonishing and opens up a whole set of new questions related to the structural evolution of the MACPF/CDC superfamily. Not all members seem to be pore-forming, suggesting that the canonical MACPF/CDC fold is flexible, both in terms of its propensity to transfold secondary structural elements and its usage.

Whereas the membrane-targeting mechanism of the MAC, that is engagement of the membrane, oligomerisation and pore formation, are reasonably well understood, secondary downstream effects, especially those elicited in host cells, are less well defined. The MAC functions in defense against bacterial infections, although deficiencies in components predominantly predispose for *Neisseria spp.* infections [115]. However, the terminal pathway of complement is also implicated in a variety of inflammatory disorders such as artherosclerosis and nephron damage [85, 94]. PFPs have been shown to activate transcription of inflammatory genes, which may result in part from direct or indirect engagement of Toll-like receptors. The inflammasome, a multiprotein intracellular signaling complex, is another host structure that is targeted by PFPs [160]. CDCs, in particular listeriolysin [42, 93] and pneumolysin [99, 182], can induce the NLRP3 inflammasome. While it is evident that large, circular MAC pores with tubular C9 can lyse cells, it is less clear whether all MAC effects require the strict formation of a giant, circular membrane β-barrel with approximately 72 β-strands. In some circumstances, effects may be more subtle. Recent studies suggest that sublytic concentrations of the MAC, akin to listeriolysin and pneumolysin, can induce the NLRP3 inflammasome via a transient calcium influx, resulting in release of the pro-inflammatory cytokine IL-1β from dendritic cells [79, 174]. Similarly, platelets treated with MAC progressively bud MAC-carrying vesicles into the extracellular medium with decreasing C9/C8 ratios, thus providing excess surface for the induction of pro-coagulatory pathways [104, 156]. Such a direct cross-talk between the effector and sensor arms of the innate immune system may shape the early immune response. In general, this interplay seems to be mediated by MACs formed under limiting component concentrations (sublytic MAC). It is tempting to speculate that arciform, toroidal membrane lesions can promote transient ion influx and budding of vesicles. It can be expected that more detailed information on the interaction of the MAC with host plasma membranes, and on the cell biological consequences of this, will allow for exciting insights into structure-function correlations of this versatile set of proteins.

Already at this point, the MAC assembly pathway is an attractive therapeutic target, as the recent introduction of the C5 antibody *eculizumab* against paroxysmal nocturnal hemoglobinuria shows [146]. An increased number of patents based on the complement pathway have been filed recently. Yet lasting therapeutic progress requires that it is paralleled by rigorous structural and functional analysis of the MAC.

Acknowledgment The Center for Chronic Immunodeficiency is supported by a Grant from the German Bundesministerium für Bildung und Forschung (BMBF 01EO0803). The author would like to thank the editors for the possibility to contribute to this volume.

References

1. Aleshin AE, DiScipio RG, Stec B, Liddington RC (2012) Crystal structure of C5b-6 suggests structural basis for priming assembly of the membrane attack complex. J Biol Chem 287:19642–19652
2. Aleshin AE, Schraufstatter IU, Stec B, Bankston LA, Liddington RC, DiScipio RG (2012) Structure of complement C6 suggests a mechanism for initiation and unidirectional, sequential assembly of membrane attack complex (MAC). J Biol Chem 287:10210–10222
3. Alouf JE (2001) Pore-forming bacterial protein toxins: an overview. Curr Top Microbiol Immunol 257:1–14
4. Amiguet P, Brunner J, Tschopp J (1985) The membrane attack complex of complement: lipid insertion of tubular and nontubular polymerized C9. Biochemistry 24:7328–7334
5. Amino R, Giovannini D, Thiberge S, Gueirard P, Boisson B, Dubremetz JF, Prevost MC, Ishino T, Yuda M, Menard R (2008) Host cell traversal is important for progression of the malaria parasite through the dermis to the liver. Cell Host Microbe 3:88–96
6. Anderluh G, Lakey JH (2008) Disparate proteins use similar architectures to damage membranes. Trends Biochem Sci 33:482–490
7. Bamford DH, Grimes JM, Stuart DI (2005) What does structure tell us about virus evolution? Curr Opin Struct Biol 15:655–663
8. Baran K, Dunstone M, Chia J, Ciccone A, Browne KA, Clarke CJP, Lukoyanova N, Saibil H, Whisstock JC, Voskoboinik I, Trapani JA (2009) The molecular basis for perforin oligomerization and transmembrane pore assembly. Immunity 30:684–695
9. Bariety J, Hinglais N, Bhakdi S, Mandet C, Rouchon M, Kazatchkine MD (1989) Immunohistochemical study of complement S protein (vitronectin) in normal and diseased human kidneys: relationship to neoantigens of the C5b-9 terminal complex. Clin Exp Immunol 75:76–81
10. Benz R, Schmid A, Wiedmer T, Sims PJ (1986) Single-channel analysis of the conductance fluctuations induced in lipid bilayer membranes by complement proteins C5b-9. J Membr Biol 94:37–45
11. Bhakdi S, Bhakdi-Lehnen B, Tranum-Jensen J (1979) Proteolytic transformation of SC5b-9 into an amphiphilic macromolecule resembling the C5b-9 membrane attack complex of complement. Immunology 37:901–912
12. Bhakdi S, Kuller G, Muhly M, Fromm S, Seibert G, Parrisius J (1987) Formation of transmural complement pores in serum-sensitive Escherichia coli. Infect Immun 55:206–210
13. Bhakdi S, Tranum-Jensen J (1984) On the cause and nature of C9-related heterogeneity of terminal complement complexes generated on target erythrocytes through the action of whole serum. J Immunol 133:1453–1463
14. Bhakdi S, Tranum-Jensen J (1986) C5b-9 assembly: average binding of one C9 molecule to C5b-8 without poly-C9 formation generates a stable transmembrane pore. J Immunol 136:2999–3005
15. Bhakdi S, Tranum-Jensen J (1991) Complement lysis: a hole is a hole. Immunol Today 12:318–320 (discussion 321)
16. Bhakdi S, Tranum-Jensen J, Klump O (1980) The terminal membrane C5b-9 complex of human complement. Evidence for the existence of multiple protease-resistant polypeptides that form the trans-membrane complement channel. J Immunol 124:2451–2457
17. Bhakdi S, Tranum-Jensen J, Sziegoleit A (1985) Mechanism of membrane damage by streptolysin-O. Infect Immun 47:52–60
18. Biesecker G, Lachmann P, Henderson R (1993) Structure of complement poly-C9 determined in projection by cryo-electron microscopy and single particle analysis. Mol Immunol 30:1369–1382

19. Bodian DL, Davis SJ, Morgan BP, Rushmere NK (1997) Mutational analysis of the active site and antibody epitopes of the complement-inhibitory glycoprotein, CD59. J Exp Med 185:507–516
20. Bork P, Beckmann G (1993) The CUB domain. A widespread module in developmentally regulated proteins. J Mol Biol 231:539–545
21. Boros T, Dourmashkin RR, Humphrey JH (1964) Lesions in erythrocyte membranes caused by immune haemolysis. Nature 202:251–252
22. Bramham J, Thai CT, Soares DC, Uhrin D, Ogata RT, Barlow PN (2005) Functional insights from the structure of the multifunctional C345C domain of C5 of complement. J Biol Chem 280:10636–10645
23. Brannen CL, Sodetz JM (2007) Incorporation of human complement C8 into the membrane attack complex is mediated by a binding site located within the C8beta MACPF domain. Mol Immunol 44:960–965
24. Bubeck D, Roversi P, Donev R, Morgan BP, Llorca O, Lea SM (2011) Structure of human complement C8, a precursor to membrane attack. J Mol Biol 405:325–330
25. Chakravarti DN, Chakravarti B, Parra CA, Müller-Eberhard HJ (1989) Structural homology of complement protein C6 with other channel-forming proteins of complement. Proc Natl Acad Sci USA 86:2799–2803
26. Chiswell B, Lovelace LL, Brannen C, Ortlund EA, Lebioda L, Sodetz JM (2007) Structural features of the ligand binding site on human complement protein C8gamma: a member of the lipocalin family. Biochim Biophys Acta 1774:637–644
27. Czajkowsky DM, Hotze EM, Shao Z, Tweten RK (2004) Vertical collapse of a cytolysin prepore moves its transmembrane beta-hairpins to the membrane. EMBO J 23:3206–3215
28. Dankert JR, Esser AF (1985) Proteolytic modification of human complement protein C9: loss of poly(C9) and circular lesion formation without impairment of function. Proc Natl Acad Sci USA 82:2128–2132
29. Dankert JR, Shiver JW, Esser AF (1985) Ninth component of complement: self-aggregation and interaction with lipids. Biochemistry 24:2754–2762
30. DiScipio RG (1981) The conversion of human complement component C5 into fragment C5b by the alternative-pathway C5 convertase. Biochem J 199:497–504
31. DiScipio RG (1992) Formation and structure of the C5b-7 complex of the lytic pathway of complement. J Biol Chem 267:17087–17094
32. DiScipio RG, Chakravarti DN, Müller-Eberhard HJ, Fey GH (1988) The structure of human complement component C7 and the C5b-7 complex. J Biol Chem 263:549–560
33. DiScipio RG, Gehring MR, Podack ER, Kan CC, Hugli TE, Fey GH (1984) Nucleotide sequence of cDNA and derived amino acid sequence of human complement component C9. Proc Natl Acad Sci USA 81:7298–7302
34. DiScipio RG, Hugli TE (1985) The architecture of complement component C9 and poly(C9). J Biol Chem 260:14802–14809
35. DiScipio RG, Hugli TE (1989) The molecular architecture of human complement component C6. J Biol Chem 264:16197–16206
36. DiScipio RG, Linton SM, Rushmere NK (1999) Function of the factor I modules (FIMS) of human complement component C6. J Biol Chem 274:31811–31818
37. DiScipio RG, Smith CA, Müller-Eberhard HJ, Hugli TE (1983) The activation of human complement component C5 by a fluid phase C5 convertase. J Biol Chem 258:10629–10636
38. Dowd KJ, Tweten RK (2012) The cholesterol-dependent cytolysin signature motif: a critical element in the allosteric pathway that couples membrane binding to pore assembly. PLoS Pathog 8:e1002787
39. Dramsi S, Cossart P (2003) Listeriolysin O-mediated calcium influx potentiates entry of Listeria monocytogenes into the human Hep-2 epithelial cell line. Infect Immun 71:3614–3618
40. Dunstone MA, Tweten RK (2012) Packing a punch: the mechanism of pore formation by cholesterol dependent cytolysins and membrane attack complex/perforin-like proteins. Curr Opin Struct Biol 22:342–349

41. Edmondson JC, Liem RK, Kuster JE, Hatten ME (1988) Astrotactin: a novel neuronal cell surface antigen that mediates neuron-astroglial interactions in cerebellar microcultures. J Cell Biol 106:505–517
42. Eitel J, Suttorp N, Opitz B (2010) Innate immune recognition and inflammasome activation in listeria monocytogenes infection. Front Microbiol 1:149
43. Enoch HG, Fleming PJ, Strittmatter P (1977) Cytochrome b5 and cytochrome b5 reductase-phospholipid vesicles. Intervesicle protein transfer and oreintation factors in protein-protein interactions. J Biol Chem 252:5656–5660
44. Esser AF (1991) Big MAC attack: complement proteins cause leaky patches. Immunol Today 12:316–318 (discussion 321)
45. Forneris F, Wu J, Gros P (2012) The modular serine proteases of the complement cascade. Curr Opin Struct Biol 22:333–341
46. Fredslund F, Jenner L, Husted LB, Nyborg J, Andersen GR, Sottrup-Jensen L (2006) The structure of bovine complement component three reveals the basis for thioester function. J Mol Biol 361:115–127
47. Fredslund F, Laursen NS, Roversi P, Jenner L, Oliveira CLP, Pedersen JS, Nunn MA, Lea SM, Discipio R, Sottrup-Jensen L, Andersen GR (2008) Structure of and influence of a tick complement inhibitor on human complement component 5. Nat Immunol 9:753–760
48. Gedde MM, Higgins DE, Tilney LG, Portnoy DA (2000) Role of listeriolysin O in cell-to-cell spread of Listeria monocytogenes. Infect Immun 68:999–1003
49. Gerard C, Gerard NP (1994) C5A anaphylatoxin and its seven transmembrane-segment receptor. Annu Rev Immunol 12:775–808
50. Gerard NP, Gerard C (1991) The chemotactic receptor for human C5a anaphylatoxin. Nature 349:614–617
51. Gilbert RJ (2005) Inactivation and activity of cholesterol-dependent cytolysins: what structural studies tell us. Structure 13:1097–1106
52. Gilbert RJ, Jiménez JL, Chen S, Tickle IJ, Rossjohn J, Parker M, Andrew PW, Saibil HR (1999) Two structural transitions in membrane pore formation by pneumolysin, the pore-forming toxin of streptococcus pneumoniae. Cell 97:647–655
53. Gilbert RJC (2010) Cholesterol-dependent cytolysins. Adv Exp Med Biol 677:56–66
54. Gilbert RJC, Mikelj M, Dalla Serra M, Froelich CJ, Anderluh G (2013) Effects of MACPF/CDC proteins on lipid membranes. Cell Mol Life Sci 70:2083–2098
55. Hadders MA, Beringer DX, Gros P (2007) Structure of C8alpha-MACPF reveals mechanism of membrane attack in complement immune defense. Science 317:1552–1554
56. Hadders MA, Bubeck D, Roversi P, Hakobyan S, Forneris F, Morgan BP, Pangburn MK, Llorca O, Lea SM, Gros P (2012) Assembly and regulation of the membrane attack complex based on structures of C5b6 and sC5b9. Cell Rep 1:200–207
57. Hadding U, Muller-Eberhard HJ (1969) The ninth component of human complement: isolation, description and mode of action. Immunology 16:719–735
58. Haefliger JA, Tschopp J, Vial N, Jenne DE (1989) Complete primary structure and functional characterization of the sixth component of the human complement system. Identification of the C5b-binding domain in complement C6. J Biol Chem 264:18041–18051
59. Hamilton KK, Zhao J, Sims PJ (1993) Interaction between apolipoproteins A-I and A-II and the membrane attack complex of complement. Affinity of the apoproteins for polymeric C9. J Biol Chem 268:3632–3638
60. Hammer CH, Nicholson A, Mayer MM (1975) On the mechanism of cytolysis by complement: evidence on insertion of C5b and C7 subunits of the C5b, 6, 7 complex into phospholipid bilayers of erythrocyte membranes. Proc Natl Acad Sci USA 72:5076–5080
61. Hobart MJ, Fernie BA, DiScipio RG (1995) Structure of the human C7 gene and comparison with the C6, C8A, C8B, and C9 genes. J Immunol 154:5188–5194
62. Hotze EM, Tweten RK (2012) Membrane assembly of the cholesterol-dependent cytolysin pore complex. Biochim Biophys Acta 1818:1028–1038

63. Huang Y, Qiao F, Abagyan R, Hazard S, Tomlinson S (2006) Defining the CD59-C9 binding interaction. J Biol Chem 281:27398–27404
64. Huang Y, Smith CA, Song H, Morgan BP, Abagyan R, Tomlinson S (2005) Insights into the human CD59 complement binding interface toward engineering new therapeutics. J Biol Chem 280:34073–34079
65. Huber-Lang M, Sarma JV, Zetoune FS, Rittirsch D, Neff TA, McGuire SR, Lambris JD, Warner RL, Flierl MA, Hoesel LM, Gebhard F, Younger JG, Drouin SM, Wetsel RA, Ward PA (2006) Generation of C5a in the absence of C3: a new complement activation pathway. Nat Med 12:682–687
66. Janssen BJ, Huizinga EG, Raaijmakers HC, Roos A, Daha MR, Nilsson-Ekdahl K, Nilsson B, Gros P (2005) Structures of complement component C3 provide insights into the function and evolution of immunity. Nature 437:505–511
67. Jenne DE, Tschopp J (1992) Clusterin: the intriguing guises of a widely expressed glycoprotein. Trends Biochem Sci 17:154–159
68. Johnson S, Brooks NJ, Smith RAG, Lea SM, Bubeck D (2013) Structural basis for recognition of the pore-forming toxin intermedilysin by human complement receptor CD59. Cell Rep 3:1369–1377
69. Joiner KA, Schmetz MA, Sanders ME, Murray TG, Hammer CH, Dourmashkin R, Frank MM (1985) Multimeric complement component C9 is necessary for killing of Escherichia coli J5 by terminal attack complex C5b-9. Proc Natl Acad Sci USA 82:4808–4812
70. Kafsack BF, Pena JD, Coppens I, Ravindran S, Boothroyd JC, Carruthers VB (2009) Rapid membrane disruption by a perforin-like protein facilitates parasite exit from host cells. Science 323:530–533
71. Kidmose RT, Laursen NS, Dobó J, Kjaer TR, Sirotkina S, Yatime L, Sottrup-Jensen L, Thiel S, Gál P, Andersen GR (2012) Structural basis for activation of the complement system by component C4 cleavage. Proc Natl Acad Sci USA 109:15425–15430
72. Kolb WP, Haxby JA, Arroyave CM, Müller-Eberhard HJ (1973) The membrane attack mechanism of complement. Reversible interactions among the five native components in free solution. J Exp Med 138:428–437
73. Kolb WP, Müller-Eberhard HJ (1974) Mode of action of human C9: adsorption of multiple C9 molecules to cell-bound C8. J Immunol 113:479–488
74. Kondos SC, Hatfaludi T, Voskoboinik I, Trapani JA, Law RHP, Whisstock JC, Dunstone MA (2010) The structure and function of mammalian membrane-attack complex/perforin-like proteins. Tissue Antigens 76:341–351
75. Kontermann R, Rauterberg EW (1989) N-deglycosylation of human complement component C9 reduces its hemolytic activity. Mol Immunol 26:1125–1132
76. LaChapelle S, Tweten RK, Hotze EM (2009) Intermedilysin-receptor interactions during assembly of the pore complex: assembly intermediates increase host cell susceptibility to complement-mediated lysis. J Biol Chem 284:12719–12726
77. Lachmann PJ, Thompson RA (1970) Reactive lysis: the complement-mediated lysis of unsensitized cells. II. The characterization of activated reactor as C56 and the participation of C8 and C9. J Exp Med 131:643–657
78. Laine RO, Esser AF (1989) Identification of the discontinuous epitope in human complement protein C9 recognized by anti-melittin antibodies. J Immunol 143:553–557
79. Laudisi F, Spreafico R, Evrard M, Hughes TR, Mandriani B, Kandasamy M, Morgan BP, Sivasankar B, Mortellaro A (2013) Cutting edge: the NLRP3 inflammasome links complement-mediated inflammation and IL-1β release. J Immunol 191:1006–1010
80. Laursen NS, Andersen KR, Braren I, Spillner E, Sottrup-Jensen L, Andersen GR (2011) Substrate recognition by complement convertases revealed in the C5-cobra venom factor complex. EMBO J 30:606–616
81. Laursen NS, Gordon N, Hermans S, Lorenz N, Jackson N, Wines B, Spillner E, Christensen JB, Jensen M, Fredslund F, Bjerre M, Sottrup-Jensen L, Fraser JD, Andersen GR (2010) Structural basis for inhibition of complement C5 by the SSL7 protein from Staphylococcus aureus. Proc Natl Acad Sci USA 107:3681–3686

82. Law RHP, Lukoyanova N, Voskoboinik I, Caradoc-Davies TT, Baran K, Dunstone MA, D'Angelo ME, Orlova EV, Coulibaly F, Verschoor S, Browne KA, Ciccone A, Kuiper MJ, Bird PI, Trapani JA, Saibil HR, Whisstock JC (2010) The structural basis for membrane binding and pore formation by lymphocyte perforin. Nature 468:447–451
83. Leto TL, Roseman MA, Holloway PW (1980) Mechanism of exchange of cytochrome b5 between phosphatidylcholine vesicles. Biochemistry 19:1911–1916
84. Letunic I, Doerks T, Bork P (2012) SMART 7: recent updates to the protein domain annotation resource. Nucleic Acids Res 40:D302–D305 (Database issue)
85. Lewis RD, Jackson CL, Morgan BP, Hughes TR (2010) The membrane attack complex of complement drives the progression of atherosclerosis in apolipoprotein E knockout mice. Mol Immunol 47:1098–1105
86. Lichtenheld MG, Olsen KJ, Lu P, Lowrey DM, Hameed A, Hengartner H, Podack ER (1988) Structure and function of human perforin. Nature 335:448–451
87. Linscott WD, Nishioka K (1963) Components of Guinea pig complement. Ii. Separation of serum fractions essential for immune hemolysis. J Exp Med 118:795–815
88. Lockert DH, Kaufman KM, Chang CP, Husler T, Sodetz JM, Sims PJ (1995) Identity of the segment of human complement C8 recognized by complement regulatory protein CD59. J Biol Chem 270:19723–19728
89. Lovelace LL, Chiswell B, Slade DJ, Sodetz JM, Lebioda L (2008) Crystal structure of complement protein C8gamma in complex with a peptide containing the C8gamma binding site on C8alpha: implications for C8gamma ligand binding. Mol Immunol 45:750–756
90. Lovelace LL, Cooper CL, Sodetz JM, Lebioda L (2011) Structure of human C8 protein provides mechanistic insight into membrane pore formation by complement. J Biol Chem 286:17585–17592
91. Lukoyanova N, Saibil HR (2008) Friend or foe: the same fold for attack and defense. Trends Immunol 29:51–53
92. Malinski JA, Nelsestuen GL (1989) Membrane permeability to macromolecules mediated by the membrane attack complex. Biochemistry 28:61–70
93. Mariathasan S, Weiss DS, Newton K, McBride J, O'Rourke K, Roose-Girma M, Lee WP, Weinrauch Y, Monack DM, Dixit VM (2006) Cryopyrin activates the inflammasome in response to toxins and ATP. Nature 440:228–232
94. Markiewski MM, Lambris JD (2007) The role of complement in inflammatory diseases from behind the scenes into the spotlight. Am J Pathol 171:715–727
95. Markiewski MM, Nilsson B, Ekdahl KN, Mollnes TE, Lambris JD (2007) Complement and coagulation: strangers or partners in crime? Trends Immunol 28:184–192
96. Martin DE, Chiu FJ, Gigli I, Muller-Eberhard HJ (1987) Killing of human melanoma cells by the membrane attack complex of human complement as a function of its molecular composition. J Clin Invest 80:226–233
97. Masson D, Tschopp J (1985) Isolation of a lytic, pore-forming protein (perforin) from cytolytic T-lymphocytes. J Biol Chem 260:9069–9072
98. Mayer MM (1972) Mechanism of cytolysis by complement. Proc Natl Acad Sci USA 69:2954–2958
99. McNeela EA, Burke A, Neill DR, Baxter C, Fernandes VE, Ferreira D, Smeaton S, El-Rachkidy R, McLoughlin RM, Mori A, Moran B, Fitzgerald KA, Tschopp J, Pétrilli V, Andrew PW, Kadioglu A, Lavelle EC (2010) Pneumolysin activates the NLRP3 inflammasome and promotes proinflammatory cytokines independently of TLR4. PLoS Pathog 6:e1001191
100. Meri S, Morgan BP, Davies A, Daniels RH, Olavesen MG, Waldmann H, Lachmann PJ (1990) Human protectin (CD59), an 18,000–20,000 MW complement lysis restricting factor, inhibits C5b-8 catalysed insertion of C9 into lipid bilayers. Immunology 71:1–9
101. Meri S, Morgan BP, Wing M, Jones J, Davies A, Podack E, Lachmann PJ (1990) Human protectin (CD59), an 18-20-kD homologous complement restriction factor, does not restrict perforin-mediated lysis. J Exp Med 172:367–370

102. Mollnes TE, Lea T, Tschopp J (1989) Activation-dependent epitopes in the terminal complement pathway. Complement Inflamm 6:223–235
103. Morgan PJ, Hyman SC, Byron O, Andrew PW, Mitchell TJ, Rowe AJ (1994) Modeling the bacterial protein toxin, pneumolysin, in its monomeric and oligomeric form. J Biol Chem 269:25315–25320
104. Moskovich O, Fishelson Z (2007) Live cell imaging of outward and inward vesiculation induced by the complement c5b-9 complex. J Biol Chem 282(41):29977–29986
105. Müller-Eberhard HJ (1986) The membrane attack complex of complement. Ann Rev Immunol 4:503–528
106. Müller-Eberhard HJ, Schreiber RD (1980) Molecular biology and chemistry of the alternative pathway of complement. Adv Immunol 29:1–53
107. Nabholz M, Tschopp J (1989) CTL-mediated cytolysis: perforin and alternative pathways? Immunol Lett 20:179–180
108. Ng SC, Rao AG, Howard OM, Sodetz JM (1987) The eighth component of human complement: evidence that it is an oligomeric serum protein assembled from products of three different genes. Biochemistry 26:5229–5233
109. Ninomiya H, Sims PJ (1992) The human complement regulatory protein CD59 binds to the alpha-chain of C8 and to the "b"domain of C9. J Biol Chem 267:13675–13680
110. Ortlund E, Parker CL, Schreck SF, Ginell S, Minor W, Sodetz JM, Lebioda L (2002) Crystal structure of human complement protein C8gamma at 1.2 Å resolution reveals a lipocalin fold and a distinct ligand binding site. Biochemistry 41:7030–7037
111. Palmer M, Harris R, Freytag C, Kehoe M, Tranum-Jensen J, Bhakdi S (1998) Assembly mechanism of the oligomeric streptolysin O pore: the early membrane lesion is lined by a free edge of the lipid membrane and is extended gradually during oligomerization. EMBO J 17:1598–1605
112. Parker CL, Sodetz JM (2002) Role of the human C8 subunits in complement-mediated bacterial killing: evidence that C8 gamma is not essential. Mol Immunol 39:453–458
113. Peitsch MC, Amiguet P, Guy R, Brunner J, Maizel JV, Tschopp J (1990) Localization and molecular modelling of the membrane-inserted domain of the ninth component of human complement and perforin. Mol Immunol 27:589–602
114. Perkins SJ, Smith KF, Nealis AS, Lachmann PJ, Harrison RA (1990) Structural homologies of component C5 of human complement with components C3 and C4 by neutron scattering. Biochemistry 29:175–180
115. Petersen BH, Lee TJ, Snyderman R, Brooks GF (1979) Neisseria meningitidis and Neisseria gonorrhoeae bacteremia associated with C6, C7, or C8 deficiency. Ann Intern Med 90:917–920
116. Phelan MM, Thai C-T, Herbert AP, Bella J, Uhrín D, Ogata RT, Barlow PN, Bramham J (2009) 1H, 15 N and 13C resonance assignment of the pair of Factor-I like modules of the complement protein C7. Biomol NMR Assign 3:49–52
117. Phelan MM, Thai C-T, Soares DC, Ogata RT, Barlow PN, Bramham J (2009) Solution structure of factor I-like modules from complement C7 reveals a pair of follistatin domains in compact pseudosymmetric arrangement. J Biol Chem 284:19637–19649
118. Podack ER (1984) Molecular composition of the tubular structure of the membrane attack complex of complement. J Biol Chem 259:8641–8647
119. Podack ER, Dennert G (1983) Assembly of two types of tubules with putative cytolytic function by cloned natural killer cells. Nature 302:442–445
120. Podack ER, Esser AF, Biesecker G, Müller-Eberhard HJ (1980) Membrane attack complex of complement: a structural analysis of its assembly. J Exp Med 151:301–313
121. Podack ER, Kolb WP, Esser AF, Müller-Eberhard HJ (1979) Structural similarities between C6 and C7 of human complement. J Immunol 123:1071–1077
122. Podack ER, Kolb WP, Muller-Eberhard HJ (1977) The SC5b-7 complex: formation, isolation, properties, and subunit composition. J Immunol 119:2024–2029
123. Podack ER, Müller-Eberhard HJ (1980) SC5b-9 complex of complement: formation of the dimeric membrane attack complex by removal of S-protein. J Immunol 124:1779–1783

124. Podack ER, Müller-Eberhard HJ, Horst H, Hoppe W (1982) Membrane attach complex of complement (MAC): three-dimensional analysis of MAC-phospholipid vesicle recombinants. J Immunol 128:2353–2357
125. Podack ER, Preissner KT, Müller-Eberhard HJ (1984) Inhibition of C9 polymerization within the SC5b-9 complex of complement by S-protein. Acta Pathol Microbiol Immunol Scand Suppl 284:89–96
126. Podack ER, Tschoop J, Müller-Eberhard HJ (1982) Molecular organization of C9 within the membrane attack complex of complement. Induction of circular C9 polymerization by the C5b-8 assembly. J Exp Med 156:268–282
127. Podack ER, Tschopp J (1982) Circular polymerization of the ninth component of complement. Ring closure of the tubular complex confers resistance to detergent dissociation and to proteolytic degradation. J Biol Chem 257:15204–15212
128. Podack ER, Tschopp J (1982) Polymerization of the ninth component of complement (C9): formation of poly(C9) with a tubular ultrastructure resembling the membrane attack complex of complement. Proc Natl Acad Sci USA 79:574–578
129. Podack ER, Tschopp J (1984) Membrane attack by complement. Mol Immunol 21:589–603
130. Polley MJ, Mueller-Eberhard HJ, Feldman JD (1971) Production of ultrastructural membrane lesions by the fifth component of complement 133:53–62
131. Ponting CP (1999) Chlamydial homologues of the MACPF (MAC/perforin) domain. Curr Biol 9:R911–R913
132. Praper T, Sonnen AF-P, Kladnik A, Andrighetti AO, Viero G, Morris KJ, Volpi E, Lunelli L, Dalla Serra M, Froelich CJ, Gilbert RJC, Anderluh G (2011) Perforin activity at membranes leads to invaginations and vesicle formation. Proc Natl Acad Sci USA 108:21016–21021
133. Preissner KP, Podack ER, Müller-Eberhard HJ (1989) SC5b-7, SC5b-8 and SC5b-9 complexes of complement: ultrastructure and localization of the S-protein (vitronectin) within the macromolecules. Eur J Immunol 19:69–75
134. Preissner KT (1991) Structure and biological role of vitronectin. Annu Rev Cell Biol 7:275–310
135. Preissner KT, Podack ER, Müller-Eberhard HJ (1985) The membrane attack complex of complement: relation of C7 to the metastable membrane binding site of the intermediate complex C5b-7. J Immunol 135:445–451
136. Preissner KT, Podack ER, Müller-Eberhard HJ (1985) Self-association of the seventh component of human complement (C7): dimerization and polymerization. J Immunol 135:452–458
137. Ramachandran R, Tweten RK, Johnson AE (2004) Membrane-dependent conformational changes initiate cholesterol-dependent cytolysin oligomerization and intersubunit beta-strand alignment. Nat Struct Mol Biol 11:697–705
138. Ramm LE, Whitlow MB, Mayer MM (1982) Transmembrane channel formation by complement: functional analysis of the number of C5b6, C7, C8, and C9 molecules required for a single channel. Proc Natl Acad Sci USA 79:4751–4755
139. Ramm LE, Whitlow MB, Mayer MM (1985) The relationship between channel size and the number of C9 molecules in the C5b-9 complex. J Immunol 134:2594–2599
140. Rawal N, Pangburn M (2001) Formation of high-affinity C5 convertases of the alternative pathway of complement. J Immunol 166:2635–2642
141. Rawal N, Pangburn MK (2003) Formation of high affinity C5 convertase of the classical pathway of complement. J Biol Chem 278:38476–38483
142. Rollins SA, Zhao J, Ninomiya H, Sims PJ (1991) Inhibition of homologous complement by CD59 is mediated by a species-selective recognition conferred through binding to C8 within C5b-8 or C9 within C5b-9. J Immunol 146:2345–2351
143. Rosado CJ, Buckle AM, Law RHP, Butcher RE, Kan W-T, Bird CH, Ung K, Browne KA, Baran K, Bashtannyk-Puhalovich TA, Faux NG, Wong W, Porter CJ, Pike RN, Ellisdon AM, Pearce MC, Bottomley SP, Emsley J, Smith AI, Rossjohn J, Hartland EL, Voskoboinik I,

Trapani JA, Bird PI, Dunstone MA, Whisstock JC (2007) A common fold mediates vertebrate defense and bacterial attack. Science 317:1548–1551
144. Rosado CJ, Kondos S, Bull TE, Kuiper MJ, Law RHP, Buckle AM, Voskoboinik I, Bird PI, Trapani JA, Whisstock JC, Dunstone MA (2008) The MACPF/CDC family of pore-forming toxins. Cell Microbiol 10:1765–1774
145. Rossjohn J, Feil SC, McKinstry WJ, Tweten RK, Parker MW (1997) Structure of a cholesterol-binding, thiol-activated cytolysin and a model of its membrane form. Cell 89:685–692
146. Rother RP, Rollins SA, Mojcik CF, Brodsky RA, Bell L (2007) Discovery and development of the complement inhibitor eculizumab for the treatment of paroxysmal nocturnal hemoglobinuria. Nat Biotechnol 25:1256–1264
147. Rubins JB, Duane PG, Clawson D, Charboneau D, Young J, Niewoehner DE (1993) Toxicity of pneumolysin to pulmonary alveolar epithelial cells. Infect Immun 61(4):1352–1358
148. Sato TK, Tweten RK, Johnson AE (2013) Disulfide-bond scanning reveals assembly state and β-strand tilt angle of the PFO β-barrel. Nat Chem Biol 9:383–389
149. Schreck SF, Plumb ME, Platteborze PL, Kaufman KM, Michelotti GA, Letson CS, Sodetz JM (1998) Expression and characterization of recombinant subunits of human complement component C8: further analysis of the function of C8 alpha and C8 gamma. J Immunol 161:311–318
150. Shatursky O, Heuck AP, Shepard LA, Rossjohn J, Parker MW, Johnson AE, Tweten RK (1999) The mechanism of membrane insertion for a cholesterol-dependent cytolysin: a novel paradigm for pore-forming toxins. Cell 99:293–299
151. Shepard LA, Heuck AP, Hamman BD, Rossjohn J, Parker MW, Ryan KR, Johnson AE, Tweten RK (1998) Identification of a membrane-spanning domain of the thiol-activated pore-forming toxin Clostridium perfringens perfringolysin O: an alpha-helical to beta-sheet transition identified by fluorescence spectroscopy. Biochemistry 37:14563–14574
152. Shinkai Y, Takio K, Okumura K (1988) Homology of perforin to the ninth component of complement (C9). Nature 334:525–527
153. Silversmith RE, Nelsestuen GL (1986) Interaction of complement proteins C5b-6 and C5b-7 with phospholipid vesicles: effects of phospholipid structural features. Biochemistry 25:7717–7725
154. Sim RB, Tsiftsoglou SA (2004) Proteases of the complement system. Biochem Soc Trans 32:21–27
155. Sims PJ (1983) Complement pores in erythrocyte membranes. Analysis of C8/C9 binding required for functional membrane damage. Biochim Biophys Acta 732:541–552
156. Sims PJ, Faioni EM, Wiedmer T, Shattil SJ (1988) Complement proteins C5b-9 cause release of membrane vesicles from the platelet surface that are enriched in the membrane receptor for coagulation factor Va and express prothrombinase activity. J Biol Chem 263:18205–18212
157. Sims PJ, Lauf PK (1980) Analysis of solute diffusion across the C5b-9 membrane lesion of complement: evidence that individual C5b-9 complexes do not function as discrete, uniform pores. J Immunol 125:2617–2625
158. Slade DJ, Lovelace LL, Chruszcz M, Minor W, Lebioda L, Sodetz JM (2008) Crystal structure of the MACPF domain of human complement protein C8 alpha in complex with the C8 gamma subunit. J Mol Biol 379:331–342
159. Smith KF, Harrison RA, Perkins SJ (1992) Molecular modeling of the domain structure of C9 of human complement by neutron and X-ray solution scattering. Biochemistry 31:754–764
160. Sonnen AF-P, Henneke P (2013) Role of pore-forming toxins in neonatal sepsis. Clin Dev Immunol 2013:608456
161. Steckel EW, Welbaum BE, Sodetz JM (1983) Evidence of direct insertion of terminal complement proteins into cell membrane bilayers during cytolysis. Labeling by a photosensitive membrane probe reveals a major role for the eighth and ninth components. J Biol Chem 258:4318–4324

162. Stevens LM, Frohnhofer HG, Klingler M, Nusslein-Volhard C (1990) Localized requirement for torso-like expression in follicle cells for development of terminal anlagen of the Drosophila embryo. Nature 346:660–663
163. Stewart JL, Kolb WP, Sodetz JM (1987) Evidence that C5b recognizes and mediates C8 incorporation into the cytolytic complex of complement. J Immunol 139:1960–1964
164. Stewart JL, Monahan JB, Brickner A, Sodetz JM (1984) Measurement of the ratio of the eighth and ninth components of human complement on complement-lysed membranes. Biochemistry 23:4016–4022
165. Stolfi RL (1968) Immune lytic transformation: a state of irreversible damage generated as a result of the reaction of the eighth component in the guinea pig complement system. J Immunol 100:46–54
166. Taylor KM, Trimby AR, Campbell AK (1997) Mutation of recombinant complement component C9 reveals the significance of the N-terminal region for polymerization. Immunology 91:20–27
167. Thai C-T, Ogata RT (2004) Complement components C5 and C7: recombinant factor I modules of C7 bind to the C345C domain of C5. J Immunol 173:4547–4552
168. Thai C-T, Ogata RT (2005) Recombinant C345C and factor I modules of complement components C5 and C7 inhibit C7 incorporation into the complement membrane attack complex. J Immunol 174:6227–6232
169. Thompson RA, Lachmann PJ (1970) Reactive lysis: the complement-mediated lysis of unsensitized cells. I. The characterization of the indicator factor and its identification as C7. J Exp Med 131:629–641
170. Tilley SJ, Orlova EV, Gilbert RJC, Andrew PW, Saibil HR (2005) Structural basis of pore formation by the bacterial toxin pneumolysin. Cell 121:247–256
171. Tomasini BR, Mosher DF (1986) On the identity of vitronectin and S-protein: immunological crossreactivity and functional studies. Blood 68:737–742
172. Torisu M, Yokoyama T, Kohler PF, Durst AL, Martineau G, Schroter G, Amemiya H, Groth CG, Starzl TE (1972) Serum complement after orthotopic transplantation of the human liver. Clin Exp Immunol 12:21–30
173. Trabue C, Pearman R, Doering T (2013) Pyogenic brain and lung abscesses due to streptococcus intermedius. J Gen Intern Med. doi:10.1007/s11606-013-2565-3
174. Triantafilou K, Hughes TR, Triantafilou M, Morgan BP (2013) The complement membrane attack complex triggers intracellular Ca2+ fluxes leading to NLRP3 inflammasome activation. J Cell Sci 126:2903–2913
175. Tschopp J (1984) Ultrastructure of the membrane attack complex of complement. Heterogeneity of the complex caused by different degree of C9 polymerization. J Biol Chem 259:7857–7863
176. Tschopp J, Engel A, Podack ER (1984) Molecular weight of poly(C9). 12 to 18 C9 molecules form the transmembrane channel of complement. J Biol Chem 259:1922–1928
177. Tschopp J, Masson D, Stanley KK (1986) Structural/functional similarity between proteins involved in complement- and cytotoxic T-lymphocyte-mediated cytolysis. Nature 322:831–834
178. Tschopp J, Podack ER (1981) Membranolysis by the ninth component of human complement. Biochem Biophys Res Commun 100:1409–1414
179. Tschopp J, Podack ER, Müller-Eberhard HJ (1982) Ultrastructure of the membrane attack complex of complement: detection of the tetramolecular C9-polymerizing complex C5b-8. Proc Natl Acad Sci USA 79:7474–7478
180. Voskoboinik I, Smyth MJ, Trapani JA (2006) Perforin-mediated target-cell death and immune homeostasis. Nat Rev Immunol 6:940–952
181. Wickham SE, Hotze EM, Farrand AJ, Polekhina G, Nero TL, Tomlinson S, Parker MW, Tweten RK (2011) Mapping the intermedilysin-human CD59 receptor interface reveals a deep correspondence with the binding site on CD59 for complement binding proteins C8alpha and C9. J Biol Chem 286:20952–20962

182. Witzenrath M, Pache F, Lorenz D, Koppe U, Gutbier B, Tabeling C, Reppe K, Meixenberger K, Dorhoi A, Ma J, Holmes A, Trendelenburg G, Heimesaat MM, Bereswill S, van der Linden M, Tschopp J, Mitchell TJ, Suttorp N, Opitz B (2011) The NLRP3 inflammasome is differentially activated by pneumolysin variants and contributes to host defense in pneumococcal pneumonia. J Immunol 187:434–440
183. Xu Q, Abdubek P, Astakhova T, Axelrod HL, Bakolitsa C, Cai X, Carlton D, Chen C, Chiu HJ, Clayton T, Das D, Deller MC, Duan L, Ellrott K, Farr CL, Feuerhelm J, Grant JC, Grzechnik A, Han GW, Jaroszewski L, Jin KK, Klock HE, Knuth MW, Kozbial P, Krishna SS, Kumar A, Lam WW, Marciano D, Miller MD, Morse AT, Nigoghossian E, Nopakun A, Okach L, Puckett C, Reyes R, Tien HJ, Trame CB, van den Bedem H, Weekes D, Wooten T, Yeh A, Zhou J, Hodgson KO, Wooley J, Elsliger MA, Deacon AM, Godzik A, Lesley SA, Wilson IA (2010) Structure of a membrane-attack complex/perforin (MACPF) family protein from the human gut symbiont Bacteroides thetaiotaomicron. Acta Cryst F66:1297–1305
184. Young JD, Hengartner H, Podack ER, Cohn ZA (1986) Purification and characterization of a cytolytic pore-forming protein from granules of cloned lymphocytes with natural killer activity. Cell 44:849–859
185. Young JD, Nathan CF, Podack ER, Palladino MA, Cohn ZA (1986) Functional channel formation associated with cytotoxic T-cell granules. Proc Natl Acad Sci USA 83:150–154
186. Zalman LS, Muller-Eberhard HJ (1990) Comparison of channels formed by poly C9, C5b-8 and the membrane attack complex of complement. Mol Immunol 27:533–537
187. Zhang X, Boyar W, Toth MJ, Wennogle L, Gonnella NC (1997) Structural definition of the C5a C terminus by two-dimensional nuclear magnetic resonance spectroscopy. Proteins 28:261–267

Part III
Functional Aspects of MACPF/CDCs

Chapter 7
Membrane Interactions and Cellular Effects of MACPF/CDC Proteins

Miša Mojca Cajnko, Miha Mikelj, Tom Turk, Marjetka Podobnik and Gregor Anderluh

Abstract The cell membrane is crucial for protection of the cell from its environment. MACPF/CDC proteins are a large superfamily known to be essential for bacterial pathogenesis and proper functioning of the immune system. The three most studied groups of MACPF/CDC proteins are cholesterol-dependant cytolysins from bacteria, the membrane attack complex of complement and human perforin. Their primary function is to form transmembrane pores in target cell membranes. The common mechanism of action comprises water-soluble monomeric proteins binding to the host cell membrane, oligomerization, and formation of a functional pore. This causes a disturbance in gradients of ions and other molecules across the membrane and can lead to cell death. Cells react to this form of attack in a complex manner. Responses can be general, like removing the perforated part of the membrane, or more specific, in many cases depending on binding of proteins to specific receptors to trigger various signalling cascades.

Keywords Cellular effects · Cholesterol-dependent cytolysins · Membrane attack complex · Perforin · Pore-formation

Abbreviations

AA	Amino acid
CDC	Cholesterol-dependent cytolysin
LLO	Listeriolysin O
MAC	Membrane attack complex
PFN	Perforin

M. M. Cajnko · M. Podobnik · G. Anderluh (✉)
Laboratory for Molecular Biology and Nanobiotechnology, National Institute of Chemistry, Hajdrihova 19, 1000 Ljubljana, Slovenia
e-mail: gregor.anderluh@ki.si

M. Mikelj · T. Turk · G. Anderluh
Department of Biology, Biotechnical Faculty, University of Ljubljana, Večna pot 111, 1000 Ljubljana, Slovenia

G. Anderluh and R. Gilbert (eds.), *MACPF/CDC Proteins - Agents of Defence, Attack and Invasion*, Subcellular Biochemistry 80, DOI: 10.1007/978-94-017-8881-6_7,

PFO	Perfringolysin O
PFP	Pore-forming proteins
PFT	Pore-forming toxin
TgPLP1	*Toxoplasma gondii* perforin-like protein 1
PLY	Pneumolysin
SLO	Streptolysin O

Introduction

Proteins of the MACPF/CDC superfamily have important roles in bacterial pathogenesis and proper functioning of the immune system [1, 2]. Although they are a large protein family, the molecular mechanism of action is well-known for only three distinct groups: Cholesterol-dependent cytolysins (CDCs), membrane attack complex (MAC) of complement and human perforin (PFN), which are collectively referred to as MACPF/CDC domain proteins. CDCs are a large group of pore-forming toxins (PFTs) that to date have been mainly identified in Gram-positive bacteria. Some of the most intensely studied and known representatives are perfringolysin O (PFO) from *Clostridium perfringens*, pneumolysin (PLY) from *Streptococcus pneumoniae*, listeriolysin O (LLO) from the foodborne pathogen *Listeria monocytogenes* and streptolysin O (SLO) from *Streptococcus pyogenes* [2, 3]. CDCs are virulence factors used by bacteria to invade host cells and tissues [3]. The MAC and PFN have important roles in the vertebrate immune system. The MAC is used for elimination of bacterial cells [4], while PFN removes unwanted cells, primarily virus-infected and tumor cells [5]. Members of all three groups act primarily through membrane interactions and formation of transmembrane pores. This chapter will review membrane interactions of some of the best known MACPF/CDC members and their effects in the cellular context. The molecular mechanism of membrane interactions and pore formation will be compared to some other well-known PFTs.

Membrane Damage by Proteins

The basic function of the cell membrane is to protect the cell from its environment. In addition, many important processes are initiated or delivered by the constituents of the membrane, like cell adhesion, ion conductivity and cell signalling. The destruction of the membrane is one of the simplest ways to harm cells, since the presence of the intact functional membrane is vital to the cell. Therefore, it is not surprising that many organisms use membrane active compounds in order to attack other organisms or to defend themselves against them. Cells are usually unable to repair extensive membrane damage, and consequently become necrotic or apoptotic and eventually die. There are a variety of different strategies that can lead to the damaging of target cell membranes, including: (i) formation of pores which can lead to colloid-osmotic lysis; (ii) enzymatic destruction of membranes; (iii) detergent-like activity which causes a solubilizing effect; and (iv) damage by direct lytic factors. Each of these mechanisms is linked to special molecules that are either able to penetrate the membrane, cleave membrane compounds, pull them out of the membrane or harm the membrane integrity and functionality in another way. Pore forming compounds are almost exclusively toxic peptides or proteins, which are expressed by the producing organism as water soluble monomers, but seek membrane surfaces and bind to them via specific receptors. Once attached to the membrane, the monomers start to oligomerize and consequently undergo large conformational changes, which generate or expose hydrophobic patches that insert into the lipid bilayer and form pores. Pores may have different characteristics (diameters, shape, permeability, selectivity, life span etc.), depending on the structure of each pore-forming compound. Such compounds represent the common toxic arsenal of many prokaryotic and eukaryotic organisms, such as many Gram-positive and Gram-negative bacteria [6–8]; Cnidaria [9] and arthropods like scorpions, spiders, wasps and bees [10]. Belonging to the second group are phospholipases that cleave phospholipids to free fatty acids and lysophospholipids that subsequently cause cell lysis. Cytolytic phospholipases can be found in many bacterial and animal venoms, i.e. those of Gram-negative bacteria, spiders and bees [11, 12]. Detergent-like compounds are less common but they are found in many marine animals like sponges, holothurians and sea stars [13]. Such molecules, which are constituted of distinct polar and nonpolar moieties, are usually denominated saponins. Many such compounds are also produced by terrestrial plants [14]. Direct lytic factors, also named cardiotoxins, are mainly three-fingered proteins found almost exclusively in some elapid snake venoms, i.e. those of cobras [15, 16]. PFTs are the most thoroughly studied molecules of the four groups and can be found in many organisms, but those most studied belong to bacteria, sea anemones and vertebrates. PFTs exhibit their toxic effect either by destruction of the membrane permeability barrier or by delivery of toxic components through the pores. Bacteria use PFTs mainly to kill other bacteria or to attack their hosts and thus survive within them. Eukaryotes such as parasites, fungi, sea anemones or plants use them in their defence mechanisms or to access nutrients. Proteins of the MAC of complement and PFN are used by the immune system of vertebrates. Members of the Bax/Bak family permeabilize mitochondria outer membranes in order to trigger apoptosis [17, 18]. Since some of these proteins are not actually behaving as toxins, a more accurate definition might be pore forming proteins (PFPs), with PFTs as the best-characterized and largest groups of PFPs.

There exists a wide variety of PFPs in terms of structural properties [8, 19]. They use a common multistep mechanism for pore formation, despite variability in the structure and properties of the final inserted pore state (see also below). Most usually this includes (i) association of a water-soluble monomeric PFP with lipid membranes (ii) oligomerization at the plane of the membrane, and (iii) conformational rearrangements of part of the polypeptide chain that allows insertion into the membrane and functional pore formation. PFPs use several distinct domains to accomplish each of these steps but there appears to be an astonishing structural conservation of domains found within different PFT families [17].

Initial Interactions with Lipid Membranes

The first step in pore formation usually starts with the interaction of the monomeric PFP and specific receptors present in the target membrane. This enables the concentration of the PFP at the specific target membrane as well as limits further diffusion of these proteins to the plane of the membrane thus raising the probability of successful oligomerization into functional pores. These receptors can be either lipids that reside in the lipid bilayers like cholesterol (typical for binding of CDCs) ([20] and see also Chap. 5) or sphingomyelin (binding of lysenin, a PFT from earthworm, and equinatoxin II from sea anemones) [21–23]. Receptors can be also proteins anchored to the lipids like glycophosphatidyl inositol-anchored protein for binding of aerolysin [24] or CD59 in the case of the CDC intermedilysin [25]. Cholesterol has been known for a long time as the specific receptor for CDCs [26, 27]; tight association to it enables the specific targeting of eukaryotic cells, with minimal harm done to the CDC-producing bacterial cell. It is now known that relatively high concentrations (35–55 %) of cholesterol are needed for efficient membrane association of CDCs in synthetic pure lipid systems [26–30] and that cholesterol exposure at the membrane is critical for efficient recognition [31, 32]. It is also known which parts of the molecule enable efficient cholesterol recognition [33–35], however, the molecular basis for it is not understood entirely. Due to these properties CDCs are now considered efficient probes for imaging of cholesterol in cells [36–38].

Variability in membrane sensing and initial association is known for members of the MACPF/CDC superfamily. Namely, while the CDC proteins show high affinity for cholesterol, enabled by the C-terminal domain folded into a β-strand rich immunoglobulin-like fold (see Chaps. 4 and 5 for details), it seems that other MACPF-domain proteins do not require any specific lipid receptor molecule. For example, PFN employs a β-strand rich C2 domain for initial membrane interaction and shows unique dependence for initial membrane binding on the presence of calcium ions [39, 40] (see discussion in Chap. 4 on the possible evolutionary relationship of the CDC C-terminal domain and the perforin C2 domain). On the other hand, MACPF proteins involved in MAC formation lack the membrane binding domain found as the C2 in PFN. The initial anchoring of the MAC proteins to the membrane is mediated by the complement subunit C7 and the full anchoring of the complex into the membrane happens after the binding of the C8α subunit. Therefore, the anchoring of complement seems to be mediated through MACPF membrane inserting regions and not through an extra lipid binding domain or receptor (see Chap. 6 for details). Yet another mode of membrane association is found in the bi-component PFP systems from fungi [41, 42]. The pore forming component, including a MACPF domain, binds various lipid membranes with low specificity and low affinity, but efficient pore formation is enabled by its association with the smaller component of the aegerolysin protein family, which possesses high specificity for cholesterol and sphingomyelin-enriched membrane lipid domains. These unique bi-component PFPs are described more thoroughly in Chap. 14.

However, for many MACPF/CDC proteins information on membrane interactions is lacking. For example, MACPF proteins from apicomplexan parasites enable parasite gliding through membranes [43]. These proteins contain β-sheet rich domains at their C-terminus that have been proposed to enable membrane interactions [44] necessary for pore formation as shown for *Toxoplasma gondii* Perforin-Like Protein 1 (TgPLP1) [45]. Furthermore, the domain at the N-terminus of the MACPF domain of TgPLP1 was also shown to promote membrane damage but was not required for pore-formation or pathogenesis. But how these domains interplay and the way in which they affect the activity of the MACPF component of the protein remains unknown in detail at present. Many other MACPF/CDC proteins have no particular domain known for membrane interactions identified within them and it remains to be determined whether a MACPF domain alone can be enough to allow membrane association and anchoring to the lipid membrane, in analogy to the complement membrane anchoring. In order to provide further insights into the membrane interactions of MACPF/CDC proteins, we have developed a screening assay that enabled us to monitor membrane interactions of some bacterial representatives.

Screening of Membrane Binding Activities of Bacterial MACPFs by Employing an In Vitro Translation System

Proteins in the large MACPF protein family mainly belong to eukaryotic species, however, around 10 % of them derive from bacteria [1]. It has been suggested that these bacterial genes coding for MACPF proteins originate from a eukaryotic source and have been acquired via horizontal gene transfer (see further discussion of MACPF/CDC evolution in Chaps. 2 and 4). They are predominantly found in Bacteroidetes, Chlamydiae and Proteobacteria [1]. We have selected seven different bacterial MACPF proteins for analysis of membrane binding activities. Some of them were chosen based on the pathogenicity of producing organism, for instance MACPF proteins from the obligate intracellular human pathogens *Chlamydophila pneumonia,* a causative agent of pneumonia, and *Chlamydia trachomatis,* which can cause trachoma and sexually transmitted diseases [46, 47]. In a similar basis we also selected the MACPF protein of the plant pathogen *Ralstonia solanacearum.* Other criteria for the selection of a MACPF protein were their size and predicted solubility, presence of additional domains and absence of transmembrane regions (Table 7.1).

Table 7.1 Description of bacterial MACPF proteins used for screening of membrane binding activities

Protein name	Designation (GenBank Acc. code)	Organism	Domain architecture	Size [kDa]	Length	Region [AA]	Predicted solubility	Synthesis	Solubility	Membrane interactions
CT153 hypothetical protein	a (AAD18329)	*Chlamydophila pneumonia*	[MACPF]-[MIR]	46	411	1–411	Insoluble	Yes	Soluble	No
C terminal region	b		[MIR]	5.3	48	364–411	Insoluble	No	/	/
MAC-PFN domain protein, partial	c (ADI88843)	*Chlamydophila pneumonia*	[MACPF]	50	456	1–456	Insoluble	Yes	Soluble	No
C-terminal region	d		–	5.5	50	407–456	Soluble	No	/	/
CT_153	e (AAC67744)	*Chlamydia trachomatis*	[MACPF]-[MIR]	90.8	810	1–810	Soluble	Yes	Soluble	No
C terminal region	f		[MIR]	7.8	69	742–810	Soluble	Yes	Soluble	No
Hypothetical protein HMPREF0765_1250	g (EEI93217)	*Sphingobacterim spiritivorum*	[MACPF]	33.9	308	1–308	Insoluble	Yes	Insoluble	/
Hypothetical protein RRSL_02063	h (EAP72888)	*Ralstonia solanacearum*	[MACPF]-[Jacalin]	55.4	519	1–519	Soluble	Yes	Soluble	No
C terminal region	i		[Jacalin]	15.4	147	373–519	Soluble	Yes	Soluble	Yes
Conserved hypothetical protein	j (EFK36569)	*Chryseobacterium gleum*	[MACPF]-[DNAQ-like]	49.5	438	23–460	Soluble	Yes	Soluble	Yes
Lipoprotein	k (ADE83751)	*Prevotella ruminicola*	[MACPF]	36.7	327	23–349	Insoluble	Yes	Soluble	No

key (Region) Part of a protein used for synthesis (numbers represent amino acids (AA) of a protein). Solubility was predicted from the sequence with PROSO II [151]. Columns synthesis, solubility and membrane interactions state whether or not proteins can be synthesised with an in vitro translation system, if they are soluble and if they interact with liposomes of different lipid compositions. (MIR) domain named after protein mannosyltransferase, inositol 1,4,5-trisphosphate receptor and ryanodine receptor. It has a β-barrel structure with a hairpin triplet and it may have a ligand tranferase function (InterPro IPR016093). (Jacalin) sugar-binding protein domain with a β-prism fold (NCBI cl03205). (DNA-Q-like) domain that belongs to DnaQ-like exonuclease superfamily, which catalyze the excision of nucleoside monophosphates at the DNA or RNA termini (NCBI: cl10012)

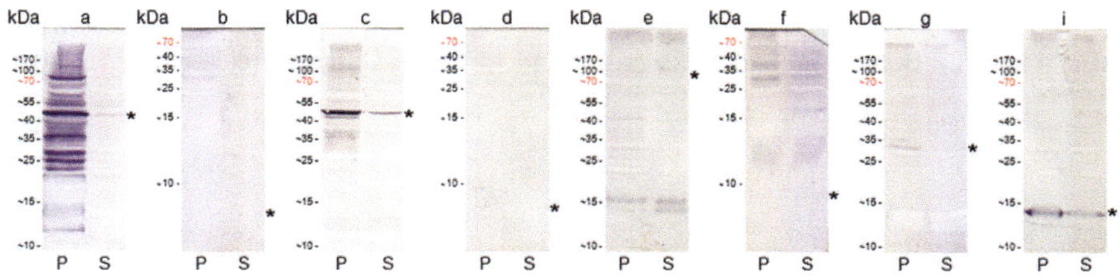

Fig. 7.1 Synthesis of selected bacterial MACPF proteins with a cell free expression system. Pellet (P) and supernatant (S) fractions were obtained after centrifugation, resolved on the SDS-PAGE gel and blotted by using penta-his primary and goat anti-mouse secondary antibody conjugated with horseradish peroxidase. The expected positions of recombinant proteins on membranes are indicated by an asterisk (*). The designations of the proteins with letters above the blots are defined in Table 7.1

Genes for selected full-length bacterial MACPF proteins without signal peptide and four DNA sequences coding for C-terminal regions of MACPF proteins from pathogenic bacteria were inserted into pET8c plasmids. Proteins with an N-terminal hexa-histidine-tag were then synthesised in a cell-free expression system. In most cases only a small proportion of the produced proteins was soluble, while the majority formed aggregates that pelleted upon centrifugation (Fig. 7.1). Results in Fig. 7.1 further show that some proteins were present only in the aggregated form or were not synthesised at all, like a conserved hypothetical protein (protein *g*) from *S. spiritivorum* and C-terminal domains (*b* and *d*) of *C. pneumonia* MACPF, respectively. In the case of protein CT_153 from *C. trachomatis* we obtained only a fraction of full-length protein, and the majority of it was degraded (Fig. 7.1e). The most soluble proteins were from *R. solanacearum*, *C. gleum* and *P. ruminicola* (*h*, *j* and *k*; data shown on Fig. 7.2).

In order to analyse membrane binding activities of bacterial MACPF proteins, produced soluble proteins were incubated with biotinylated liposomes and recovered by affinity purification with streptavidin magnetic beads. Pellet fractions containing liposomes with bound proteins and supernatant fractions with unbound proteins were analysed by immunoblotting (Fig. 7.2). Tested bacterial MACPF proteins did not exhibit significant binding to liposomes in comparison with non-specific binding to streptavidin magnetic beads (lanes Buffer in Fig. 7.2). Exceptions were conserved hypothetical MACPF protein (protein *j*) from *C. gleum* and Jacalin domain (i), the C-terminal part of hypothetical protein RRSL_02063 (protein *h*) from *R. solanacearum*. In both cases a modest difference was observed in the amount of protein found in pellet fractions between the samples that were incubated with liposomes or buffer alone (Fig. 7.2).

In conclusion, bacterial MACPF proteins that we tested for interactions with lipid membranes were found difficult to produce in soluble form as the majority of proteins formed insoluble aggregates. This may be a common outcome for many MACPF-domain containing proteins, since they are large and complex proteins with many hydrophobic regions. The membrane association of the soluble protein examples was not significant; however, nonetheless we observed membrane

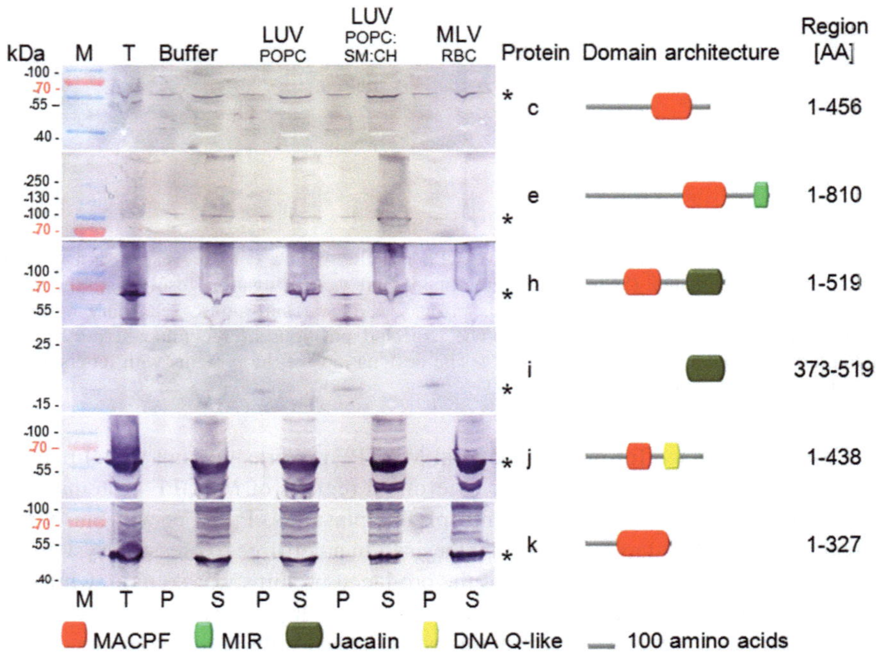

Fig. 7.2 Binding of selected bacterial MACPF proteins to biotinylated liposomes of different lipid composition. Membranes were blotted with penta-his primary and goat anti-mouse secondary antibody conjugated with horseradish peroxidase. (T) Total amount of produced protein (LUV) large unilamellar vesicles (MLV-RBC) multilamellar vesicles composed of erythrocyte membranes (M) marker (P) pellet and (S) supernatant fractions after affinity purification with streptavidin magnetic beads. The expected positions of recombinant proteins on membranes are indicated by an asterisk (*). The designations of the proteins with letters defined in Table 7.1

binding in some cases. Additional experiments are needed to verify these results and to get insights into membrane sensing; that is, which region of a protein is important for interaction and what is the nature of the receptor molecule on the lipid membrane. Also further effort will be needed to produce more proteins in their soluble form in order to determine whether C-terminal or any other additional domains are responsible for interactions with membranes or if the MACPF itself enables membrane interactions.

Pore Properties

Membrane binding is followed by oligomerisation on the plane of the membrane and consequently functional pores are formed. PFPs can be divided into two classes based on the secondary structure of the lipid bilayer spanning segment of

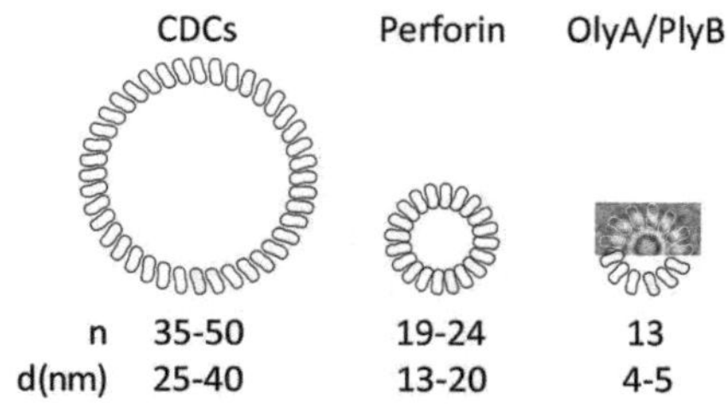

Fig. 7.3 Pore sizes formed by some of the MACPF/CDC proteins. The approximate dimensions of pores are presented for CDCs, PFN and a MACPF protein from fungi (pleurotolysin B, PlyB), which needs a smaller component for efficient membrane association (in this case ostreolysin A, OlyA). The number of monomers that build the pore and approximate diameters of pores (d) are also stated. Schemes of pores are composed of 40 monomers in the case of CDCs, 20 for PFN and 13 for OlyA/PlyB pore. OlyA/PlyB pore is overlayed on an actual electron-microscopy image (courtesy of Peter Maček)

the pore. The α-PFP class contains proteins which build the membrane integrated part of the pore with α-helical elements, including bacterial pore-forming colicins, as well as members of the actinoporin family from sea anemones like equinatoxin II, and diphtheria toxin translocation domain. Pores built by β-PFPs are rich in β-sheet and are organised in a β-barrel shape in the pore state. They include proteins from the aerolysin family, α-toxin family and members of the MACP/CDC superfamily [19, 48]. The stoichiometry of the oligomeric pore assembly greatly varies between PFPs, which is consequently reflected in different pore size. For example, small pores built of 3–4 monomers of 2 nm in diameter are formed by α-PFP equinatoxin II [49] and pores of a similar size are formed by a heptameric β-PFP α-toxin [50]. Some degree of variability of the pore size has been noticed for these smaller pores, as in the case of two-component staphylococcal γ-hemolysin, showing hexa-, hepta- or octameric structures [51–53], and the pores of cytolysin A have been found in dodecameric and tridecamric stoichiometry [54, 55]. What determines the stoichiometry of these pores is not so well understood.

Significantly larger oligomers are formed by members of the MACPF/CDC superfamily. CDCs tend to oligomerize with a large number of monomeric units varying from 35 to 50, resulting in pore sizes between 25 and 40 nm, forming the largest protein pores known to date [20, 56] (Fig. 7.3). The complete pores of PFN were found to be smaller, assembled by 19–24 monomers with the diameter of the pores having an average size of 13–20 nm [57, 58]. The MAC also form pores of similar size to those of PFN [59, 60]. However, not all MACPF/CDCs form large pores. For example, recent work on pleurotolysin, a bi-component lytic PFP from fungi, provided an insight into its pore size and stoichiometry [41]. Pleurotolysin forms significantly smaller pores than CDCs or PFN, composed of 13 monomers and 4 nm in diameter [41, 42, 61]. It therefore appears that there is a large variability in pore sizes that may be formed by the MACPF/CDC proteins, despite their having pore forming domains similar in size. It is hard to understand the underlying molecular basis for such variability in the absence of high resolution

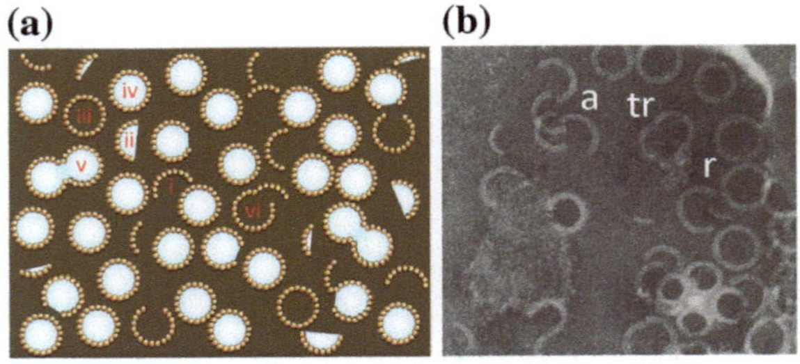

Fig. 7.4 Formation of pores of a variable size by MACPF/CDC superfamily members at the target membrane. **a** A scheme showing different oligomeric structures at the target membrane (*dark green*). Each monomer of PFP is shown as an *orange circle*; (i) represents a small oligomer attached to the surface of the membrane but not yet inserted, while (ii) represents an inserted *arc* oligomer, which together with the membrane forms a pore (the *blue* surface indicates that the oligomer is inserted and the pore is open for the conductance of the molecules). (iii) depicts a fully formed proteinaceous pre-pore and (iv) a respective full pure inserted into the membrane (*full rings*). (v) *twinned* or *clustered* pores, as recently observed in dynamic equilibrium for PFN [58]. (vi) a cluster of *arcs* before insertion into the membrane. **b** Electron micrographs of lipid layers with PFO, reprinted with permission from [150]. Pore elements described in (**a**) are highlighted by white letters; *arcs* (**a**), *full rings* (r), *twinned rings* (tr)

structural information on MACPF/CDC pores. Interestingly, despite their size, the large pores of MACPF/CDC proteins on plasma membranes are more efficiently resealed than the small pores formed by PFPs like aerolysin [62, 63]. A prepore intermediate is quite often formed in the case of MACPF/CDC proteins, having been observed for CDCs PLY [56] and PFO [64] and for human PFN [58].

Interestingly, for many members of the MACPF/CDC superfamily it has been shown that diverse pore structures exist at the surface of the membrane, as determined by biophysical methods like atomic force microscopy, electron microscopy and planar lipid membrane conductance studies (Fig. 7.4). These methods showed that the pores can exist as full proteinaceous channels and also as arcs, where a part of the pore is lined by the lipid bilayer, with the size, shape and conductivity of the pores dependent on the lipid composition of the target membrane [1, 65]. For example, examination of the pore properties of human PFN has recently revealed that distinctively different pores are formed in different composition bilayers, with variable diameters and noise levels [58]. Pores partially lined with the membrane components have also been reported for other PFPs, like actinoporins [66, 67], colicins [68, 69] or the proapoptotic proteins Bax and Bak, which form stable toroidal pores tunable in size in a concentration dependent manner [70–74]. Toroidal lipid structures generated by these proteins are made so that the protein is located within lipids around the pore circumference, which is enabled by the amphipathic nature of the inserted helices [73, 75]. Such an arrangement is not possible in the MACPF/CDC pores due to the presence of β-sheets in the lumen of the pore. Consequently, arcs are formed, where one side of the pore is lined by the protein and the other by the lipid headgroups (Fig. 7.4) (reviewed in [1]). In agreement with such structural arrangements of protein and membrane lipids, an increased flip-flop of lipids in the membrane was shown to be induced by PFN [76]. Formation of different sizes of pores was also reported for LLO [77]. It was shown that initially LLO, released by the intracellular pathogen *Listeria monocytogenes*, forms smaller pores that allow exchange of ions and small molecules between *Listeria* containing phagosomes and the cytoplasm of macrophages. After several minutes these small pores expand into larger pores that allow transport of proteins and finally the release of bacteria into the cytoplasm [77, 78]. The initial small perforations inhibit vacuole fusion with lysosomes and thus protect the intravacuolar bacterium from the harsh environment. The smaller size of pores could be attributed to arcs growing into full pores of defined radius with time. As shown in the mutational studies of SLO, these arcs are able to perforate the pore and conduct molecules through the lipid bilayer [79]. In this work, Bhakdi and colleagues showed that arc-shaped oligomers form pores of reduced functional diameter conducting only ions and small molecules, but not large dextran molecules that can only pass through fully proteinaceous pores of SLO. Thus, the formation of arcs might trigger different signalling mechanisms and cellular responses compared to the effects upon formation of the full pore. However, the role of arcs in the cellular response is not well understood and the biological implication of various sizes of pores remains to be shown.

Other Effects on Membranes

Apart from pore formation in lipid membranes a variety of other effects have also been observed upon interaction of MACPF/CDC proteins with lipid membranes. In general, large-scale changes in membrane morphology may be induced by the binding of proteins to lipid bilayers. Interactions of proteins within one monolayer of the bilayer membrane creates tension that is relieved by membrane bending leading to formation of invaginations or intraluminal vesicles [80–84]. PFPs initially interact with lipid membranes by inserting relatively large domains into one of the monolayers and this may directly induce changes in membrane shape. Indeed, it was recently observed that human PFN can induce invaginations and formation of internal vesicles in a giant unilamellar vesicle model system and on Jurkat cells. Effects on cells were ATP-independent and other PFPs that were used for comparisons (lysenin and PLY) were not able to induce these effects [85]. Induction of invaginations may aid rapid endocytosis that has been observed upon PFN interactions with the target cell membranes [86].

In another example, a CDC member PLY was shown to be able to induce a range of different activities on large unilamellar vesicles. By using freeze-fracture electron microscopy and NMR it was shown that PLY can induce, apart from pore formation, also formation of ring-like structures composed of 30–50 PLY monomers and up to 1,400 lipids. It was also noticed that many liposomes were aggregated upon PLY addition [87]. It thus seems that PLY can induce a multitude of effects including pore formation in lipid membranes, destabilisation of membranes through extraction of lipids into oligomeric complexes in solution and aggregation of liposomes. As proposed, some of the observed structures may be important for observed biological effects connected to non-cytolytic mechanisms, such as activation of the complement system through direct interaction of PLY and components of the immune system [87].

Effects on Cells

While the structure and function of PFPs have been extensively studied, less is known about their effects on cells. Exposure to high lytic concentrations of PFT can lead to cell death, but recent studies have shown that physiological concentrations are rarely this high, giving the cells an opportunity to respond [6]. Pore formation can trigger a wide variety of events that depend on cell type, exposure time, concentration and purpose of the PFT. Not all PFTs are meant to lyse cells. Some, like LLO, have many different functions ranging from enabling the bacteria to escape from endosomes to modulating the host immune response [88]. Although most CDCs act from the outside of the cell, their effects can be seen well beyond the plasma membrane. It is becoming clear that the role of these proteins is not merely to perforate a target cell, but also to act as potent regulators of signalling and immunity. Some of the most common cellular responses described to date are listed in Table 7.2 and schematically presented in Fig. 7.5.

But what triggers all these effects? Is it an ion flux through the pores or is it merely the binding of the protein to its receptor? It has been found that some PFPs affect cells in a pore-independent manner. CDCs are putative ligands for the host pattern-recognition molecule Toll-like receptor-4 independently of their pore forming activity [89–91]. A non-hemolytic PLY toxoid has proven equally efficient in triggering an immune response compared with its native form [89]. More on PLY and TLR receptor signalling can be found in Chap. 8. Another example of pore-independent signalling is spontaneous aggregation of lipid rafts via oligomerization of the PFT monomers on the membrane, as shown for LLO-induced signalling cascades within a host cell [92, 93]. Pore-independent action of CDCs was thought to be the path for histone modifications observed [94], however, it was later shown that pore formation was indeed needed for dephoshorylation of histones [95].

Once a stable pore is opened on the plasma membrane, the consequences can be dire. Leakage of ions along their concentration gradients can severely disrupt the physiology of the cell. Along with ions other molecules like ATP and even some cytosolic enzymes can pass through large CDC pores [96]. These leakages can damage the cell to the point of no return, finally resulting in necrosis or apoptosis [97–99]. The primary effect of the membrane pore formation is the change in the concentration of ions in the target cell cytosol. In the case of PFTs two ions have been shown to have a major role, potassium and calcium. One of the major roles of

Table 7.2 Cellular effects of MACPF/CDCs

Cellular effect	Protein	Protein
Endocytosis	PFN	(Keefe et al. [101]; Thiery et al. [127]; Praper et al. [85]; Lopez et al. [125])
	LLO	(Vadia et al. [103])
	SLO	(Idone et al. [100]; Corrotte et al. [124])
	MAC	(Morgan et al. [126]; Moskovich and Fishelson [139])
Histone modifications	LLO, PLY, PFO	(Hamon et al. [94])
MAPK activation	LLO	(Gonzalez et al. [63])
	PLY	(Ratner et al. [112]; Aguilar et al. [106])
	SLO	(Ratner et al. [112]; Kloft et al. [109])
	MAC	(Rus et al. [113]; Niculescu et al. [110]; Peng et al. [111]; Aoudjit et al. [107]; Gancz et al. [108])
Activation of a quiescent-like state (lipid droplet formation, activation of autophagy, inhibition of protein synthesis)	LLO, SLO	(Kloft et al. [141]; Gonzalez et al. [63])
Impairment of SUMOylation	LLO, PLY, PFO	(Ribet et al. [140])
Induction of cytokine expression via NF-κB and/or NLRP3 inflammasome	LLO	(Nishibori et al. [119]; Kayal et al. [116]; Tsuchiya et al. [120]; Meixenberger et al. [118]; Gonzalez et al. [63])
	PLY	(Houldsworth et al. [115]; Fickl et al. [105]; McNeela et al. [152])
	SLO	(Walev et al. [122])
	Tetanolysin	(Chu et al. [114])
	MAC	(Kilgore et al. [117]; Viedt et al. [121])
Caspase-7 activation	LLO, PFO	(Cassidy et al. [130])
Activation of the unfolded protein response	LLO	(Pillich et al. [153])
Inhibition of PLC mediated NADPH oxidase inhibition	LLO	(Lam et al. [154])
Induction of chloride secretion	LLO	(Richter et al. [155])
Impaired barrier function	LLO	(Richter et al. [155])
	PLY	(Zysk et al. [156]; Lucas et al. [157])
Release of Ca^{2+} from intracellular stores	LLO	(Gekara et al. [104])
Disruption of mitochondrial dynamics	LLO	(Stavru et al. [158])
	PLY	(Braun et al. [159])
	MAC	(Papadimitriou et al. [132])
Induction of iNOS	PLY	(Braun et al. [160])
Apoptosis	PLY	(Braun et al. [161]; Srivastava et al. [91])
	LLO	(Carrero et al. [162])
	PFN	(Keefe et al. [101])
Ciliary slowing	PLY	(Steinfort et al. [163]; Rayner et al. [164])

(continued)

Table 7.2 (continued)

Cellular effect	Protein	Protein
Shedding of receptors	SLO	(Walev et al. [165])
Ectocytosis	SLO	(Keyel et al. [137])
	MAC	(Stein and Luzio [138])
Inhibition of apoptosis and cell cycle activation	MAC	(Rus et al. [113, 134]; Soane et al. [136]; Rus et al. [135]; Cudrici et al. [133]; Tegla et al. [4])

Some of the effects of various members of MACPF/CDCs superfamily are listed. The list is by no means complete. For more information on cellular effects of PLY see Chap. 8, of LLO Chap. 9 and of PFN Chap. 10

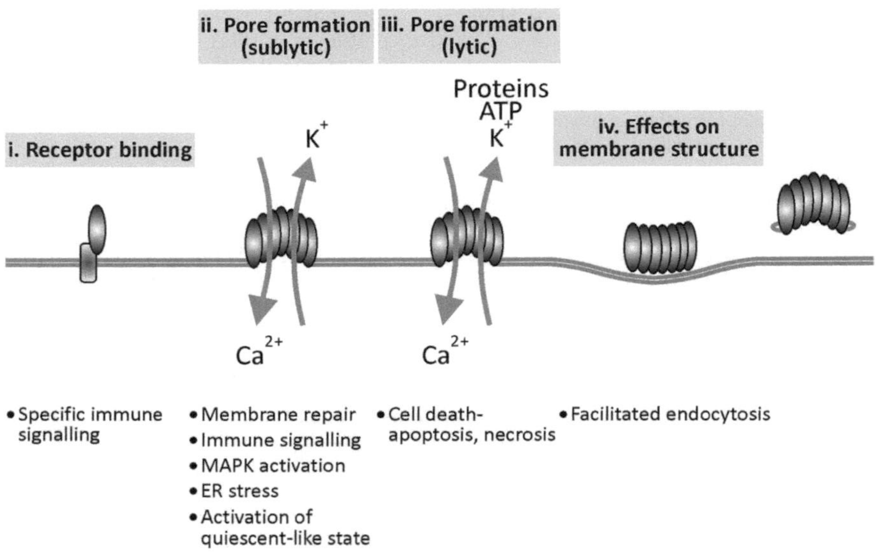

Fig. 7.5 Various effects induced in cells after interaction of MACPF/CDCs with lipid membranes. For more information and representative papers see the text and Table 7.2

calcium ions is triggering of endocytosis and the membrane repair processes mentioned below. When a hole is formed on the plasma membrane, either a stable protein pore formed by PFP or a mechanical lesion, calcium influx triggers rapid endocytosis [100, 101]. This process is used by certain intracellular pathogens like *L. monocytogenes* for efficient invasion [102, 103]. In addition, LLO affects calcium signalling also by release of calcium ions from intracellular stores, which can lead to stress of the endoplasmic reticulum [104]. Another effect that was linked to calcium influx due to pore formation by a CDC is NF-κB activation and IL-8 production [105]. In contrast to calcium, potassium ions are pushed out of the cells upon pore formation. The efflux of potassium ions has been shown to result in activation of the inflammasome, histone modifications, MAPK activation, activation of autophagy, arrest in protein synthesis and lipid droplet formation [63, 95]. It is, however, hard to say whether a cellular effect is the result of only one type of signalling, either specific protein-receptor interaction or ion fluxes through pores and it is quite possible in many instances to involve both. For example, valinomycin, a potassium ionophore, triggers a much lower histone dephosphorylation response than listeriolysin O, suggesting that potassium efflux is not the only trigger for this effect [95].

Certain cellular responses have a protective role, while others work in favour of the attacking pathogen secreting the toxin. But before cells can respond to the action of any of these PFP, they must first detect their presence. CDCs from pathogenic bacteria like LLO, PLY and SLO and also MAC proteins have been shown to activate phosphorylation of certain MAP kinases like p38, JNK and ERK at sublytic concentrations, which is a way of sensing osmotic stress [63, 106–113]. This, along with other signalling pathways, including activation of protein complexes like NF-κB and the NLRP3 inflammasome, results in cytokine production and leads to an effective immune response [63, 105, 114–122]. Activation of certain MAPKs by sublytic doses of MAC proteins also protects some cells from complement-mediated injury by desensitizing them and making them resistant to lytic complement doses [107, 123].

Immune response, membrane repair and apoptosis are just some of the protective cellular responses, important at the cellular level or for the whole organism. Induction of endocytosis of protein pores is a way of repairing an injured plasma membrane. This effect was documented for MAC proteins, LLO, SLO as well as PFN [76, 100, 101, 103, 124–127], however in the case of PFN this effect may serve a very different purpose. When pore formation by perforin triggers endocytosis, granzymes are endocytosed along with perforin pores leading to apoptotic death of the target cell [86, 101, 125, 127]. More on this subject is presented in Chap. 10. Here apoptosis has a protective role, since it prevents pathogen invasion through the organism [91] and protects against tumors and virus infected cells [128, 129]. With regards to apoptosis, one unlikely protector has surfaced. Caspase-7 is known as one of the effector caspases in apoptosis, but recent research has shown that it can also have a cytoprotective role against attack by certain CDCs like LLO and PFO [130]. When it comes to MAC proteins, characterisation of cell death is more complex. It exhibits features of apoptosis, like DNA fragmentation as well as features of necrosis, like swelling and mitochondrial abnormalities [131, 132]. Another interesting feature about the effects of MAC proteins is that sublytic concentrations of these complexes can altogether inhibit cell death and induce cell proliferation [4, 113, 133–136].

Endocytosis of protein pores can also have less desirable effects on cells. It can cause leakage of lysosomal contents into the cytoplasm and even promote bacterial invasion and thus work in favour of the pathogen, as is the case with LLO [88]. Here, endocytosis is designed to promote cellular damage and not repair. Some cells have developed an alternative way of dealing with the problem. Ectocytosis, a process of vesicle shedding, has been shown to be the mechanism of eliminating

SLO and MAC pores from the plasma membrane surface as a mode of cell protection [137, 138]. Also, this process has shown itself to be of a physical nature, since it was seen not only in living but also in chemically fixed cells [137]. In the case of MAC and membrane repair, both endocytosis and ectocytosis can occur in the same cell type, just to a different extent [139].

Not all cellular effects result in a protective response from the host cell. With the help of certain CDCs, pathogens can modulate cell responses to their own advantage. For example, LLO, PFO and PLY cause histone modifications which correlate with reduced transcriptional activity of a subset of host genes, including key immunity genes [94]. The same three CDCs also affect a fundamental eukaryotic post-translational modification process provided by the SUMOylation machinery, triggering the degradation of Ubc9, the E2 SUMO enzyme [140]. Since attack by CDCs is an energy consuming process resembling starvation, some cells have been found to enter a quiescent-like state, which involves formation of lipid droplets, arrest in protein synthesis and autophagy [63, 141].

There are many similarities in cell responses when CDCs are compared to some other PFTs. There are, however, also some differences that are probably due to specific binding to receptors of target cells. These include effects such as endocytosis, MAPK activation, cytokine production via NF-κB and NLRP3 inflammasome signalling, caspase-7 activation, lipid droplet formation, activation of autophagy and histone modifications, as reported for *Staphylococcus aureus* α-toxin, *Vibrio cholerae* cytolysin, *Escherichia coli* α-hemolysin and *Aeromonas* sp. aerolysin [63, 95, 109, 112, 130, 141–146]. Binding of the α-PFP equinatoxin II to the membrane of the target cell resulted in membrane blebbing, actin cytoskeleton reorganization and inhibition of endocytosis [147]. Recovery of intracellular potassium and ATP levels in cells treated with different PFTs is another very interesting process for comparison. For example, the difference in pore sizes between LLO and aerolysin is huge. Aerolysin pores are approximately 1.5 nm in diameter [8, 148] in comparison to LLO pores that can exceed 35 nm. Yet it was shown that cells treated with LLO replenish their ATP and potassium levels in a significantly shorter time than cells treated with aerolysin. There appears to be an inverse correlation between pore size and the ability of cells to repair their plasma membrane. This would indicate a difference in signalling, or that pore number or stability at the membrane surface are more important for cell recovery than the pore size [63], which may be the case for pores formed at the interface of lipids and proteins, i.e. arcs, and may be inherently less stable. Another possible explanation for this phenomenon could be the amount of calcium ions that pass through a pore since the influx of calcium and the subsequent rise in its intracellular concentration has been linked to membrane repair [100, 101]. Examination of the effect that terminal MAC complexes have on cellular calcium levels revealed that these complexes increased intracellular calcium concentrations in proportion with the known functional pore size of the terminal complement complex in the plasma membrane, the smaller the complex the smaller the influx of calcium [149].

Conclusions

Cellular effects of PFPs are diverse and complex and reflect their mode of interactions with cellular membranes. The responses of cells range from general ones, which are linked to pore formation and dissipation of ionic and other gradients across the membranes, to more specific ones that result from a protein binding to its specific lipidic or protein receptor. One indicator of this is comparison of MACPF/CDCs to other PFTs, which reveals that despite their differences in receptor binding and mode of action, some effects like membrane repair and immune signalling are shared by many different PFPs. On the other hand, there are many differences in cellular responses between members of the MACPF/CDC superfamily, even when they bind to the same receptor like in the case of CDC proteins. Differences between the effects that pore-forming proteins have on cells can be attributed to many things like cell type and the purpose of the protein in question. It is clear that cellular effects are complex and go beyond simply responding to membrane perforation. Further research into the role of pore size on cellular responses and other activities of PFPs on lipid membranes is clearly needed.

Acknowledgements We would like to thank Slovenian Research Agency for a financial support.

References

1. Gilbert RJ, Mikelj M, Dalla Serra M, Froelich CJ, Anderluh G (2013) Effects of MACPF/CDC proteins on lipid membranes. Cell Mol Life Sci 70:2083–2098
2. Rosado CJ, Kondos S, Bull TE, Kuiper MJ, Law RHP, Buckle AM, Voskoboinik I, Bird PI, Trapani JA, Whisstock JC, Dunstone MA (2008) The MACPF/CDC family of pore-forming toxins. Cell Microbiol 10:1765–1774
3. Tweten RK (2005) Cholesterol-dependent cytolysins, a family of versatile pore-forming toxins. Infect Immun 73:6199–6209
4. Tegla CA, Cudrici C, Patel S, Trippe R 3rd, Rus V, Niculescu F, Rus H (2011) Membrane attack by complement: the assembly and biology of terminal complement complexes. Immunol Res 51:45–60
5. Voskoboinik I, Smyth MJ, Trapani JA (2006) Perforin-mediated target-cell death and immune homeostasis. Nat Rev Immunol 6:940–952
6. Aroian R, van der Goot FG (2007) Pore-forming toxins and cellular non-immune defenses (CNIDs). Curr Opin Microbiol 10:57–61
7. Knapp O, Stiles BG, Popoff MR (2010) The aerolysin-like toxin family of cytolytic, pore-forming toxins. Open Toxinol J 3:53–68
8. Parker MW, Feil SC (2005) Pore-forming protein toxins: from structure to function. Prog Biophys Mol Biol 88:91–142
9. Anderluh G, Maček P (2002) Cytolytic peptide and protein toxins from sea anemones (Anthozoa: Actiniaria). Toxicon 40:111–124
10. Kuhn-Nentwig L (2003) Antimicrobial and cytolytic peptides of venomous arthropods. Cell Mol Life Sci 60:2651–2668
11. Istivan TS, Coloe PJ (2006) Phospholipase A in gram-negative bacteria and its role in pathogenesis. Microbiology 152:1263–1274

12. Shimuta K, Ohnishi M, Iyoda S, Gotoh N, Koizumi N, Watanabe H (2009) The hemolytic and cytolytic activities of *Serratia marcescens* phospholipase A (PhlA) depend on lysophospholipid production by PhlA. BMC Microbiol 9:261
13. Caulier G, Van Dyck S, Gerbaux P, Eeckhaut I, Flammang P (2011) Review of saponin diversity in sea cucumbers belonging to the family Holothuriidae. SPC Beche Mer Inf Bull 31:48–54
14. Podolak I, Galanty A, Sobolewska D (2010) Saponins as cytotoxic agents: a review. Phytochem Rev 9:425–474
15. Dufton MJ, Hider RC (1991) The structure and pharmacology of elapid cytotoxins. In: Harvey AL (ed) Snake toxins. Pergamon Press, New York, pp 259–302
16. Wu WG, Tjong SC, Wu PL, Kuo JH, Wu K (2010) Role of heparan sulfates and glycosphingolipids in the pore formation of basic polypeptides of cobra cardiotoxin. Adv Exp Med Biol 677:143–149
17. Anderluh G, Lakey JH (2008) Disparate proteins use similar architectures to damage membranes. Trends Biochem Sci 33:482–490
18. Bischofberger M, Iacovache I, van der Goot FG (2012) Pathogenic pore-forming proteins: function and host response. Cell Host Microbe 12:266–275
19. Gouaux E (1997) Channel-forming toxins: tales of transformation. Curr Opin Struct Biol 7:566–573
20. Hotze EM, Tweten RK (2012) Membrane assembly of the cholesterol-dependent cytolysin pore complex. Biochim Biophys Acta 1818:1028–1038
21. Bakrač B, Gutierrez-Aguirre I, Podlesek Z, Sonnen AF, Gilbert RJ, Maček P, Lakey JH, Anderluh G (2008) Molecular determinants of sphingomyelin specificity of a eukaryotic pore-forming toxin. J Biol Chem 283:18665–18677
22. De Colibus L, Sonnen AF, Morris KJ, Siebert CA, Abrusci P, Plitzko J, Hodnik V, Leippe M, Volpi E, Anderluh G, Gilbert RJ (2012) Structures of lysenin reveal a shared evolutionary origin for pore-forming proteins and its mode of sphingomyelin recognition. Structure 20:1498–1507
23. Yamaji A, Sekizawa Y, Emoto K, Sakuraba H, Inoue K, Kobayashi H, Umeda M (1998) Lysenin, a novel sphingomyelin-specific binding protein. J Biol Chem 273:5300–5306
24. Abrami L, Fivaz M, Glauser PE, Parton RG, van der Goot FG (1998) A pore-forming toxin interacts with a GPI-anchored protein and causes vacuolation of the endoplasmic reticulum. J Cell Biol 140:525–540
25. Giddings KS, Zhao J, Sims PJ, Tweten RK (2004) Human CD59 is a receptor for the cholesterol-dependent cytolysin intermedilysin. Nat Struct Mol Biol 11:1173–1178
26. Alving CR, Habig WH, Urban KA, Hardegree MC (1979) Cholesterol-dependent tetanolysin damage to liposomes. Biochim Biophys Acta 551:224–228
27. Rosenqvist E, Michaelsen TE, Vistnes AI (1980) Effect of streptolysin O and digitonin on egg lecithin/cholesterol vesicles. Biochim Biophys Acta 600:91–102
28. Bavdek A, Gekara NO, Priselac D, Gutierrez Aguirre I, Darji A, Chakraborty T, Maček P, Lakey JH, Weiss S, Anderluh G (2007) Sterol and pH interdependence in the binding, oligomerization, and pore formation of Listeriolysin O. Biochemistry 46:4425–4437
29. Heuck AP, Hotze EM, Tweten RK, Johnson AE (2000) Mechanism of membrane insertion of a multimeric beta-barrel protein: perfringolysin O creates a pore using ordered and coupled conformational changes. Mol Cell 6:1233–1242
30. Ohno-Iwashita Y, Iwamoto M, Ando S, Iwashita S (1992) Effect of lipidic factors on membrane cholesterol topology–mode of binding of theta-toxin to cholesterol in liposomes. Biochim Biophys Acta 1109:81–90
31. Flanagan JJ, Tweten RK, Johnson AE, Heuck AP (2009) Cholesterol exposure at the membrane surface is necessary and sufficient to trigger perfringolysin O binding. Biochemistry 48:3977–3987
32. Moe PC, Heuck AP (2010) Phospholipid hydrolysis caused by *Clostridium perfringens* alpha-toxin facilitates the targeting of perfringolysin O to membrane bilayers. Biochemistry 49:9498–9507

33. Farrand AJ, LaChapelle S, Hotze EM, Johnson AE, Tweten RK (2010) Only two amino acids are essential for cytolytic toxin recognition of cholesterol at the membrane surface. Proc Natl Acad Sci USA 107:4341–4346
34. Johnson BB, Moe PC, Wang D, Rossi K, Trigatti BL, Heuck AP (2012) Modifications in perfringolysin O domain 4 alter the cholesterol concentration threshold required for binding. Biochemistry 51:3373–3382
35. Soltani CE, Hotze EM, Johnson AE, Tweten RK (2007) Structural elements of the cholesterol-dependent cytolysins that are responsible for their cholesterol-sensitive membrane interactions. Proc Natl Acad Sci USA 104:20226–20231
36. Ohno-Iwashita Y, Shimada Y, Waheed AA, Hayashi M, Inomata M, Nakamura M, Maruya M, Iwashita S (2004) Perfringolysin O, a cholesterol-binding cytolysin, as a probe for lipid rafts. Anaerobe 10:125–134
37. Pocognoni CA, De Blas GA, Heuck AP, Belmonte SA, Mayorga LS (2013) Perfringolysin O as a useful tool to study human sperm physiology. Fertil Steril 99:99–106
38. Waheed AA, Shimada Y, Heijnen HFG, Nakamura M, Inomata M, Hayashi M, Iwashita S, Slot JW, Ohno-Iwashita Y (2001) Selective binding of perfringolysin O derivative to cholesterol-rich membrane microdomains (rafts). Proc Natl Acad Sci USA 98:4926–4931
39. Praper T, Beseničar MP, Istinič H, Podlesek Z, Metkar SS, Froelich CJ, Anderluh G (2010) Human perforin permeabilizing activity, but not binding to lipid membranes, is affected by pH. Mol Immunol 47:2492–2504
40. Voskoboinik I, Thia MC, Fletcher J, Ciccone A, Browne K, Smyth MJ, Trapani JA (2005) Calcium-dependent plasma membrane binding and cell lysis by perforin are mediated through its C2 domain: a critical role for aspartate residues 429, 435, 483, and 485 but not 491. J Biol Chem 280:8426–8434
41. Ota K, Leonardi A, Mikelj M, Skočaj M, Wohlschlager T, Kunzler M, Aebi M, Narat M, Križaj I, Anderluh G, Sepčić K, Maček P (2013) Membrane cholesterol and sphingomyelin, and ostreolysin A are obligatory for pore-formation by a MACPF/CDC-like pore-forming protein, pleurotolysin B. Biochimie 95:1855–1864
42. Tomita T, Noguchi K, Mimuro H, Ukaji F, Ito K, Sugawara-Tomita N, Hashimoto Y (2004) Pleurotolysin, a novel sphingomyelin-specific two-component cytolysin from the edible mushroom *Pleurotus ostreatus*, assembles into a transmembrane pore complex. J Biol Chem 279:26975–26982
43. Kafsack BF, Pena JD, Coppens I, Ravindran S, Boothroyd JC, Carruthers VB (2009) Rapid membrane disruption by a perforin-like protein facilitates parasite exit from host cells. Science 323:530–533
44. Kafsack BF, Carruthers VB (2010) Apicomplexan perforin-like proteins. Commun Integr Biol 3:18–23
45. Roiko MS, Carruthers VB (2013) Functional dissection of toxoplasma gondii perforin-like protein 1 reveals a dual domain mode of membrane binding for cytolysis and parasite egress. J Biol Chem 288:8712–8725
46. Ponting CP (1999) Chlamydial homologues of the MACPF (MAC/perforin) domain. Curr Biol 9:R911–R913
47. Taylor LD, Nelson DE, Dorward DW, Whitmire WM, Caldwell HD (2010) Biological characterization of *Chlamydia trachomatis* plasticity zone MACPF domain family protein CT153. Infect Immun 78:2691–2699
48. Iacovache I, Bischofberger M, van der Goot FG (2010) Structure and assembly of pore-forming proteins. Curr Opin Struct Biol 20:241–246
49. Belmonte G, Pederzolli C, Maček P, Menestrina G (1993) Pore formation by the sea anemone cytolysin equinatoxin II in red blood cells and model lipid membranes. J Membr Biol 131:11–22
50. Song L, Hobaugh MR, Shustak C, Cheley S, Bayley H, Gouaux JE (1996) Structure of staphylococcal alpha-hemolysin, a heptameric transmembrane pore. Science 274:1859–1866

51. Comai M, Dalla Serra M, Coraiola M, Werner S, Colin DA, Monteil H, Prevost G, Menestrina G (2002) Protein engineering modulates the transport properties and ion selectivity of the pores formed by staphylococcal gamma-haemolysins in lipid membranes. Mol Microbiol 44:1251–1267
52. Sugawara-Tomita N, Tomita T, Kamio Y (2002) Stochastic assembly of two-component staphylococcal gamma-hemolysin into heteroheptameric transmembrane pores with alternate subunit arrangements in ratios of 3:4 and 4:3. J Bacteriol 184:4747–4756
53. Yamashita K, Kawai Y, Tanaka Y, Hirano N, Kaneko J, Tomita N, Ohta M, Kamio Y, Yao M, Tanaka I (2011) Crystal structure of the octameric pore of staphylococcal gamma-hemolysin reveals the beta-barrel pore formation mechanism by two components. Proc Natl Acad Sci USA 108:17314–17319
54. Eifler N, Vetsch M, Gregorini M, Ringler P, Chami M, Philippsen A, Fritz A, Muller SA, Glockshuber R, Engel A, Grauschopf U (2006) Cytotoxin ClyA from *Escherichia coli* assembles to a 13-meric pore independent of its redox-state. EMBO J 25:2652–2661
55. Mueller M, Grauschopf U, Maier T, Glockshuber R, Ban N (2009) The structure of a cytolytic alpha-helical toxin pore reveals its assembly mechanism. Nature 459:726–730
56. Tilley SJ, Orlova EV, Gilbert RJ, Andrew PW, Saibil HR (2005) Structural basis of pore formation by the bacterial toxin pneumolysin. Cell 121:247–256
57. Law RH, Lukoyanova N, Voskoboinik I, Caradoc-Davies TT, Baran K, Dunstone MA, D'Angelo ME, Orlova EV, Coulibaly F, Verschoor S, Browne KA, Ciccone A, Kuiper MJ, Bird PI, Trapani JA, Saibil HR, Whisstock JC (2010) The structural basis for membrane binding and pore formation by lymphocyte perforin. Nature 468:447–451
58. Praper T, Sonnen A, Viero G, Kladnik A, Froelich CJ, Anderluh G, Dalla Serra M, Gilbert RJ (2011) Human perforin employs different avenues to damage membranes. J Biol Chem 286:2946–2955
59. Aleshin AE, Schraufstatter IU, Stec B, Bankston LA, Liddington RC, DiScipio RG (2012) Structure of complement C6 suggests a mechanism for initiation and unidirectional, sequential assembly of membrane attack complex (MAC). J Biol Chem 287:10210–10222
60. Tschopp J, Masson D, Stanley KK (1986) Structural/functional similarity between proteins involved in complement- and cytotoxic T-lymphocyte-mediated cytolysis. Nature 322: 831–834
61. Sepčić K, Berne S, Potrich C, Turk T, Maček P, Menestrina G (2003) Interaction of ostreolysin, a cytolytic protein from the edible mushroom *Pleurotus ostreatus*, with lipid membranes and modulation by lysophospholipids. Eur J Biochem 270:1199–1210
62. Bischofberger M, Gonzalez MR, van der Goot FG (2009) Membrane injury by pore-forming proteins. Curr Opin Cell Biol 21:589–595
63. Gonzalez MR, Bischofberger M, Freche B, Ho S, Parton RG, van der Goot FG (2011) Pore-forming toxins induce multiple cellular responses promoting survival. Cell Microbiol 13:1026–1043
64. Czajkowsky DM, Hotze EM, Shao Z, Tweten RK (2004) Vertical collapse of a cytolysin prepore moves its transmembrane beta-hairpins to the membrane. EMBO J 23:3206–3215
65. Marchioretto M, Podobnik M, Dalla Serra M, Anderluh G (2013) What planar lipid membranes tell us about the pore-forming activity of cholesterol-dependent cytolysins. Biophys Chem 182:64–70
66. Anderluh G, Dalla Serra M, Viero G, Guella G, Maček P, Menestrina G (2003) Pore formation by equinatoxin II, a eukaryotic protein toxin, occurs by induction of nonlamellar lipid structures. J Biol Chem 278:45216–45223
67. Valcarcel CA, Dalla Serra M, Potrich C, Bernhart I, Tejuca M, Martinez D, Pazos F, Lanio ME, Menestrina G (2001) Effects of lipid composition on membrane permeabilization by sticholysin I and II, two cytolysins of the sea anemone *Stichodactyla helianthus*. Biophys J 80:2761–2774
68. Sobko AA, Kotova EA, Antonenko YN, Zakharov SD, Cramer WA (2004) Effect of lipids with different spontaneous curvature on the channel activity of colicin E1: evidence in favor of a toroidal pore. FEBS Lett 576:205–210

69. Sobko AA, Kotova EA, Antonenko YN, Zakharov SD, Cramer WA (2006) Lipid dependence of the channel properties of a colicin E1-lipid toroidal pore. J Biol Chem 281: 14408–14416
70. Basanez G, Sharpe JC, Galanis J, Brandt TB, Hardwick JM, Zimmerberg J (2002) Bax-type apoptotic proteins porate pure lipid bilayers through a mechanism sensitive to intrinsic monolayer curvature. J Biol Chem 277:49360–49365
71. Bleicken S, Landeta O, Landajuela A, Basanez G, Garcia-Saez AJ (2013) Proapoptotic bax and bak proteins form stable protein-permeable pores of tunable size. J Biol Chem 288:33241–33252
72. Epand RF, Martinou JC, Montessuit S, Epand RM, Yip CM (2002) Direct evidence for membrane pore formation by the apoptotic protein bax. Biochem Biophys Res Commun 298:744–749
73. Qian S, Wang W, Yang L, Huang HW (2008) Structure of transmembrane pore induced by bax-derived peptide: evidence for lipidic pores. Proc Natl Acad Sci USA 105:17379–17383
74. Xu XP, Zhai D, Kim E, Swift M, Reed JC, Volkmann N, Hanein D (2013) Three-dimensional structure of bax-mediated pores in membrane bilayers. Cell Death Dis 4:e683
75. Westphal D, Dewson G, Czabotar PE, Kluck RM (2011) Molecular biology of bax and bak activation and action. Biochim Biophys Acta 1813:521–531
76. Metkar SS, Wang B, Catalan E, Anderluh G, Gilbert RJ, Pardo J, Froelich CJ (2011) Perforin rapidly induces plasma membrane phospholipid flip-flop. PLoS One 6:e24286
77. Birmingham CL, Canadien V, Kaniuk NA, Steinberg BE, Higgins DE, Brumell JH (2008) Listeriolysin O allows listeria monocytogenes replication in macrophage vacuoles. Nature 451:350–354
78. Shaughnessy LM, Hoppe AD, Christensen KA, Swanson JA (2006) Membrane perforations inhibit lysosome fusion by altering pH and calcium in Listeria monocytogenes vacuoles. Cell Microbiol 8:781–792
79. Palmer M, Harris R, Freytag C, Kehoe M, Tranum-Jensen J, Bhakdi S (1998) Assembly mechanism of the oligomeric streptolysin O pore: the early membrane lesion is lined by a free edge of the lipid membrane and is extended gradually during oligomerization. EMBO J 17:1598–1605
80. Ewers H, Romer W, Smith AE, Bacia K, Dmitrieff S, Chai W, Mancini R, Kartenbeck J, Chambon V, Berland L, Oppenheim A, Schwarzmann G, Feizi T, Schwille P, Sens P, Helenius A, Johannes L (2010) GM1 structure determines SV40-induced membrane invagination and infection. Nat Cell Biol 12:11–18
81. McMahon HT, Gallop JL (2005) Membrane curvature and mechanisms of dynamic cell membrane remodelling. Nature 438:590–596
82. Romer W, Berland L, Chambon V, Gaus K, Windschiegl B, Tenza D, Aly MR, Fraisier V, Florent JC, Perrais D, Lamaze C, Raposo G, Steinem C, Sens P, Bassereau P, Johannes L (2007) Shiga toxin induces tubular membrane invaginations for its uptake into cells. Nature 450:670–675
83. Wollert T, Wunder C, Lippincott-Schwartz J, Hurley JH (2009) Membrane scission by the ESCRT-III complex. Nature 458:172–177
84. Zimmerberg J, Kozlov MM (2006) How proteins produce cellular membrane curvature. Nat Rev Mol Cell Biol 7:9–19
85. Praper T, Sonnen AF, Kladnik A, Andrighetti AO, Viero G, Morris KJ, Volpi E, Lunelli L, Dalla Serra M, Froelich CJ, Gilbert RJ, Anderluh G (2011) Perforin activity at membranes leads to invaginations and vesicle formation. Proc Natl Acad Sci USA 108:21016–21021
86. Thiery J, Keefe D, Boulant S, Boucrot E, Walch M, Martinvalet D, Goping IS, Bleackley RC, Kirchhausen T, Lieberman J (2011) Perforin pores in the endosomal membrane trigger the release of endocytosed granzyme B into the cytosol of target cells. Nat Immunol 12:770–777
87. Bonev BB, Gilbert RJ, Andrew PW, Byron O, Watts A (2001) Structural analysis of the protein/lipid complexes associated with pore formation by the bacterial toxin pneumolysin. J Biol Chem 276:5714–5719

88. Hamon MA, Ribet D, Stavru F, Cossart P (2012) Listeriolysin O: the Swiss army knife of Listeria. Trends Microbiol 20:360–368
89. Malley R, Henneke P, Morse SC, Cieslewicz MJ, Lipsitch M, Thompson CM, Kurt-Jones E, Paton JC, Wessels MR, Golenbock DT (2003) Recognition of pneumolysin by Toll-like receptor 4 confers resistance to pneumococcal infection. Proc Natl Acad Sci USA 100:1966–1971
90. Park JM, Ng VH, Maeda S, Rest RF, Karin M (2004) Anthrolysin O and other gram-positive cytolysins are toll-like receptor 4 agonists. J Exp Med 200:1647–1655
91. Srivastava A, Henneke P, Visintin A, Morse SC, Martin V, Watkins C, Paton JC, Wessels MR, Golenbock DT, Malley R (2005) The apoptotic response to pneumolysin is toll-like receptor 4 dependent and protects against pneumococcal disease. Infect Immun 73:6479–6487
92. Gekara NO, Jacobs T, Chakraborty T, Weiss S (2005) The cholesterol-dependent cytolysin listeriolysin O aggregates rafts via oligomerization. Cell Microbiol 7:1345–1356
93. Gekara NO, Weiss S (2004) Lipid rafts clustering and signalling by listeriolysin O. Biochem Soc Trans 32:712–714
94. Hamon MA, Batsche E, Regnault B, Tham TN, Seveau S, Muchardt C, Cossart P (2007) Histone modifications induced by a family of bacterial toxins. Proc Natl Acad Sci USA 104:13467–13472
95. Hamon MA, Cossart P (2011) K+ efflux is required for histone H3 dephosphorylation by *Listeria monocytogenes* listeriolysin O and other pore-forming toxins. Infect Immun 79:2839–2846
96. Walev I, Bhakdi SC, Hofmann F, Djonder N, Valeva A, Aktories K, Bhakdi S (2001) Delivery of proteins into living cells by reversible membrane permeabilization with streptolysin-O. Proc Natl Acad Sci USA 98:3185–3190
97. Bhakdi S, Tranum-Jensen J (1991) Alpha-toxin of *Staphylococcus aureus*. Microbiol Rev 55:733–751
98. Kennedy CL, Smith DJ, Lyras D, Chakravorty A, Rood JI (2009) Programmed cellular necrosis mediated by the pore-forming alpha-toxin from *Clostridium septicum*. PLoS Pathog 5:e1000516
99. Nelson KL, Brodsky RA, Buckley JT (1999) Channels formed by sub-nanomolar concentrations of the toxin aerolysin trigger apoptosis of T lymphomas. Cell Microbiol 1:69–74
100. Idone V, Tam C, Goss JW, Toomre D, Pypaert M, Andrews NW (2008) Repair of injured plasma membrane by rapid Ca2+-dependent endocytosis. J Cell Biol 180:905–914
101. Keefe D, Shi L, Feske S, Massol R, Navarro F, Kirchhausen T, Lieberman J (2005) Perforin triggers a plasma membrane-repair response that facilitates CTL induction of apoptosis. Immunity 23:249–262
102. Dramsi S, Cossart P (2003) Listeriolysin O-mediated calcium influx potentiates entry of *Listeria monocytogenes* into the human Hep-2 epithelial cell line. Infect Immun 71:3614–3618
103. Vadia S, Arnett E, Haghighat AC, Wilson-Kubalek EM, Tweten RK, Seveau S (2011) The pore-forming toxin listeriolysin O mediates a novel entry pathway of *Listeria monocytogenes* into human hepatocytes. PLoS Pathog 7:e1002356
104. Gekara NO, Groebe L, Viegas N, Weiss S (2008) *Listeria monocytogenes* desensitizes immune cells to subsequent Ca2+ signaling via listeriolysin O-induced depletion of intracellular Ca2+ stores. Infect Immun 76:857–862
105. Fickl H, Cockeran R, Steel HC, Feldman C, Cowan G, Mitchell TJ, Anderson R (2005) Pneumolysin-mediated activation of NFkappaB in human neutrophils is antagonized by docosahexaenoic acid. Clin Exp Immunol 140:274–281
106. Aguilar JL, Kulkarni R, Randis TM, Soman S, Kikuchi A, Yin Y, Ratner AJ (2009) Phosphatase-dependent regulation of epithelial mitogen-activated protein kinase responses to toxin-induced membrane pores. PLoS One 4:e8076

107. Aoudjit L, Stanciu M, Li H, Lemay S, Takano T (2003) p38 mitogen-activated protein kinase protects glomerular epithelial cells from complement-mediated cell injury. Am J Physiol Ren Physiol 285:F765–F774
108. Gancz D, Donin N, Fishelson Z (2009) Involvement of the c-jun N-terminal kinases JNK1 and JNK2 in complement-mediated cell death. Mol Immunol 47:310–317
109. Kloft N, Busch T, Neukirch C, Weis S, Boukhallouk F, Bobkiewicz W, Cibis I, Bhakdi S, Husmann M (2009) Pore-forming toxins activate MAPK p38 by causing loss of cellular potassium. Biochem Biophys Res Comm 385:503–506
110. Niculescu F, Badea T, Rus H (1999) Sublytic C5b-9 induces proliferation of human aortic smooth muscle cells: role of mitogen activated protein kinase and phosphatidylinositol 3-kinase. Atherosclerosis 142:47–56
111. Peng H, Takano T, Papillon J, Bijian K, Khadir A, Cybulsky AV (2002) Complement activates the c-Jun N-terminal kinase/stress-activated protein kinase in glomerular epithelial cells. J Immunol 169:2594–2601
112. Ratner AJ, Hippe KR, Aguilar JL, Bender MH, Nelson AL, Weiser JN (2006) Epithelial cells are sensitive detectors of bacterial pore-forming toxins. J Biol Chem 281:12994–12998
113. Rus H, Niculescu F, Badea T, Shin ML (1997) Terminal complement complexes induce cell cycle entry in oligodendrocytes through mitogen activated protein kinase pathway. Immunopharmacology 38:177–187
114. Chu J, Thomas LM, Watkins SC, Franchi L, Nunez G, Salter RD (2009) Cholesterol-dependent cytolysins induce rapid release of mature IL-1beta from murine macrophages in a NLRP3 inflammasome and cathepsin B-dependent manner. J Leukoc Biol 86:1227–1238
115. Houldsworth S, Andrew PW, Mitchell TJ (1994) Pneumolysin stimulates production of tumor necrosis factor alpha and interleukin-1 beta by human mononuclear phagocytes. Infect Immun 62:1501–1503
116. Kayal S, Lilienbaum A, Join-Lambert O, Li X, Israel A, Berche P (2002) Listeriolysin O secreted by *Listeria monocytogenes* induces NF-kappaB signalling by activating the IkappaB kinase complex. Mol Microbiol 44:1407–1419
117. Kilgore KS, Schmid E, Shanley TP, Flory CM, Maheswari V, Tramontini NL, Cohen H, Ward PA, Friedl HP, Warren JS (1997) Sublytic concentrations of the membrane attack complex of complement induce endothelial interleukin-8 and monocyte chemoattractant protein-1 through nuclear factor-kappa B activation. Am J Pathol 150:2019–2031
118. Meixenberger K, Pache F, Eitel J, Schmeck B, Hippenstiel S, Slevogt H, N'Guessan P, Witzenrath M, Netea MG, Chakraborty T, Suttorp N, Opitz B (2010) *Listeria monocytogenes*-infected human peripheral blood mononuclear cells produce IL-1beta, depending on listeriolysin O and NLRP3. J Immunol 184:922–930
119. Nishibori T, Xiong H, Kawamura I, Arakawa M, Mitsuyama M (1996) Induction of cytokine gene expression by listeriolysin O and roles of macrophages and NK cells. Infect Immun 64:3188–3195
120. Tsuchiya K, Kawamura I, Takahashi A, Nomura T, Kohda C, Mitsuyama M (2005) Listeriolysin O-induced membrane permeation mediates persistent interleukin-6 production in Caco-2 cells during *Listeria monocytogenes* infection in vitro. Infect Immun 73:3869–3877
121. Viedt C, Hansch GM, Brandes RP, Kubler W, Kreuzer J (2000) The terminal complement complex C5b-9 stimulates interleukin-6 production in human smooth muscle cells through activation of transcription factors NF-kappa B and AP-1. FASEB J 14:2370–2372
122. Walev I, Hombach M, Bobkiewicz W, Fenske D, Bhakdi S, Husmann M (2002) Resealing of large transmembrane pores produced by streptolysin O in nucleated cells is accompanied by NF-kappaB activation and downstream events. FASEB J 16:237–239
123. Kraus S, Seger R, Fishelson Z (2001) Involvement of the ERK mitogen-activated protein kinase in cell resistance to complement-mediated lysis. Clin Exp Immunol 123:366–374
124. Corrotte M, Fernandes MC, Tam C, Andrews NW (2012) Toxin pores endocytosed during plasma membrane repair traffic into the lumen of MVBs for degradation. Traffic 13:483–494
125. Lopez JA, Susanto O, Jenkins MR, Lukoyanova N, Sutton VR, Law RH, Johnston A, Bird CH, Bird PI, Whisstock JC, Trapani JA, Saibil HR, Voskoboinik I (2013) Perforin forms

transient pores on the target cell plasma membrane to facilitate rapid access of granzymes during killer cell attack. Blood 121:2659–2668

126. Morgan BP, Dankert JR, Esser AF (1987) Recovery of human neutrophils from complement attack: removal of the membrane attack complex by endocytosis and exocytosis. J Immunol 138:246–253
127. Thiery J, Keefe D, Saffarian S, Martinvalet D, Walch M, Boucrot E, Kirchhausen T, Lieberman J (2010) Perforin activates clathrin- and dynamin-dependent endocytosis, which is required for plasma membrane repair and delivery of granzyme B for granzyme-mediated apoptosis. Blood 115:1582–1593
128. Lieberman J (2003) The ABCs of granule-mediated cytotoxicity: new weapons in the arsenal. Nat Rev Immunol 3:361–370
129. Russell JH, Ley TJ (2002) Lymphocyte-mediated cytotoxicity. Annu Rev Immunol 20:323–370
130. Cassidy SK, Hagar JA, Kanneganti TD, Franchi L, Nunez G, O'Riordan MX (2012) Membrane damage during *Listeria monocytogenes* infection triggers a caspase-7 dependent cytoprotective response. PLoS Pathog 8:e1002628
131. Cragg MS, Howatt WJ, Bloodworth L, Anderson VA, Morgan BP, Glennie MJ (2000) Complement mediated cell death is associated with DNA fragmentation. Cell Death Diff 7:48–58
132. Papadimitriou JC, Ramm LE, Drachenberg CB, Trump BF, Shin ML (1991) Quantitative analysis of adenine nucleotides during the prelytic phase of cell death mediated by C5b-9. J Immunol 147:212–217
133. Cudrici C, Niculescu F, Jensen T, Zafranskaia E, Fosbrink M, Rus V, Shin ML, Rus H (2006) C5b-9 terminal complex protects oligodendrocytes from apoptotic cell death by inhibiting caspase-8 processing and up-regulating FLIP. J Immunol 176:3173–3180
134. Rus HG, Niculescu F, Shin ML (1996) Sublytic complement attack induces cell cycle in oligodendrocytes. J Immunol 156:4892–4900
135. Rus HG, Niculescu FI, Shin ML (2001) Role of the C5b-9 complement complex in cell cycle and apoptosis. Immunol Rev 180:49–55
136. Soane L, Rus H, Niculescu F, Shin ML (1999) Inhibition of oligodendrocyte apoptosis by sublytic C5b-9 is associated with enhanced synthesis of bcl-2 and mediated by inhibition of caspase-3 activation. J Immunol 163:6132–6138
137. Keyel PA, Loultcheva L, Roth R, Salter RD, Watkins SC, Yokoyama WM, Heuser JE (2011) Streptolysin O clearance through sequestration into blebs that bud passively from the plasma membrane. J Cell Sci 124:2414–2423
138. Stein JM, Luzio JP (1991) Ectocytosis caused by sublytic autologous complement attack on human neutrophils. The sorting of endogenous plasma-membrane proteins and lipids into shed vesicles. Biochem J 274:381–386
139. Moskovich O, Fishelson Z (2007) Live cell imaging of outward and inward vesiculation induced by the complement c5b-9 complex. J Biol Chem 282:29977–29986
140. Ribet D, Hamon M, Gouin E, Nahori MA, Impens F, Neyret-Kahn H, Gevaert K, Vandekerckhove J, Dejean A, Cossart P (2010) *Listeria monocytogenes* impairs SUMOylation for efficient infection. Nature 464:1192–1195
141. Kloft N, Neukirch C, Bobkiewicz W, Veerachato G, Busch T, von Hoven G, Boller K, Husmann M (2010) Pro-autophagic signal induction by bacterial pore-forming toxins. Med Microbiol Immunol 199:299–309
142. Bucker R, Krug SM, Rosenthal R, Gunzel D, Fromm A, Zeitz M, Chakraborty T, Fromm M, Epple HJ, Schulzke JD (2011) Aerolysin from *Aeromonas hydrophila* perturbs tight junction integrity and cell lesion repair in intestinal epithelial HT-29/B6 cells. J Infect Dis 204:1283–1292
143. Dragneva Y, Anuradha CD, Valeva A, Hoffmann A, Bhakdi S, Husmann M (2001) Subcytocidal attack by staphylococcal alpha-toxin activates NF-kappaB and induces interleukin-8 production. Infect Immun 69:2630–2635

144. Gurcel L, Abrami L, Girardin S, Tschopp J, van der Goot FG (2006) Caspase-1 activation of lipid metabolic pathways in response to bacterial pore-forming toxins promotes cell survival. Cell 126:1135–1145
145. Gutierrez MG, Saka HA, Chinen I, Zoppino FC, Yoshimori T, Bocco JL, Colombo MI (2007) Protective role of autophagy against *Vibrio cholerae* cytolysin, a pore-forming toxin from *V. cholerae*. Proc Natl Acad Sci USA 104:1829–1834
146. Husmann M, Beckmann E, Boller K, Kloft N, Tenzer S, Bobkiewicz W, Neukirch C, Bayley H, Bhakdi S (2009) Elimination of a bacterial pore-forming toxin by sequential endocytosis and exocytosis. FEBS Lett 583:337–344
147. Garcia-Saez AJ, Buschhorn SB, Keller H, Anderluh G, Simons K, Schwille P (2011) Oligomerization and pore formation by equinatoxin II inhibit endocytosis and lead to plasma membrane reorganization. J Biol Chem 286:37768–37777
148. Degiacomi MT, Iacovache I, Pernot L, Chami M, Kudryashev M, Stahlberg H, van der Goot FG, Dal Peraro M (2013) Molecular assembly of the aerolysin pore reveals a swirling membrane-insertion mechanism. Nat Chem Biol 9:623–629
149. Carney DF, Hammer CH, Shin ML (1986) Elimination of terminal complement complexes in the plasma membrane of nucleated cells: influence of extracellular Ca^{2+} and association with cellular Ca^{2+}. J Immunol 137:263–270
150. Olofsson A, Hebert H, Thelestam M (1993) The projection structure of perfringolysin O (*Clostridium perfringens* θ-toxin). FEBS Lett 319:125–127
151. Smialowski P, Doose G, Torkler P, Kaufmann S, Frishman D (2012) PROSO II–a new method for protein solubility prediction. FEBS J 279:2192–2200
152. McNeela EA, Burke A, Neill DR, Baxter C, Fernandes VE, Ferreira D, Smeaton S, El-Rachkidy R, McLoughlin RM, Mori A, Moran B, Fitzgerald KA, Tschopp J, Petrilli V, Andrew PW, Kadioglu A, Lavelle EC (2010) Pneumolysin activates the NLRP3 inflammasome and promotes proinflammatory cytokines independently of TLR4. PLoS Pathog 6:e1001191
153. Pillich H, Loose M, Zimmer KP, Chakraborty T (2012) Activation of the unfolded protein response by *Listeria monocytogenes*. Cell Microbiol 14:949–964
154. Lam GY, Fattouh R, Muise AM, Grinstein S, Higgins DE, Brumell JH (2011) Listeriolysin O suppresses phospholipase C-mediated activation of the microbicidal NADPH oxidase to promote *Listeria monocytogenes* infection. Cell Host Microbe 10:627–634
155. Richter JF, Gitter AH, Gunzel D, Weiss S, Mohamed W, Chakraborty T, Fromm M, Schulzke JD (2009) Listeriolysin O affects barrier function and induces chloride secretion in HT-29/B6 colon epithelial cells. Am J Physiol Gastrointest Liver Physiol 296:G1350–G1359
156. Zysk G, Schneider-Wald BK, Hwang JH, Bejo L, Kim KS, Mitchell TJ, Hakenbeck R, Heinz HP (2001) Pneumolysin is the main inducer of cytotoxicity to brain microvascular endothelial cells caused by *Streptococcus pneumoniae*. Infect Immun 69:845–852
157. Lucas R, Yang G, Gorshkov BA, Zemskov EA, Sridhar S, Umapathy NS, Jezierska-Drutel A, Alieva IB, Leustik M et al (2012) Protein kinase C-alpha and arginase I mediate pneumolysin-induced pulmonary endothelial hyperpermeability. Am J Respir Cell Mol Biol 47:445–453
158. Stavru F, Bouillaud F, Sartori A, Ricquier D, Cossart P (2011) *Listeria monocytogenes* transiently alters mitochondrial dynamics during infection. Proc Natl Acad Sci USA 108:3612–3617
159. Braun JS, Hoffmann O, Schickhaus M, Freyer D, Dagand E, Bermpohl D, Mitchell TJ, Bechmann I, Weber JR (2007) Pneumolysin causes neuronal cell death through mitochondrial damage. Infect Immun 75:4245–4254
160. Braun JS, Novak R, Gao G, Murray PJ, Shenep JL (1999) Pneumolysin, a protein toxin of *Streptococcus pneumoniae*, induces nitric oxide production from macrophages. Infect Immun 67:3750–3756

161. Braun JS, Sublett JE, Freyer D, Mitchell TJ, Cleveland JL, Tuomanen EI, Weber JR (2002) Pneumococcal pneumolysin and H(2)O(2) mediate brain cell apoptosis during meningitis. J Clin Invest 109:19–27
162. Carrero JA, Calderon B, Unanue ER (2004) Listeriolysin O from *Listeria monocytogenes* is a lymphocyte apoptogenic molecule. J Immunol 172:4866–4874
163. Steinfort C, Wilson R, Mitchell T, Feldman C, Rutman A, Todd H, Sykes D, Walker J, Saunders K, Andrew PW et al (1989) Effect of *Streptococcus pneumoniae* on human respiratory epithelium in vitro. Infect Immun 57:2006–2013
164. Rayner CF, Jackson AD, Rutman A, Dewar A, Mitchell TJ, Andrew PW, Cole PJ, Wilson R (1995) Interaction of pneumolysin-sufficient and -deficient isogenic variants of *Streptococcus pneumoniae* with human respiratory mucosa. Infect Immun 63:442–447
165. Walev I, Vollmer P, Palmer M, Bhakdi S, Rose-John S (1996) Pore-forming toxins trigger shedding of receptors for interleukin 6 and lipopolysaccharide. Proc Natl Acad Sci USA 93:7882–7887

Chapter 8
The Biology of Pneumolysin

Tim J. Mitchell and Catherine E. Dalziel

Abstract Cholesterol dependent cytolysins are important in the ability of some bacteria to cause disease in man and animals. Pneumolysin (PLY) plays a key role in the diseases caused by *Streptococcus pneumoniae* (the pneumococcus). This chapter describes the role of PLY in some of the key process in disease. These include induction of cell death by pore formation and toxin-induced apoptosis as well as more subtle effects on gene expression of host cells including epigenetic effects of the toxin. The use of bacterial mutants that either do not express the toxin or express altered versions in biological systems is described. Use of isolated tissue and whole animal systems to dissect the structure/function relationships of the toxin as well as the role played by different activities in the pathogenesis of infection are described. The role of PLY in meningitis and the associated deafness is discussed as well as the role of the toxin in promoting increased lung permeability and inflammation during pneumococcal pneumonia. Different clinical strains of the pneumococcus produce different forms of PLY and the impact of this on disease caused by these strains is discussed. Finally, the impact of this knowledge on the development of treatment and prevention strategies for pneumococcal disease is discussed.

Keywords *Streptococcus pneumoniae* · Pneumolysin · Pathogenesis of pneumococcal infection · Meningitis · Human respiratory epithelium · Cell death

T. J. Mitchell (✉)
Institute of Microbiology and Infection, University of Birmingham,
Birmingham B15-2TT, UK
e-mail: t.j.mitchell@bham.ac.uk

T. J. Mitchell
School of Immunity and Infection, College of Medical and Dental Sciences,
University of Birmingham, Birmingham B5-2TT, UK

C. E. Dalziel
Institute of Infection, Immunity and Inflammation, University of Glasgow,
Glasgow G12-8TA, UK

G. Anderluh and R. Gilbert (eds.), *MACPF/CDC Proteins - Agents of Defence, Attack and Invasion*, Subcellular Biochemistry 80, DOI: 10.1007/978-94-017-8881-6_8,

Abbreviations

AIF	Apoptosis-inducing factor
CDC	Cholesterol dependent cytolysin
CFU	Colony forming units
CRP	C-reactive protein
CYLD	Deubiquitinating enzyme cylindromatosis
D1–3	Domains 1–3
HBMEC	Human brain microvascular endothelial cells
ICAM-1	Intracellular adhesion molecule-1
MAC	Membrane attack complex
NALP3	NOD-like receptor family pyrin domain containing 3
p38-MAPK	p38 mitogen-activated protein kinase
PAI-1	Plasminogen activator-1
PI3K	Phosphoinositide-3-kinase
PKC	Protein kinase-C
PLY	Pneumolysin
ROCK	Rho-associated kinase
SDM	Site-directed mutagenesis
TLO	Tetanolysin

Introduction

Pneumolysin (PLY) is a 53 kDa protein composed of 471 amino acids [75] and consists of four distinct functional domains. In practice PLY can be physically separated into two peptide fragments, domain 4 and domains 1–3 as the polypeptide backbone weaves in and out of domains 1–3 making them physically inseparable (domain 1 (residues 6–21, 58–147, 198–243, 319–342), domain 2 (residues 22–57, 343–359) and domain 3 (residues 148–197, 244–318)). Structural domains 1–3 (D1–3) form the N-terminal part of the molecule while domain 4 (residues 360–469) forms the important C-terminal part of the protein. Domains 1–3 are connected to domain 4 via domain 2, which is considered as the "neck" of the molecule. Membrane binding of the intact toxin is mediated through domain 4 and isolated domain 4 is capable of binding to membranes and competing with full-length toxin. Domain 4 also contains the undecapeptide sequence that is largely conserved in all members of the cholesterol dependent cytolysin (CDC) family. As a member of the CDC family of toxins, PLY is unique in that it is not actively secreted from the bacterium, as it lacks a typical signal secretion leader sequence. Instead of depending on a type II secretion system for bacterial release the accepted dogma is that PLY escapes from the cell on lysis of the pneumococcus, either by autolysis or the action of lytic antibiotics. However, there have been some reports of active secretion of the PLY protein [6, 56]. When discussing the biology of PLY it is worth remembering that there are several sequence variants of the protein and this may affect the biological activity of the toxin. For example, serotype 1 pneumococci belonging to multi-locus sequence type ST306 express a version of the toxin that is not lytic [34]. This has effects on the interaction of these bacteria with host cells and tissues.

PLY and Cell Death

Cholesterol dependent cytolysins form large pores in all eukaryotic cells that have cholesterol in their membranes. The mechanism of pore formation has been investigated by cryo-electron microscopy [74]. Following oligomer complex formation a conformational change occurs in PLY structure; first, two α-helical bundles in domain three change into two β-hairpin structures capable of piercing the membrane, then domain two collapses reducing the height of the overall complex by around 40 Å. The dramatic collapse of the complex brings the β-hairpins close enough to the membrane to pierce it resulting in a large pore. The pores formed can be up to 350 Å in diameter with each pore containing as many as 50 PLY monomers. The formation of pores by PLY frequently results in host cell death as membrane integrity is destroyed.

The sensitivity of mammalian cells to PLY as well as other CDCs is dependent on both dose and cell type. The sensitivity of human monocytes and epithelial cells to PLY are different with monocytes being more resistant to the toxin. Moreover, the resistance of monocytes to the toxin was further increased by treatment with γ-interferon [29]. At low doses some cell types may be able to survive cytolytic attack by either shedding or internalizing damaged areas of their membranes. Chu et al. [14] have shown that murine macrophages are capable of recovering from cytolytic damage after exposure to low doses of tetanolysin (TLO, a CDC). At low doses the macrophage is able to remove the damaged membrane by endocytosis, essentially "pinching" out the damaged area. Although these studies do not all examine PLY the conserved structures and functions of some CDCs suggest that host survival mechanisms may have adapted to respond similarly to membrane attack by this group of cytolysins.

Interaction of PLY with the Immune System

Complement

The classical complement pathway is usually activated by the aggregation of antibodies on a target pathogen; it results in specific targeting of complement components to the pathogen surface resulting in opsonisation or the formation of the membrane attack complex (MAC) which results in cytolysis of the target. Like other Gram-positive bacteria *S. pneumoniae* is resistant to the formation of membrane attack complexes (MAC) by complement, and its polysaccharide capsule provides some protection against opsonisation with C3b and subsequent phagocytosis by immune cells [33, 80]. However, it has been shown that complement is essential for pneumococcal clearance and that both the classical and alternative pathways are required and that they initiate and determine the magnitude of C3b deposition, respectively [80]. Complement activation in the lungs during early infection is important in preventing bacterial spread and invasive disease. In fact complement has an essential role in the prevention of septicaemia due to bacterial spread from the lungs to the blood [39, 80]. PLY plays a central role in protecting the pneumococcus from complement attack and aiding its spread to other tissues/organs. PLY is able to activate the classical complement pathway, even in the absence of PLY specific antibody [49]. Although this may seem counterintuitive, this mechanism can result in depletion of complement in the host and induce inflammation aiding bacterial survival and spread. Patients with active pneumococcal infection have reduced serum complement levels and PLY-treated serum has a reduced opsonic activity [2, 25].

The complement activating ability of PLY was once believed to be due to sequence homology in the fourth domain to human CRP (C-reactive protein) but it is now known to be due to structural homology to the Fc region of IgG [65]. Site-directed mutagenesis (SDM) has been employed to show that the complement activating ability of PLY is dependent on amino acids Tyr384 and Asp385 as amino acid substitutions at these positions caused significant reduction or abolishment of complement activation, respectively. The ability of PLY to inhibit complement activation may be important when the availability of complement to opsonize bacteria is limited as occurs in the lung, or in the blood or lungs of rats with decreased levels of complement in a model of liver cirrhosis [1, 57].

Inflammatory Properties of PLY

Although at high concentrations PLY lyses all cells that contain cholesterol in their membranes, at sub-lytic concentrations the toxin can have marked effects on the production of immune regulatory molecules. PLY activates expression of a large number of genes in eukaryotic cells [64]. Microarray analysis using a THP-1 monocyte cell line exposed to wild type pneumococci or a PLY-deficient mutant showed that more than 140 genes were up regulated by PLY in the context of infection with *S. pneumoniae*. Up-regulated genes included those for chemokines and cytokines including those involved in recruitment of inflammatory cells such as IL-8, MIP-1β and MCP-3. Molecules such as lysozyme, caspases (4 and 6), cathepsin E and mannose binding lectin were up-regulated. In a study using purified PLY protein alone it was shown that intracellular adhesion molecule-1 (ICAM-1) was up-regulated. The effect on ICAM-1 expression appears to be related to the pore-forming activity of the toxin, as a single amino acid substitution in the toxin that reduces lytic activity by 99 % reduces ability to stimulate ICAM-1 expression. Since ICAM-1 is an integral component of the immunological synapse between T lymphocyte and antigen presenting cell, increased ICAM-1 expression on antigen presenting cells could enhance adaptive responses. In this context, it is interesting to note that PLY has been shown to enhance immune responses to other proteins when they are genetically coupled to the toxin [21]. The role of PLY in altering eukaryotic gene transcription also involves modification of epigenetic regulation [27]. One of the epigenetic changes induced by PLY is histone H3 dephosphorylation and this is independent of the pore forming ability of the toxin. PLY has also been shown decrease the level of SUMOylation in mammalian cells by increasing the degradation of Ubc9 (E2 SUMO enzyme) needed for the SUMOylation process [63]. Targets of Ubc9 are involved in transcriptional regulation. These effects of PLY (and other members of the CDC family) are dependent on the pore forming activity of the toxin. During pneumococcal lung infection in mice the deubiquitinating enzyme cylindromatosis (CYLD) in involved in pathology and poor outcome [42]. This study demonstrates that infection of mice with pneumococci or administration of PLY induces a mortality rate higher than that seen in *Cyld*$^{-/-}$ mice. The knockout mice had less microvascular leak and less haemorrhage. Lack of CYLD is associated with increased expression of prothrombic protein plasminogen activator-1 (PAI-1), which is a member of the serine protease inhibitor (serpin) family. PAI-1 deficiency results in enhanced lung haemorrhage due to pneumococcus of PLY. CYLD negatively regulates phosphorylation of p38MAPK and pMKK3 as *Cyld*$^{-/-}$ mice show enhanced phosphorylation of these proteins. Pharmacologic inhibition of the p38MAPK-MKK3 reduces APAI-1 expression and enhances mortality in *S. pneumoniae*-infected *Cyld*$^{-/-}$ mice.

The toxin is very pro-inflammatory and stimulates the production of cytokines including TNF, interleukin-1 and IL-6 [11]. The production of these cytokines may play a role in pro-inflammatory disease as well as regulation of the immune response to the pneumococcus. The toxin stimulates nitric oxide production from macrophages. Activation of mouse macrophages by PLY is dependent on IFN as cells with an inactivated IFN signalling pathway are unable to produce NO in response to the toxin [11]. Increases in transcription of the genes for COX-2 in response to PLY may also be involved in increased prostaglandin synthesis. Activation of phospholipase A in pulmonary epithelial cells [68] could result in the degradation of membrane phospholipids which could contribute directly to lung damage by release of free fatty acids and lysophosphatides, while release of arachidonic acid could promote chemotaxis and respiratory burst of neutrophils [5, 18]. Neutrophil transendothelial migration in vitro to the pneumococcus has been shown to be dependent on PLY [50]. The toxin causes increased production of superoxide from neutrophils as well as increased production of elastase, expression of B2 integrins and increased production of prostaglandin E2 and leukotriene B4 [16, 17]. Synthesis and release of matrix metalloproteinase 8 and 9 is also increased by treatment of neutrophils with PLY. As well as promoting release of pro-inflammatory molecules the toxin can also promote cell recruitment to sites of infection. For example PLY promotes the production of IL-8 from neutrophils [15] and epithelial [59] and endothelial cells [45]. The toxin is therefore very pro-inflammatory. It seems somewhat of a paradox that production of the toxin may promote the demise of the organism by promoting inflammation. However, it may be that the inflammation induced is not effective in clearing the organism and PLY has been shown to reduce killing of pneumococci by neutrophils in vitro [23]. It should also be remembered that the pneumococcus is mostly found as a commensal colonising the nasopharynx of humans. It is possible the PLY plays a role in colonization of the host and only causes overt inflammatory disease when the organism accesses normally 'sterile' sites such as the lower respiratory tract or meninges.

The cytotoxic effect of PLY on the respiratory epithelium causes cytoplasmic blebbing, mitochondrial swelling and cell death [72]. PLY helps pneumococci to move from alveoli to interstitium and into the blood stream. This is achieved by breaking the epithelial tight junctions and increasing alveolar permeability, inflammation and thus aiding its spread in the blood (bacteraemia) [37, 52, 66]. In addition to this PLY also impedes ciliary bronchial epithelial cell clearance and helps in propagation of the organism in the lower respiratory tract. This effect is dependent on hemolytic activity as a mutant lacking haemolytic activity (W433F) showed reduced effect. Thus Ply perturbs host defences in the respiratory tract, which helps in proliferation and invasion of the bacterium [22, 62, 72].

There are at least 16 different naturally-occurring variants of PLY including allele 5 PLY which is expressed in specific strains of serotypes 1 and 8 pneumococci [34, 40, 43]. It has been demonstrated that both human and murine mononuclear cells exposed to *S. pneumoniae* that express fully lytic toxin produce IL-1β and the production of this cytokine depended on the NOD-like receptor family, pyrin domain containing 3 (NALP3) inflammasome [79]. Strains expressing the non-haemolytic allele five of the toxin did not stimulate IL-1β production. NLRP3 activation was beneficial for mice during pneumonia caused by pneumococcal strains expressing fully active toxin due to cytokine production and maintenance of the pulmonary micro-vascular barrier. Thus, polymorphisms in the PLY protein may substantially affect recognition of bacteria by the innate immune system.

The Interaction of PLY with Pattern Recognition Receptors

PLY interacts with the innate immune system through TLR-4 [47]. Production of pro-inflammatory cytokines such as TNF and IL-6 by macrophages is dependent on the presence of TLR-4 and Myd88. Lack of TLR4 increases the rate of infection following colonization of mice. This supports the view mentioned above that activation of some inflammatory pathways plays a role in the normal colonisation process. In models of nasopharyngeal colonisation lack of PLY impedes bacterial clearance [69]. PLY has been shown to physically interact with TLR4 in a solid phase binding assay and mediates signalling in macrophages and epithelial cells [71]. Other studies have suggested that TLR4 plays only a small role in the immune response to PLY [10, 48]. However, the inflammatory response to PLY does appear to involve both TLR2 and TLR4 [19]. In this study PLY caused a classic inflammatory response when instilled into the respiratory tract. This response included neutrophil influx, production of pro-inflammatory cytokines and changes in epithelial barrier permeability. The inflammation observed at high toxin doses in control animals showed phagocytic cell influx, increase in cytokines (Il-1β, Il-6 and TNF-α and the chemokine KC). There was also an increase in total protein levels in the lung showing a change in permeability. In TLR-4 deficient mice there were fewer neutrophils, less Il-6, Il-1β and KC and less of a change in permeability. TLR2 knockout mice showed a reduced level of Il-6, KC and protein. PLY also promotes the production of IFN-γ and IL-17A in the lungs during pneumonia in a TLR4 dependent manner [48].

Effects of PLY on Cells and Tissues

At sub-lytic concentration PLY has been shown to have several effects on eukaryotic cell signalling. The toxin induces early membrane depolarization and micropores form in the plasma membrane. These micropores lead to calcium influx and activation of rac and rho GTPases, including rac-1 and the rho-associated kinase (ROCK) [31]. These changes cause cytoskeletal rearrangement in the intoxicated cell. The changes in cell morphology have been attributed to PLY-induced changes to the actin cytoskeleton. At sublytic concentrations rearrangement of the actin cytoskeleton and microtubule stabilization (as indicated by increased acetylation and bundling) can be observed in astrocytes [24, 32]. Although these effects were seen in the absence of observable pores, pore-forming competency was required and non-lytic mutants did not induce these cytoskeletal changes. Cholesterol was also required for microtubule stabilization as these effects were abolished in cholesterol-depleted cells and when using cholesterol-treated PLY. This suggested that PLY must be bound to the cell surface in order to induce these effects. Recently, the mechanism of PLY-dependent cytoskeletal remodelling has been further elucidated, it has been shown that PLY has the ability to directly bind actin in vitro. Domains 1–3 and full length PLY bind strongly to actin, the non-lytic deletion mutant Δ6PLY has comparatively reduced binding activity suggesting that the refolding and/or conformational changes that occur in D1–3 during pore formation are required for actin binding [30].

Sublytic concentrations of PLY also activate p38 mitogen-activated protein kinase (p38-MAPK) due to osmotic stress [59]. Cells protected from osmotic stress by use of high molecular weight dextran do not activate p38-MAPK in response to PLY suggesting that osmosensing is a general mechanism for activating immune responses [59]. Activation of p38MAPK may protect tissues from damage as described above in relation to CYLD. Macrophages exposed to sublytic concentrations of PLY show activation of phosphoinositide-3-kinase (PI3K). Activation of PI3K is required for recruitment of inflammatory macrophages to the site of infection.

Streptococcus pneumoniae has been reported to induce apoptosis in a number of cell types including neutrophils, macrophages and neuronal cells [12, 20, 73, 81]. Zysk et al. [81] report that PLY is responsible for mainly necrosis in human neutrophils and that apoptosis is induced by heat killed bacteria. Dockrell et al. [20] used a PLY-negative mutant of the pneumococcus to show that apoptosis of infected macrophages was reduced in the absence of PLY.

Streptococcus pneumoniae is a major cause of bacterial meningitis and a crucial interaction is therefore between the bacterium and the blood brain barrier. Using brain endothelial cells it has been shown that most of the damage to the blood brain barrier is due to the action of PLY [82]. The damage induced by the toxin was dependent on active protein synthesis, tyrosine phosphorylation and the activation of caspases [82].

The most detailed studies of the role of PLY in apoptosis have been done using neuronal tissues. PLY has been shown to induce an apoptosis-inducing factor (AIF)—dependent form of apoptosis [12]. Both PLY and bacterial hydrogen peroxide were shown to be important in the induction of apoptosis of human microglial cells and ultra structural analysis of the cells following the blocking of PLY and hydrogen peroxide activity revealed reduced mitochondrial damage. The effects of PLY in this system were dependent on the pore forming activity and PLY-induced increases in intracellular calcium levels were shown to trigger the release of AIF from mitochondria in rat neurones. Chelation of calcium blocked the release of AIF and showed that this event is mediated by increases in intracellular calcium. The importance of calcium levels in PLY-induced apoptosis has also been demonstrated in neuroblastoma cells [73]. The changes in intracellular calcium levels were shown to be dependent on pore formation by the toxin and p38MAPK was also shown to play a role in cell death.

The spread of the pneumococcus from the nasopharynx to the eyes, ears and lungs and subsequent spread to the blood and brain is facilitated by PLY. The inflammation induced by immune responses to the toxin no doubt contributes to loss of integrity in epithelial and endothelial barriers as seen in other disease states but PLY itself directly induces changes in cellular morphology and function that disrupt these protective barriers and mechanisms. In fact, it has been shown histologically that damage to endothelial cells precedes the infiltration of inflammatory cells [67]. In a model of the human respiratory mucosa, isogenic PLY positive and negative pneumococci rapidly adhere to mucus and both caused significant damage to the ultrastructure of the epithelium over 48 h; damage includes extrusion of cells and cells debris. Pneumococci did not adhere directly to ciliated cells, however, though PLY positive pneumococci caused significant damage to the ciliated epithelium; reduced ciliary beat or ciliary stasis and disorganization of cilia was observed. In addition, PLY-positive pneumococci caused the separation of epithelial tight junctions and increased pneumococcal adherence was seen on the edges of separated cells [61]. Reduced ciliary beat in the respiratory tract may prevent pneumococcal clearance via the mucociliary elevator and the breakdown of tight junctions allows pneumococci to pass through epithelial and endothelial barriers. The damage to ciliated cells during infection not only aids the spread of infection but also can cause permanent damage to the host; an example of such damage is the sensorineural hearing loss that is associated with pneumococcal meningitis. PLY is responsible for loss of hearing due to damage caused to cochlear hair cells; the damage can be severe ranging from stereocilia disorganization as seen on ciliated epithelial cells, to total cell loss mediated by calcium influx [9, 76].

In a model of the blood brain barrier using Human Brain Microvascular Endothelial Cells (HBMEC), rapid PLY-dependent rounding and detachment of HBMEC was observed [82]. As seen in other ciliated cells PLY has the ability to slow ciliary beat in the ependymal cells of the rat brain; this was shown with both purified PLY and penicillin-lysed pneumococci [28]. In addition ciliary sloughing and cytoplasmic extrusions were noted following PLY treatment of ciliated ependymal cells; these effects of PLY could be blocked with anti-PLY antibodies [28].

Pulmonary permeability edema, which is a major complication of pneumonia is characterised by endothelial hyper-permeability and can occur several days after antibiotic therapy when the tissues are sterile. The occurrence of hyper-permeability correlates with the presence of PLY [77]. PLY increases permeability in endothelial monolayers by reducing stable and dynamic microtubule content and by modulating VE-cadherin expression [44]. These effects are preceded by activation of protein kinase-C (PKC), perturbation of the RhoA/Rac1 balance and increased phosphorylation of myosin light chain. The expression of arginase is also increased by PLY treatment. The use of a PKC inhibitor reduces PLY-induced activation of arginase and reduces PLY-induced hyper-permeability in mice. The role of arginase in the PLY-induced change in endothelial permeability is supported by the finding that arginase knockout mice are less susceptible to PLY-induced capillary leak. As well as changing endothelial barrier permeability PLY can alter local systemic blood pressure and perfusion. When isolated mouse lungs were exposed intravascularly to PLY a dose dependent increase in vascular resistance was observed. Detection of the toxin by immunohistochemistry showed that it was localised in pulmonary arterial vessels, which displayed vasoconstriction [77]. In a later study by the Witzenrath group [78] it was shown that the dose dependent increase in pulmonary vascular resistance was associated with an increase in levels of platelet-activating factor in the lung. The pressure response was reduced in lungs from PAF-receptor deficient mice and after pharmacological blockade of the PAF receptor. Analysis of cell signalling pathways suggested increased pressure involved phosphatidylcholine-specific phospholipase C, protein kinase C and the Rho kinase system. PLY-induced microvascular leakage was reduced in PAF-receptor knockout mice. PLY therefore plays a very important role in pulmonary vascular hypertension and microvascular leakage and the PAF-receptor and its downstream signalling pathways may provide a target for clinical intervention during pneumococcal-induced respiratory failure.

Role of PLY in Intracellular Delivery

PLY can allow the intracellular delivery of other microbial products into the cytoplasm of host cells. Ratner et al. [60] showed that simultaneous colonisation of the mouse upper airway with *Streptococcus pneumoniae* and *Haemophilus influenzae* leads to synergistic production of chemokines at the mucosal surface and vigorous recruitment of neutrophils. These neutrophils mediate the outcome of the competition between these two species to establish colonisation [46]. Co-colonization resulted in the clearance of the pneumococci by attracted neutrophils. PLY and the host factor NFkB were show to be essential for this response. Fragments of *H. influenzae* peptidoglycan were shown to enter cells through PLY pores, leading to Nod1-mediated signalling and activation of NFkB [58]. PLY can also mediate the entry of pneumococcal DNA into respiratory epithelial and dendritic cells where it then stimulates type 1 interferon signalling [53].

Role of PLY in Pathogenesis of Infection

There are a large range of studies, mostly using murine models of infection that demonstrate the important role that PLY plays in pathogenesis. Early studies using PLY-deficient mutants of the pneumococcus demonstrated decreased colonisation of the nasopharynx, increased bacterial clearance from the lung and prolonged survival of animals following infection [8]. When instilled into the lung PLY-negative versions of the pneumococcus induce much less inflammation [13]. Cell recruitment into the lung during infection with PLY-ve pneumococci is delayed in comparison to infection with wild-type bacteria [36]. Neutrophil responses were the most affected and the redistribution of T and B lymphocytes in and around inflamed bronchioles was also delayed and reduced in the absence of PLY. The PLY-negative mutant also causes bacteraemia later than the wild type organism following respiratory infection.

The role of PLY in bacteraemia is probably to protect the pneumococcus from the host response. In a model of bacteraemia wild type bacteria grew exponentially until bacterial numbers as high as 10^{10} colony forming units (CFU) per ml of blood were reached, when the animals died [7]. When PLY-negative bacteria were used the initial growth rate of bacteria in the blood was the same until numbers of about 10^{7} CFU/ml were reached when the increase in bacterial numbers stopped and remained constant for several days. Interestingly, if PLY-negative bacteria were mixed with wild type organisms they demonstrated wild-type characteristics, suggesting PLY mediates its effects at a distance.

PLY plays a crucial role in hearing loss during experimental meningitis [76]. Using a guinea pig model of meningitis it was shown that a PLY-negative strain of the pneumococcus induced much less damage to the cochlea than the wild type strain despite inducing an equivalent amount of inflammation. It is interesting to note that inflammation was not reduced in the brain as use of the same mutants in the lung model showed much reduced inflammation in the absence of the toxin [13]. This presumably reflects the presence of different inflammatory receptors in brain and lung. The extent of damage to the cochlea was reflected in the amount of hearing loss in infected animals. Thus, in animal models at least, PLY plays a major role in meningitis associated hearing loss. This could be an important point of clinical intervention to prevent this common complication.

The virulence of PLY^- mutants has also been studied in eye infection models. *S. pneumoniae* is one of the three main species causing bacterial keratitis [4]. PLY-negative bacteria used in models of ocular infection show greatly reduced virulence [35, 51]. Immunization with PLY protects against endophthalmitis [70].

The Use of PLY as a Vaccine

It was first shown in 1983 that PLY can protect mice from challenge with *S. pneumoniae* [54]. The toxin has been shown to confer protection against challenge in a serotype independent manner [3]. PLY has been used as a carrier protein for pneumococcal polysaccharide [55]. Highly protective anti-pneumococcal responses could be induced in infant mice by immunization with the conjugate during gestation or early infancy. Non-toxic versions of PLY have been developed for use in vaccines. One such version of the toxin was created by deletion of a single amino acid (Ala146) [41]. This mutant protein was unable to form pores in cell membranes or stimulate the in vivo inflammatory effects associated with native PLY treatment. Mice vaccinated with the mutant PLY generated high levels of neutralizing antibodies and were significantly protected from challenge with virulent pneumococci. It has also been demonstrated that PLY can act as a powerful mucosal adjuvant for genetically coupled proteins [21]. Intranasal administration of mice with pneumococcal PsaA coupled to PLY resulted in high levels of serum IgG and mucosal IgA to PsaA. Fusion of PLY to PspA has also shown to elicit protective responses in animals [26]. Another toxoid version of PLY was designed by using structural approaches to select appropriate mutations and led to the development of PlyD1, which contains three amino acid substitutions (T65C, G293C and C428A). PlyD1 has been used in phase 1 clinical trials in humans [38] where all dose levels used in adults were shown to be safe and immunogenic. Vaccination with PlyD1 generated functional neutralising antibodies. Thus, the first stage of using PLY as a vaccine has been achieved.

Summary

PLY has a range of biological functions that contribute to the disease process during pneumococcal infection. Some of these effects are intimately linked with the ability of the toxin to undergo the elegant process of changing from a monomer in solution to a large oligomeric structure inserted into the membrane of host cells. The toxin can also mediate some actions, such as acting as a mucosal adjuvant, in the absence of the pore formation. The toxin interacts with a number of host cell components including membrane cholesterol, actin and Toll-like receptors. These interactions mediate changes in cell signalling and behaviour. As well as acting as a protective antigen in its own right PLY can act as a modulator of the immune system and affect response of the host to other bacterial products. This modulation may be at the level of cell signalling or may be through allowing access of other molecules to intracellular receptors via the PLY pore. There is an ever-increasing amount of knowledge concerning the biological effects of PLY and other pore forming proteins. This knowledge is being used to design new potential methods to treat or prevent pneumococcal disease. This is highlighted by the first Phase 1 trials of PLY as a vaccine in humans.

References

1. Alcantara RB, Preheim LC, Gentry MJ (1999) Role of pneumolysin's complement-activating activity during pneumococcal bacteremia in cirrhotic rats. Infect Immun 67:2862–2866
2. Alcantara RB, Preheim LC, Gentry-Nielsen MJ (2001) Pneumolysin-induced complement depletion during experimental pneumococcal bacteremia. Infect Immun 69:3569–3575
3. Alexander JE, Lock RA, Peeters CC, Poolman JT, Andrew PW, Mitchell TJ, Hansman D, Paton JC (1994) Immunization of mice with pneumolysin toxoid confers a significant degree of protection against at least nine serotypes of *Streptococcus pneumoniae*. Infect Immun 62:5683–5688
4. Asbell P, Stenson S (1982) Ulcerative keratitis. Survey of 30 years' laboratory experience. Arch Ophthalmol 100:77–80
5. Badwey JA, Curnutte JT, Robinson JM, Berde CB, Karnovsky MJ, Karnovsky ML (1984) Effects of free fatty acids on release of superoxide and on change of shape by human neutrophils. Reversibility by albumin. J Biol Chem 259:7870–7877
6. Balachandran P, Hollingshead SK, Paton JC, Briles DE (2001) The autolytic enzyme LytA of *Streptococcus pneumoniae* is not responsible for releasing pneumolysin. J Bacteriol 183:3108–3116
7. Benton KA, Everson MP, Briles DE (1995) A pneumolysin-negative mutant of *Streptococcus pneumoniae* causes chronic bacteremia rather than acute sepsis in mice. Infect Immun 63:448–455
8. Berry AM, Yother J, Briles DE, Hansman D, Paton JC (1989) Reduced virulence of a defined pneumolysin-negative mutant of *Streptococcus pneumoniae*. Infect Immun 57:2037–2042
9. Beurg M, Hafidi A, Skinner L, Cowan G, Hondarrague Y, Mitchell TJ, Dulon D (2005) The mechanism of pneumolysin-induced cochlear hair cell death in the rat. J Physiol 568:211–227
10. Branger J, Knapp S, Weijer S, Leemans JC, Pater JM, Speelman P, Florquin S, van der Poll T (2004) Role of Toll-like receptor 4 in gram-positive and gram-negative pneumonia in mice. Infect Immun 72:788–794
11. Braun JS, Novak R, Gao G, Murray PJ, Shenep JL (1999) Pneumolysin, a protein toxin of *Streptococcus pneumoniae*, induces nitric oxide production from macrophages. Infect Immun 67:3750–3756
12. Braun JS, Sublett JE, Freyer D, Mitchell TJ, Cleveland JL, Tuomanen EI, Weber JR (2002) Pneumococcal pneumolysin and H(2)O(2) mediate brain cell apoptosis during meningitis. J Clin Invest 109:19–27

13. Canvin JR, Marvin AP, Sivakumaran M, Paton JC, Boulnois GJ, Andrew PW, Mitchell TJ (1995) The role of pneumolysin and autolysin in the pathology of pneumonia and septicemia in mice infected with a type 2 pneumococcus. J Infect Dis 172:119–123
14. Chu J, Thomas LM, Watkins SC, Franchi L, Nunez G, Salter RD (2009) Cholesterol-dependent cytolysins induce rapid release of mature IL-1beta from murine macrophages in a NLRP3 inflammasome and cathepsin B-dependent manner. J Leukoc Biol 86:1227–1238
15. Cockeran R, Durandt C, Feldman C, Mitchell TJ, Anderson R (2002) Pneumolysin activates the synthesis and release of interleukin-8 by human neutrophils in vitro. J Infect Dis 186:562–565
16. Cockeran R, Steel HC, Mitchell TJ, Feldman C, Anderson R (2001) Pneumolysin potentiates production of prostaglandin E(2) and leukotriene B(4) by human neutrophils. Infect Immun 69:3494–3496
17. Cockeran R, Theron AJ, Steel HC, Matlola NM, Mitchell TJ, Feldman C, Anderson R (2001) Proinflammatory interactions of pneumolysin with human neutrophils. J Infect Dis 183:604–611
18. Curnutte JT, Badwey JM, Robinson JM, Karnovsky MJ, Karnovsky ML (1984) Studies on the mechanism of superoxide release from human neutrophils stimulated with arachidonate. J Biol Chem 259:11851–11857
19. Dessing MC, Hirst RA, de Vos AF, van der Poll T (2009) Role of Toll-like receptors 2 and 4 in pulmonary inflammation and injury induced by pneumolysin in mice. PLoS One 4:e7993
20. Dockrell DH, Lee M, Lynch DH, Read RC (2001) Immune-mediated phagocytosis and killing of *Streptococcus pneumoniae* are associated with direct and bystander macrophage apoptosis. J Infect Dis 184:713–722
21. Douce G, Ross K, Cowan G, Ma J, Mitchell TJ (2010) Novel mucosal vaccines generated by genetic conjugation of heterologous proteins to pneumolysin (PLY) from *Streptococcus pneumoniae*. Vaccine 28:3231–3237
22. Feldman C, Mitchell TJ, Andrew PW, Boulnois GJ, Read RC, Todd HC, Cole PJ, Wilson R (1990) The effect of *Streptococcus pneumoniae* pneumolysin on human respiratory epithelium in vitro. Microb Pathog 9:275–284
23. Ferrante A, RowanKelly B, Paton JC (1984) Inhibition of in vitro human lymphocyte response by the pneumococcal toxin pneumolysin. Infect Immun 46:585–589
24. Fortsch C, Hupp S, Ma J, Mitchell TJ, Maier E, Benz R, Iliev AI (2011) Changes in astrocyte shape induced by sublytic concentrations of the cholesterol-dependent cytolysin pneumolysin still require pore-forming capacity. Toxins (Basel) 3:43–62
25. Giebink GS, Dee TH, Kim Y, Quie PG (1980) Alterations in serum opsonic activity and complement levels in pneumococcal disease. Infect Immun 29:1062–1066
26. Goulart C, da Silva TR, Rodriguez D, Politano WR, Leite LC, Darrieux M (2013) Characterization of protective immune responses induced by pneumococcal surface protein A in fusion with pneumolysin derivatives. PLoS One 8:e59605
27. Hamon MA, Batsche E, Regnault B, Tham TN, Seveau S, Muchardt C, Cossart P (2007) Histone modifications induced by a family of bacterial toxins. Proc Natl Acad Sci USA 104:13467–13472
28. Hirst RA, Mohammed BJ, Mitchell TJ, Andrew PW, O'Callaghan C (2004) *Streptococcus pneumoniae*-induced inhibition of rat ependymal cilia is attenuated by antipneumolysin antibody. Infect Immun 72:6694–6698
29. Hirst RA, Yesilkaya H, Clitheroe E, Rutman A, Dufty N, Mitchell TJ, O'Callaghan C, Andrew PW (2002) Sensitivities of human monocytes and epithelial cells to pneumolysin are different. Infect Immun 70:1017–1022
30. Hupp S, Fortsch C, Wippel C, Ma J, Mitchell TJ, Iliev AI (2013) Direct transmembrane interaction between actin and the pore-competent, cholesterol-dependent cytolysin pneumolysin. J Mol Biol 425:636–646
31. Iliev AI, Djannatian JR, Nau R, Mitchell TJ, Wouters FS (2007) Cholesterol-dependent actin remodeling via RhoA and Rac1 activation by the *Streptococcus pneumoniae* toxin pneumolysin. Proc Natl Acad Sci USA 104:2897–2902

32. Iliev AI, Djannatian JR, Opazo F, Gerber J, Nau R, Mitchell TJ, Wouters FS (2009) Rapid microtubule bundling and stabilization by the *Streptococcus pneumoniae* neurotoxin pneumolysin in a cholesterol-dependent, non-lytic and Src-kinase dependent manner inhibits intracellular trafficking. Mol Microbiol 71:461–477
33. Jarva H, Jokiranta TS, Wurzner R, Meri S (2003) Complement resistance mechanisms of streptococci. Mol Immunol 40:95–107
34. Jefferies JM, Johnston CH, Kirkham LA, Cowan GJ, Ross KS, Smith A, Clarke SC, Brueggemann AB, George RC, Pichon B, Pluschke G, Pfluger V, Mitchell TJ (2007) Presence of nonhemolytic pneumolysin in serotypes of *Streptococcus pneumoniae* associated with disease outbreaks. J Infect Dis 196:936–944
35. Johnson MK, Hobden JA, Hagenah M, OCallaghan RJ, Hill JM, Chen S (1990) The role of pneumolysin in ocular infections with *Streptococcus pneumoniae*. Curr Eye Res 9:1107–1114
36. Kadioglu A, Gingles NA, Grattan K, Kerr A, Mitchell TJ, Andrew PW (2000) Host cellular immune response to pneumococcal lung infection in mice. Infect Immun 68:492–501
37. Kadioglu A, Taylor S, Iannelli F, Pozzi G, Mitchell TJ, Andrew PW (2002) Upper and lower respiratory tract infection by *Streptococcus pneumoniae* is affected by pneumolysin deficiency and differences in capsule type. Infect Immun 70:2886–2890
38. Kamtchoua T, Bologa M, Hopfer R, Neveu D, Hu B, Sheng X, Corde N, Pouzet C, Zimmermann G, Gurunathan S (2013) Safety and immunogenicity of the pneumococcal pneumolysin derivative PlyD1 in a single-antigen protein vaccine candidate in adults. Vaccine 31:327–333
39. Kerr AR, Paterson GK, Riboldi-Tunnicliffe A, Mitchell TJ (2005) Innate immune defense against pneumococcal pneumonia requires pulmonary complement component C3. Infect Immun 73:4245–4252
40. Kirkham LA, Jefferies JM, Kerr AR, Jing Y, Clarke SC, Smith A, Mitchell TJ (2006) Identification of invasive serotype 1 pneumococcal isolates that express nonhemolytic pneumolysin. J Clin Microbiol 44:151–159
41. Kirkham LA, Kerr AR, Douce GR, Paterson GK, Dilts DA, Liu DF, Mitchell TJ (2006) Construction and immunological characterization of a novel nontoxic protective pneumolysin mutant for use in future pneumococcal vaccines. Infect Immun 74:586–593
42. Lim JH, Stirling B, Derry J, Koga T, Jono H, Woo CH, Xu H, Bourne P, Ha UH, Ishinaga H, Andalibi A, Feng XH, Zhu H, Huang Y, Zhang W, Weng X, Yan C, Yin Z, Briles DE, Davis RJ, Flavell RA, Li JD (2007) Tumor suppressor CYLD regulates acute lung injury in lethal *Streptococcus pneumoniae* infections. Immunity 27:349–360
43. Lock RA, Zhang QY, Berry AM, Paton JC (1996) Sequence variation in the *Streptococcus pneumoniae* pneumolysin gene affecting haemolytic activity and electrophoretc mobility of the toxin. Microb Pathogen 21:71–83
44. Lucas R, Yang G, Gorshkov BA, Zemskov EA, Sridhar S, Umapathy NS, Jezierska-Drutel A, Alieva IB, Leustik M, Hossain H, Fischer B, Catravas JD, Verin AD, Pittet JF, Caldwell RB, Mitchell TJ, Cederbaum SD, Fulton DJ, Matthay MA, Caldwell RW, Romero MJ, Chakraborty T (2012) Protein kinase C-alpha and arginase I mediate pneumolysin-induced pulmonary endothelial hyperpermeability. Am J Respir Cell Mol Biol 47:445–453
45. Luttge M, Fulde M, Talay SR, Nerlich A, Rohde M, Preissner KT, Hammerschmidt S, Steinert M, Mitchell TJ, Chhatwal GS, Bergmann S (2012) *Streptococcus pneumoniae* induces exocytosis of Weibel-Palade bodies in pulmonary endothelial cells. Cell Microbiol 14:210–225
46. Lysenko ES, Ratner AJ, Nelson AL, Weiser JN (2005) The role of innate immune responses in the outcome of interspecies competition for colonization of mucosal surfaces. PLoS Pathog 1:e1
47. Malley R, Henneke P, Morse SC, Cieslewicz MJ, Lipsitch M, Thompson CM, Kurt-Jones E, Paton JC, Wessels MR, Golenbock DT (2003) Recognition of pneumolysin by Toll-like receptor 4 confers resistance to pneumococcal infection. Proc Natl Acad Sci USA 100:1966–1971

48. McNeela EA, Burke A, Neill DR, Baxter C, Fernandes VE, Ferreira D, Smeaton S, El-Rachkidy R, McLoughlin RM, Mori A, Moran B, Fitzgerald KA, Tschopp J, Petrilli V, Andrew PW, Kadioglu A, Lavelle EC (2010) Pneumolysin activates the NLRP3 inflammasome and promotes proinflammatory cytokines independently of TLR4. PLoS Pathog 6:e1001191
49. Mitchell TJ, Andrew PW, Saunders FK, Smith AN, Boulnois GJ (1991) Complement activation and antibody binding by pneumolysin via a region of the toxin homologous to a human acute-phase protein. Mol Microbiol 5:1883–1888
50. Moreland JG, Bailey G (2006) Neutrophil transendothelial migration in vitro to *Streptococcus pneumoniae* is pneumolysin dependent. Am J Physiol Lung Cell Mol Physiol 290:L833–L840
51. Norcross EW, Sanders ME, Moore QC 3rd, Marquart ME (2011) Pathogenesis of a clinical ocular strain of *Streptococcus pneumoniae* and the interaction of pneumolysin with corneal cells. J Bacteriol Parasitol 2:108
52. Orihuela CJ, Gao GL, Francis KP, Yu J, Tuomanen EI (2004) Tissue-specific contributions of pneumococcal virulence factors to pathogenesis. J Infect Dis 190(9):1661–1669
53. Parker D, Martin FJ, Soong G, Harfenist BS, Aguilar JL, Ratner AJ, Fitzgerald KA, Schindler C, Prince A (2011) *Streptococcus pneumoniae* DNA initiates type I interferon signaling in the respiratory tract. MBio 2:e00016–e00011
54. Paton JC, Lock RA, Hansman DJ (1983) Effect of immunization with pneumolysin on survival time of mice challenged with *Streptococcus pneumoniae*. Infect Immun 40:548–552
55. Paton JC, Lock RA, Lee CJ, Li JP, Berry AM, Mitchell TJ, Andrew PW, Hansman D, Boulnois GJ (1991) Purification and immunogenicity of genetically obtained pneumolysin toxoids and their conjugation to *Streptococcus pneumoniae* type 19F polysaccharide. Infect Immun 59:2297–2304
56. Price KE, Greene NG, Camilli A (2012) Export requirements of pneumolysin in *Streptococcus pneumoniae*. J Bacteriol 194:3651–3660
57. Propst-Graham KL, Preheim LC, Vander Top EA, Snitily MU, Gentry-Nielsen MJ (2007) Cirrhosis-induced defects in innate pulmonary defenses against *Streptococcus pneumoniae*. BMC Microbiol 7:94
58. Ratner AJ, Aguilar JL, Shchepetov M, Lysenko ES, Weiser JN (2007) Nod1 mediates cytoplasmic sensing of combinations of extracellular bacteria. Cell Microbiol 9:1343–1351
59. Ratner AJ, Hippe KR, Aguilar JL, Bender MH, Nelson AL, Weiser JN (2006) Epithelial cells are sensitive detectors of bacterial pore-forming toxins. J Biol Chem 281:12994–12998
60. Ratner AJ, Lysenko ES, Paul MN, Weiser JN (2005) Synergistic proinflammatory responses induced by polymicrobial colonization of epithelial surfaces. Proc Natl Acad Sci USA 102:3429–3434
61. Rayner CF, Jackson AD, Rutman A, Dewar A, Mitchell TJ, Andrew PW, Cole PJ, Wilson R (1995) Interaction of pneumolysin-sufficient and -deficient isogenic variants of *Streptococcus pneumoniae* with human respiratory mucosa. Infect Immun 63:442–447
62. Rayner CFJ, Jackson AD, Rutman A, Dewar A, Mitchell TJ, Andrew PW, Cole PJ, Wilson R (1995) Interaction of Pneumolysin-sufficient and -deficient isogenic variants of *Streptococcus pneumoniae* with human respiratory mucosa. Infect Immun 63:442–447
63. Ribet D, Hamon M, Gouin E, Nahori MA, Impens F, Neyret-Kahn H, Gevaert K, Vandekerckhove J, Dejean A, Cossart P (2010) *Listeria monocytogenes* impairs SUMOylation for efficient infection. Nature 464:1192–1195
64. Rogers PD, Thornton J, Barker KS, McDaniel DO, Sacks GS, Swiatlo E, McDaniel LS (2003) Pneumolysin-dependent and -independent gene expression identified by cDNA microarray analysis of THP-1 human mononuclear cells stimulated by *Streptococcus pneumoniae*. Infect Immun 71:2087–2094
65. Rossjohn J, Gilbert RJ, Crane D, Morgan PJ, Mitchell TJ, Rowe AJ, Andrew PW, Paton JC, Tweten RK, Parker MW (1998) The molecular mechanism of pneumolysin, a virulence factor from *Streptococcus pneumoniae*. J Mol Biol 284(2):449–461

66. Rubins JB, Charboneau D, Paton JC, Mitchell TJ, Andrew PW, Janoff EN (1995) Dual function of pneumolysin in the early pathogenesis of murine pneumococcal pneumonia. J Clin Invest 95:142–150
67. Rubins JB, Duane PG, Charboneau D, Janoff EN (1992) Toxicity of pneumolysin to pulmonary endothelial cells in vitro. Infect Immun 60:1740–1746
68. Rubins JB, Mitchell TJ, Andrew PW, Niewoehner DE (1994) Pneumolysin activates phospholipase A in pulmonary artery endothelial cells. Infect Immun 62:3829–3836
69. Rubins JB, Paddock AH, Charboneau D, Berry AM, Paton JC, Janoff EN (1998) Pneumolysin in pneumococcal adherence and colonization. Microb Pathog 25:337–342
70. Sanders ME, Norcross EW, Moore QC 3rd, Fratkin J, Thompson H, Marquart ME (2010) Immunization with pneumolysin protects against both retinal and global damage caused by *Streptococcus pneumoniae* endophthalmitis. J Ocul Pharmacol Ther 26:571–577
71. Srivastava A, Henneke P, Visintin A, Morse SC, Martin V, Watkins C, Paton JC, Wessels MR, Golenbock DT, Malley R (2005) The apoptotic response to pneumolysin is Toll-like receptor 4 dependent and protects against pneumococcal disease. Infect Immun 73:6479–6487
72. Steinfort C, Wilson R, Mitchell T, Feldman C, Rutman A, Todd H, Sykes D, Walker J, Saunders K, Andrew PW, Boulnois GJ, Cole PJ (1989) Effect of *Streptococcus pneumoniae* on human respiratory epithelium in vitro. Infect Immun 57:2006–2013
73. Stringaris AK, Geisenhainer J, Bergmann F, Balshusemann C, Lee U, Zysk G, Mitchell TJ, Keller BU, Kuhnt U, Gerber J, Spreer A, Bahr M, Michel U, Nau R (2002) Neurotoxicity of pneumolysin, a major pneumococcal virulence factor, involves calcium influx and depends on activation of p38 mitogen-activated protein kinase. Neurobiol Dis 11:355–368
74. Tilley SJ, Orlova EV, Gilbert RJ, Andrew PW, Saibil HR (2005) Structural basis of pore formation by the bacterial toxin pneumolysin. Cell 121:247–256
75. Walker JA, Allen RL, Falmagne P, Johnson MK, Boulnois GJ (1987) Molecular cloning, characterization, and complete nucleotide sequence of the gene for pneumolysin, the sulfhydryl-activated toxin of *Streptococcus pneumoniae*. Infect Immun 55:1184–1189
76. Winter AJ, Comis SD, Osborne MP, Tarlow MJ, Stephen J, Andrew PW, Hill J, Mitchell TJ (1997) A role for pneumolysin but not neuraminidase in the hearing loss and cochlear damage induced by experimental pneumococcal meningitis in guinea pigs. Infect Immun 65:4411–4418
77. Witzenrath M, Gutbier B, Hocke AC, Schmeck B, Hippenstiel S, Berger K, Mitchell TJ, de los Toyos JR, Rosseau S, Suttorp N, Schutte H (2006) Role of pneumolysin for the development of acute lung injury in pneumococcal pneumonia. Crit Care Med 34:1947–1954
78. Witzenrath M, Gutbier B, Owen JS, Schmeck B, Mitchell TJ, Mayer K, Thomas MJ, Ishii S, Rosseau S, Suttorp N, Schutte H (2007) Role of platelet-activating factor in pneumolysin-induced acute lung injury. Crit Care Med 35:1756–1762
79. Witzenrath M, Pache F, Lorenz D, Koppe U, Gutbier B, Tabeling C, Reppe K, Meixenberger K, Dorhoi A, Ma J, Holmes A, Trendelenburg G, Heimesaat MM, Bereswill S, van der Linden M, Tschopp J, Mitchell TJ, Suttorp N, Opitz B (2011) The NLRP3 inflammasome is differentially activated by pneumolysin variants and contributes to host defense in pneumococcal pneumonia. J Immunol 187:434–440
80. Yuste J, Botto M, Paton JC, Holden DW, Brown JS (2005) Additive inhibition of complement deposition by pneumolysin and PspA facilitates *Streptococcus pneumoniae* septicemia. J Immunol 175:1813–1819
81. Zysk G, Bejo L, Schneider-Wald BK, Nau R, Heinz H (2000) Induction of necrosis and apoptosis of neutrophil granulocytes by *Streptococcus pneumoniae*. Clin Exp Immunol 122:61–66
82. Zysk G, Schneider-Wald BK, Hwang JH, Bejo L, Kim KS, Mitchell TJ, Hakenbeck R, Heinz HP (2001) Pneumolysin is the main inducer of cytotoxicity to brain microvascular endothelial cells caused by *Streptococcus pneumoniae*. Infect Immun 69:845–852

Chapter 9
Multifaceted Activity of Listeriolysin O, the Cholesterol-Dependent Cytolysin of *Listeria monocytogenes*

Stephanie Seveau

Abstract The cholesterol-dependent cytolysins (CDCs) are a large family of pore-forming toxins that are produced by numerous Gram-positive bacterial pathogens. These toxins are released in the extracellular environment as water-soluble monomers or dimers that bind to cholesterol-rich membranes and assemble into large pore complexes. Depending upon their concentration, the nature of the host cell and membrane (cytoplasmic or intracellular) they target, the CDCs can elicit many different cellular responses. Among the CDCs, listeriolysin O (LLO), which is a major virulence factor of the facultative intracellular pathogen *Listeria monocytogenes*, is involved in several stages of the intracellular lifecycle of the bacterium and displays unique characteristics. It has long been known that following *L. monocytogenes* internalization into host cells, LLO disrupts the internalization vacuole, enabling the bacterium to replicate into the host cell cytosol. LLO is then used by cytosolic bacteria to spread from cell to cell, avoiding bacterial exposure to the extracellular environment. Although LLO is continuously produced during the intracellular lifecycle of *L. monocytogenes*, several processes limit its toxicity to ensure the survival of infected cells. It was previously thought that LLO activity was limited to mediating vacuolar escape during bacterial entry and cell to cell spreading. This concept has been challenged by compelling evidence suggesting that LLO secreted by extracellular *L. monocytogenes* perforates the host cell plasma membrane, triggering important host cell responses. This chapter provides an overview of the well-established intracellular activity of LLO and the multiple roles attributed to LLO secreted by extracellular *L. monocytogenes*.

Keywords Bacterial toxin · Cholesterol-dependent cytolysin · Infectious disease · *Listeria monocytogenes* · Listeriolysin O

S. Seveau (✉)
Department of Microbiology, Department of Microbial Infection and Immunity, The Ohio State University, 484 West, 12th Avenue, Columbus, OH 43210-1292, USA
e-mail: seveau.1@osu.edu

G. Anderluh and R. Gilbert (eds.), *MACPF/CDC Proteins - Agents of Defence, Attack and Invasion*, Subcellular Biochemistry 80, DOI: 10.1007/978-94-017-8881-6_9,

Abbreviations

ALO	Anthrolysin O
ASM	Acid sphingomyelinase
CDC	Cholesterol-dependent cytolysin
CFTR	Cystic fibrosis transmembrane conductance regulator
ER	Endoplasmic reticulum
ERK	Extracellular-signal-regulated kinase
GILT	Gamma-interferon inducible lysosomal thiol reductase
HGF	Hepatocyte growth factor receptor
HNP1	Human neutrophil peptide
ILY	Intermedilysin O
InlA	Internalin A
InlB	Internalin B
IPTG	Isopropyl β-D-1-thiogalactopyranoside
JNK	C-Jun N-terminal kinase
LIPI-1	*Listeria* pathogenicity island
LLO	Listeriolysin O
MAPK	Mitogen-activated protein kinase
MHC	Major histocompatibility complex
Mpl	Metalloprotease
NLRs	NOD-like receptors
PERK	Double-stranded RNA activated protein kinase (PKR)-like ER kinase
PC-PLC	Phosphatidylcholine-specific phospholipase
PI-PLC	Phosphatidylinositol-specific phospholipase
PFO	Perfringolysin O
PLY	Pneumolysin
PrfA	Positive regulatory factor A
ROS	Reactive oxygen species
SLO	Streptolysin O
SUMO	Small ubiquitin-like modifier
UPR	Unfolded protein response
UTR	Untranslated region

Listeriolysin O is a Major Virulence Factor of *L. monocytogenes*

Listeriosis

The Gram-positive, facultative anaerobe *Listeria monocytogenes* is the causative agent of listeriosis, a life-threatening disease associated with a very high rate of mortality in humans (20–30 %) and numerous other vertebrate species [1, 2]. This bacterium was isolated from diseased rabbits in 1926 by E. G. D. Murray and was recognized as the cause of a severe human foodborne illness in the early 1980s [3–5]. *L. monocytogenes* is ubiquitous in the environment, where it is found in soils, water, and plants, and frequently contaminates a large variety of raw and processed foods. The versatility of this organism comes from its ability to grow at a wide range of temperatures (1–45 °C) and pH (4.4–9.6), at high concentrations of salts (up to 10 % NaCl), and to resist the harsh environment of the animal gut [6–9]. It is estimated that a brief intestinal carriage of *L. monocytogenes*, below 4 days, occurs at least twice a year in healthy adults [10]. Although healthy individuals typically remain asymptomatic, a self-limiting flu-like illness and gastroenteritis may develop [2, 11, 12]. *L. monocytogenes* is a greater concern for several high-risk populations, in which it causes invasive infections and crosses the blood-brain or placental barriers [13–18]. In immunocompromised individuals, mainly the elderly, *L. monocytogenes* can cause bacteremia, meningitis, encephalitis, liver abscesses, and cardiac infections. Women are about twenty times more susceptible to listeriosis during pregnancy. While the mother may only exhibit mild symptoms, infection has devastating consequences for the developing fetus, resulting in miscarriages, preterm birth, still birth, or severe infection of the newborn [16]. Listeriosis is generally treated with ampicillin or amoxicillin, sometimes in combination with gentamicin [19]. However, late diagnosis combined with the immunodeficiency of the listeriosis patients and the high virulence of the bacterium likely explains the elevated rate of morbidity and mortality despite treatment [20].

Listeriolysin O Plays a Critical Role in the **L. monocytogenes** *Intracellular Lifecycle*

L. monocytogenes is a facultative intracellular pathogen that infects professional phagocytes and cells that are normally nonphagocytic in multiple organs: the intestines, spleen, liver, heart, brain, and placenta. The *L. monocytogenes* intracellular lifecycle is critical for pathogenesis since *L. monocytogenes* strains that are unable to infect host cells cannot cause disease. Major efforts have been devoted to the discovery of *L. monocytogenes* virulence factors and virulence mechanisms that orchestrate host cell invasion. Throughout the 1980s and 1990s, advancement of molecular biology techniques such as transposon mutagenesis, cloning, and sequencing led to the identification of a number of virulence genes. These genes are clustered on the *Listeria* Pathogenicity island-1 (LIPI-1) and the *inlAB* operon on the bacterial chromosome [21, 22]. Elucidating the role of these genes and discovering additional virulence genes is still the object of extensive studies [23, 24, 25].

The first step of the *L. monocytogenes* intracellular lifecycle is the entry of the pathogen into a host cell (Fig. 9.1). *L. monocytogenes* is phagocytosed with high

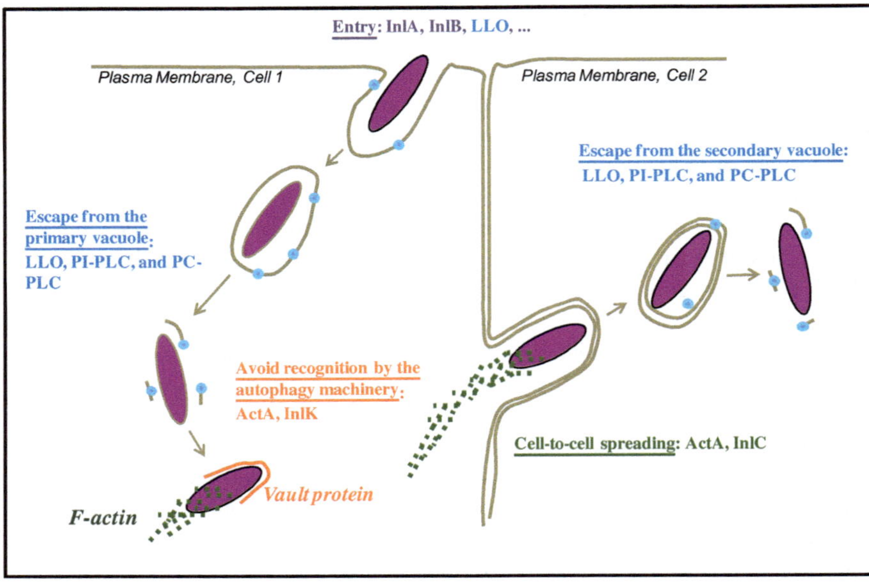

Fig. 9.1 Depiction of the canonical intracellular lifecycle of *L. monocytogenes. L. monocytogenes* expresses virulence factors that orchestrate the various stages of its intracellular lifecycle [33, 34, 90, 188]. The surface proteins InlA and InlB facilitate *L. monocytogenes* internalization into normally nonphagocytic cells. Additional virulence factors, such as the secreted CDC toxin LLO, also mediate *L. monocytogenes* entry into host cells (cell 1) [33, 89]. Secreted in the primary vacuole, LLO, PI-PLC, and PC-PLC target and disrupt the vacuolar membrane to release the bacterium into the cytosol. In the cytosol, *L. monocytogenes* replicates, while avoiding autophagy recognition by recruiting the host proteins F-actin and the major vault protein, through interaction with the bacterial surface proteins ActA and InlK, respectively [116]. ActA recruits the host F-actin polymerization machinery at one pole of the bacterium. F-actin-mediated propulsion of the bacterium leads to the formation of an intercellular protrusion; this process is facilitated by the bacterial protein InlC that decreases the plasma membrane surface tension [189]. The resulting intercellular protrusion is ingested by the adjacent cell (cell 2). LLO and the phospholipases (PI-PLC and PC-PLC) disrupt the newly formed secondary vacuole to release the bacterium into the cytosol, where it repeats its intracellular lifecycle (Host proteins are italicized)

efficiency by professional phagocytes, which express multiple phagocytic receptors such as complement, immunoglobulin, and scavenger receptors. This is in contrast to normally nonphagocytic cells that ingest *L. monocytogenes* with a lower efficiency. *L. monocytogenes* produces several virulence factors to promote its attachment to normally nonphagocytic cells and activate its internalization [26]. In particular, the surface proteins internalin (InlA) and InlB, encoded by the *inlAB* operon, specifically bind to their respective host cell receptors, E-cadherin and the hepatocyte growth factor receptor (HGF-Rc/c-Met) to stimulate *L. monocytogenes* internalization [27–34]. Following internalization into phagocytic or nonphagocytic cells, the bacterium is located into an endosome, called the primary vacuole. This vacuole is rapidly disrupted by the secreted pore-forming toxin listeriolysin

O (LLO) encoded by *hly* on LIPI-1. LLO was initially identified as a hemolytic factor [35, 36], its role in host cell invasion was discovered later by performing electron microscopy analysis of macrophages and epithelial cells incubated with wild type or LLO-deficient *L. monocytogenes.* At an early stage of infection, wild type bacteria were located within a vacuole, and were then observed to proliferate in the cytosol. In contrast, *L. monocytogenes* strains in which *hly* was either interrupted by the insertion of a transposon or deleted, remained trapped in the vacuole, unable to divide [26, 37]. LLO-deficient bacteria were also nonvirulent in vivo, revealing the essential role of this toxin and the bacterial intracellular lifecycle in pathogenesis [37–41]. Additional studies showed that the *L. monocytogenes* surface protein ActA (encoded by *actA* on LIPI-1) is enriched at one pole of the bacterium, where it recruits the host cell F-actin polymerization machinery to generate a propulsive actin comet tail [42, 43]. F-actin polymerization randomly propels *L. monocytogenes* until it forms an extracellular protrusion that can invade an adjacent cell [44, 45]. Once the adjacent cell has ingested the protrusion, *L. monocytogenes* is located into a secondary vacuole made of two membranes that originate from the donor and recipient cells. LLO is also required for the disruption of the secondary vacuole, releasing the bacterium into the cytosol of the newly infected cell. Cell-to-cell spreading is a very efficient process that propagates the bacterium within tissues, while avoiding exposure to the many antimicrobial molecules and phagocytes present in the extracellular environment [46, 47]. In addition to LLO, two bacterial phospholipases, PI-PLC and PC-PLC respectively encoded by *plcA* and *plcB* on LIPI-1, facilitate *L. monocytogenes* escape from the primary and secondary vacuoles [48]. As further developed in the next paragraphs, recent studies suggest that LLO plays additional roles during the intracellular lifecycle of the bacterium. For example, LLO can also act as an invasin that stimulates *L. monocytogenes* internalization and affects the transcriptional activity of infected cells [49, 50].

Regulation of Listeriolysin O Expression and Activity

Listeriolysin O Expression

Transcriptional regulation of *hly,* the 1590-base pair gene coding for LLO, is predominantly controlled by the positive regulatory factor A (PrfA) [51–53]. PrfA binds to a 14-base pair palindromic sequence upstream of the promoter region to positively regulate the transcription of the major virulence genes of *L. monocytogenes,* which form the PrfA regulon [52, 54]. PrfA expression and activity are regulated at the transcriptional, translational, and post-translational levels [52]. Among the several factors that control PrfA expression or activity, temperature, carbon source, bacterial growth phase, and environmental stresses play an important role [55–57]. Of particular significance, the *prfA* mRNA functions as a thermosensor that selectively allows the translation of PrfA at 37 °C once the bacterium is within its animal host. The *prfA* mRNA 5' untranslated region (UTR) adopts a hairpin structure that masks the ribosome binding site at temperatures below 30 °C, whereas this structure melts at the temperature of the host, 37 °C, leading to PrfA synthesis and transcription of *hly* and other virulence genes [58, 59]. Additionally, the *hly* promoter is more active when the bacterium is within host cells. This is independent of PrfA and requires the *hly* 5′ UTR, but the underlying mechanism remains unknown [60, 61]. Finally, the 5′ coding region of the *hly* mRNA is also thought to control LLO translation in the cytosol [62].

The open reading frame of *hly* codes for a preprotein of 529 residues with an amino-terminal secretion signal sequence typical of those found in Gram-positive bacteria [63]. Upon secretion by the general secretory pathway, the signal sequence is cleaved by a signal peptidase releasing a mature protein of 504 residues. Due to the central role LLO plays during pathogenesis, molecular strategies preventing its synthesis, secretion, and/or activity are of great medical interest. For example, human antimicrobial peptides that belong to the α-defensin family, such as HNP1, prevent both LLO release by the bacterium and LLO activity via unknown mechanisms [64, 65]. Once these mechanisms are solved, the development of defensin-like molecules could constitute promising therapeutic tools against *L. monocytogenes* and other CDC-producing pathogens [66].

Mechanism of Assembly of the CDC Pore Complex

LLO crystallization and X-ray crystallographic analysis were recently carried out; however, detailed structural determination is not yet established [67]. Structural information and models for pore formation are available for several other CDCs including PFO (perfringolysin O, produced by *Clostridium perfringens*) [68, 69]. Based upon the high degree of identity between their amino acid sequences (28–99 %), CDCs likely share a similar three-dimensional organization into four domains (referred to as D1 to D4) (Fig. 9.2) and mechanism of pore formation [70–73]. These toxins are produced as water soluble monomers or dimers that bind to the lipid bilayer in an upright position through three hydrophobic loops—L1 to L3—located at the base of the carboxy-terminal domain (D4) and form an oligomeric pore complex. It has been proposed that CDCs could form pores of different sizes [74]. In particular, studies using the patch-clamp technique or measuring the traversal of various membrane impermeant dyes across LLO pores, suggested that the size and/or shape of LLO pores formed in mammalian membranes could vary [75, 76]. However, the model for CDC pore formation that currently prevails proposes that CDCs assemble into a large ring-shaped pore oligomer. Whether there is some variation in size and/or shape of the pores among the CDC family members remain to be established.

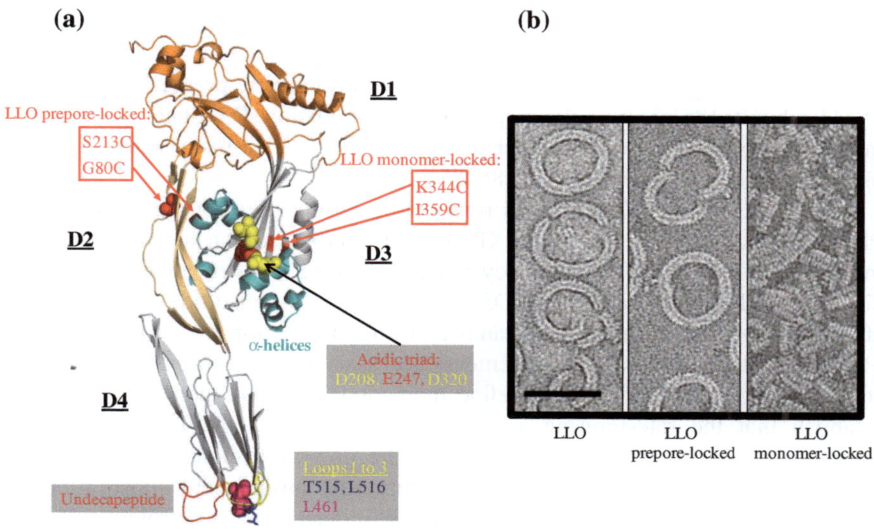

Fig. 9.2 Model for LLO pore formation. **a** Model for LLO ribbon structure based upon PFO crystal structure (PBD#1PFO) generated with Pymol [69]. LLO, similar to other CDCs, is thought to be organized into four domains (D1–D4). D4 is involved in binding to the lipid bilayer through three hydrophobic loops (L1–L3) located at the base of D4. L1 contains the cholesterol-binding motif T515/L516. Leucine 461 controls the LLO oligomerization rate. The undecapeptide sequence, which contains the unique cysteine, controls conformational remodeling in D3 during toxin oligomerization. The α-helices in D3 convert into two *β*-strands per monomer to form the *β* barrel pore. The acidic triad in D3 is responsible for LLO denaturation at neutral pH (at the temperature of the host). Amino acid substitutions in LLO monomer-locked and LLO prepore-locked variants are indicated by the *red boxes*. Dr. Eusondia Arnett (Ohio State University, USA) helped designing the LLO model. **b** Transmission electron microscopy of the LLO pore formed on lipid droplets. The images represent oligomers of native LLO, LLO prepore-locked, and LLO monomer-locked variants (scale bar = 50 nm) [89]. The transmission electron microscopy images were captured by Dr. Elisabeth M. Wilson-Kubalek (The Scripps Research Institute, USA)

Cholesterol is the receptor for most CDCs and plays an essential role in initiating the structural transitions required for formation of the pore complex [77, 78]. A threonine-leucine pair conserved in all CDCs and located in L1, mediates direct binding of PFO, SLO (streptolysin O, secreted by *Streptococcus pyogenes*), and PLY (pneumolysin, produced by *Streptococcus pneumoniae*) to cholesterol [79]. We recently confirmed that the corresponding amino acid pair (Thr515/Leu516 in LLO) is required for LLO binding to cholesterol and pore formation (our unpublished observations). To date, cholesterol is the only known mammalian cell receptor for LLO, but the existence of additional lipid and/or protein receptors cannot be excluded [80, 81]. Also, preloading LLO with cholesterol does not abrogate its binding to biological membranes and a few CDCs utilize the surface protein CD59 as initial receptor before binding to cholesterol [81–84]. Following binding to cholesterol, the two proximal hydrophobic loops, L2 and L3, insert into the lipid bilayer.

The undecapeptide signature sequence of CDCs –ECTGLAWEWWR- at the tip of D4, is then critical for stable monomer–monomer interactions, which transmit structural changes to D3, leading to the formation of stable prepore oligomers of 35–44 subunits [85]. During the prepore-to-pore transition, D3 α-helices convert of into two amphipathic β-hairpins per toxin monomer, which insert across the lipid bilayer to form a β-barrel pore [71, 86, 87].

Most studies on CDC structural remodeling during pore-formation have used PFO. For example, substitution of Gly57 and Ser190, (which are in close proximity in the folded protein) with cysteine residues generated an intramolecular disulfide bridge between D2 and D3. Formation of this disulfide bridge did not affect PFO binding to host membranes and oligomerization into a prepore complex, but halted structural rearrangements necessary for pore formation [88]. This construct allowed the authors to refine their model of CDC pore formation and establish that the rate-limiting step in the formation of the PFO pore is the assembly of the prepore complex and not the prepore-to-pore transition at the temperature of the animal host [88]. Importantly, amino acid substitutions at corresponding locations in LLO (G80C/S213C), determined by protein sequence alignment, also arrested LLO at the prepore stage [77, 89]. Also, amino acid substitution into cysteines at similar locations in LLO (K344C/I359C) and PFO blocked formation of the prepore complex in both toxins [88, 89]. Therefore, LLO and PFO adopt a similar fold and undergo similar intra- and inter-molecular remodeling to form a pore complex.

Regulation of LLO Activity by pH, Oxidation, and Cholesterol

Several physicochemical parameters control the formation of LLO pores, among which are the temperature and pH [35]. Indeed, the formation of LLO pores is pH-sensitive at the temperature of the host (37 °C), LLO being more active at acidic pH [90]. Therefore, LLO activity is likely increased in acidic environments, such as the acidifying phagosome. The pH sensitivity of LLO is due to unfolding of the D3 transmembrane beta-hairpins at 37°C and neutral pH leading to LLO aggregation [91, 92]. Denaturation involves three acidic residues located in D3 (Asp208, Glu247, and Asp320) that act as a pH sensor. At neutral pH, the combination of charge repulsions between acidic residues and thermal fluctuations leads to unfurling of the twin α-helical bundles in D3, exposing hydrophobic residues and causing LLO aggregation [92]. Denaturation operates on a relatively long timescale, since it takes several minutes to inactivate LLO in physiological buffer at 37 °C (at pH 7.4) [79, 89]. However, if bound to a lipid bilayer, LLO appears protected from denaturation and remains active [89]. The slow pH-dependent inactivation of LLO has two important biological implications. First, LLO secreted by extracellular *L. monocytogenes* can still perforate host cells at physiological temperature and neutral pH, affecting the biology of the host cell from the extracellular compartment [80]. Second, the lifetime of LLO in extracellular fluids is limited, lessening its cytotoxicity [49, 89, 93, 94].

The efficiency of LLO pore formation also involves the amino acid Leu461, which exchange with a threonine generates a LLO variant that retains full activity at neutral pH [95]. This observation led to the initial conclusion that Leu461 was responsible for the pH-dependent activity of LLO. However, the LLO L461T variant was later shown to display similar rates of pH- and temperature-dependent denaturation, but an increased rate of hemolysis, suggesting that L461 rather controls the rate of LLO oligomerization [92].

CDCs were initially named the sulfhydryl-activated hemolysins for their ability to cause hemolysis in their reduced state, while they were inactive when oxidized [35]. Indeed, a unique cysteine in the undecapeptide sequence in D4 acts as a redox switch that turns off the activity of the oxidized toxin [35]. However, a cysteine-free LLO variant (LLO C484A) is still able to form pores, albeit with a lower efficiency [96]. The biological significance of this redox switch remains uncertain. A study proposed that *L. monocytogenes* exploits the gamma-interferon inducible lysosomal thiol reductase (GILT), located in lysosomes of antigen-presenting cells, to maintain LLO in its active reduced state [97].

It has also long been known that CDCs are inactivated by preloading with cholesterol prior to their binding to host membranes are inactive [35, 84]. This inactivation is attributed to premature and irreversible structural transitions in D3 of the CDC monomer. This property has been used as an experimental approach to establish the role of pore-formation in CDC activity. However, preloading CDCs with cholesterol raises several concerns. First, cholesterol insolubility in aqueous buffer may lead to uncontrolled formation of toxin/cholesterol aggregates, decreasing toxin binding to host cells in addition to inhibiting pore formation. Also, the proportion of CDC associated to cholesterol is difficult to establish and some residual pore-forming activity may persist. As an alternative approach, amino acid variations in the LLO undecapeptide sequence were proposed to decrease the LLO pore-forming activity; however it is unclear if binding and/or pore-formation are affected and these variants retain substantial capacity to form pores [96]. To unambiguously dissect the contribution of each stage of the LLO pore formation, i.e. binding as a monomer, formation of the prepore oligomer, and prepore-to-pore transition, in LLO function, two full length recombinant LLO variants were constructed (LLO monomer-locked and LLO-prepore-locked). These variants retain the membrane binding capability of native LLO, but the LLO monomer-locked variant fails to oligomerize into a prepore complex, while the LLO prepore-locked variant forms a prepore complex that cannot undergo the prepore-to-pore conversion (Fig. 9.2) [89]. These variants are of great value to establish the structure-function relationship of LLO.

Listeriolysin O is an Intracellular Pore-Forming Toxin with Reduced Cytotoxicity

LLO Facilitates L. monocytogenes *Escape from the Endocytic Vacuoles*

An essential function of LLO is to mediate *L. monocytogenes* escape from the primary and secondary vacuoles (Fig. 9.1). This activity is facilitated by three additional virulence factors: phosphatidylinositol-specific (PI-PLC) phospholipase, phosphatidylcholine-specific (PC-PLC) phospholipase, and a metalloprotease (Mpl) that activates PC-PLC, encoded respectively by *plcA*, *plcB,* and *mpl* on LIPI-1 [90].

Escape from the Primary Vacuole

LLO is critical for *L. monocytogenes* escape from the primary endocytic vacuole in most cells, including professional phagocytes and nonphagocytic cells [26]. However, there are exceptions, since LLO-deficient *L. monocytogenes* can reach the cytosol of some nonphagocytic human cell lines [98]. In those cells, PC-PLC plays a major role in vacuolar escape in the absence of LLO [99, 100]. On the contrary, in cells that require LLO for vacuolar escape, PI-PLC, but not PC-PLC, facilitates escape [48, 101]. Therefore, the process of vacuolar escape is controlled by both the *L. monocytogenes* virulence factors and host cells. The molecular basis for such variation in the mechanism of escape in different cell types or species is not yet understood. Also, how LLO alone or in cooperation with the bacterial phospholipases dismantles the primary endocytic vacuole is not fully explained. Many studies have addressed this question leading to multiple, non-mutually exclusive, hypotheses. Overall, LLO and the phospholipases delay the maturation of the vacuole and decrease its toxicity toward the bacterium, while compromising the integrity of the membrane. These virulence factors even affect host cell signaling before the bacterium is fully internalized. Indeed, extracellular LLO and PI-PLC induce the translocation of the host protein kinase C β isoforms (PKCβ) from the cytosol to endosomal membranes, which increases the efficiency of escape, possibly by altering endosomal recycling during early stages of phagosomal maturation [102]. The process of vacuolar escape takes 15–30 min in murine macrophages [76] and requires early maturation of the phagosome, such as its acidification and some low level of fusion with lysosomes. Phagosomal acidification may increase LLO pore-forming activity [95, 103]. Perforation of the phagosome by LLO then perturbs the intraphagosomal pH and calcium concentration, which in turn decreases the rate of phagosome fusion with lysosomes, giving time to the bacterium to fully disrupt the phagosomal membrane without being harmed by the toxic lysosomal contents [76, 104]. Indeed, LLO delays the acquisition of the late endosomal and lysosomal marker LAMP-1 [104]. The host thiol reductase GILT, which accumulates in the phagosome of macrophages, facilitates *L. monocytogenes* phagosomal escape in a cell culture model and is also important for *L. monocytogenes* virulence in mice. The direct role of GILT in maintaining LLO in its reduced active state is supported by the fact that purified GILT activates the hemolytic activity of recombinant LLO [97]. Beyond Ca^{2+} and H^{+} ions, the intraphagosomal concentration of Cl^{-} also affects the efficiency of vacuolar escape. Interestingly, the cystic fibrosis transmembrane conductance regulator (CFTR), which increases Cl^{-} concentration in the phagosome, facilitates bacterial escape in vitro and potentiates *L. monocytogenes* virulence in mice [105]. This activity might be explained by the fact that high Cl^{-} concentrations enhance the rate of LLO oligomerization, thereby facilitating pore-formation [105]. Recently, the *L. monocytogenes* PLCs were shown to stimulate the production of reactive oxygen species (ROS), which is unfavorable to *L. monocytogenes* intracellular survival. However, LLO counteracts this activity, importantly decreasing the toxicity of the phagosome [106]. When LLO and the phospholipases cooperate for escape, LLO may translocate the PLCs to grant them access to both sides of the lipid bilayer for more efficient digestion of their target phospholipids. A role for LLO in the translocation of the PLCs has not yet been shown, but another CDC, SLO was shown to translocate the virulence factor nicotinamide adenine dinucleotide-glycohydrolase into the host cell cytosol [107]. In conclusion, in addition to physical and biochemical damages inflicted to the phagosomal membrane, LLO and the PLCs also modify the maturation of the phagosome in multiple ways, which collectively leads to the release of the bacterium into the cytosol.

The Efficiency of Escape is Cell-Type Dependent and Varies with the Cell Activation State

L. monocytogenes does not escape from all phagosomes. The efficiency of escape depends upon the nature and activation status of the host cell, and can vary from phagosome to phagosome within the same cell. For example, LLO and the PLCs are less efficient in activated than resting macrophages and are unable to protect *L. monocytogenes* from intracellular killing in human neutrophils [108, 109]. In these examples, it appears that host cells can use various strategies to limit or even prevent vacuolar escape. These include exacerbated bactericidal activity of the phagosome, direct inhibition or degradation of LLO, or increased physical resistance of the endolysosomal compartment. In activated murine macrophages, the rapid production of reactive oxygen and nitrogen intermediates likely damage bacteria, limiting their ability to secrete the virulence factors required for escape [110]. Also, lysosomal cathepsin-D degrades LLO in murine fibroblasts and macrophages, limiting the efficiency of *L. monocytogenes* escape in cultured cells, and impacting the course of infection in a murine model of listeriosis [111]. A similar enzymatic process was recently proposed to exist in neutrophils in which the matrix metalloproteinase-8, stored in secondary granules, degrades LLO [109].

Also, the antimicrobial peptides, α-defensins, stored at high concentrations in neutrophil primary granules, inhibit LLO release from the bacterium in addition to blocking its activity [64, 65]. Paradoxically, during *L. monocytogenes* phagocytosis by neutrophils, LLO triggers its own inactivation by stimulating the exocytosis of secondary and primary granules [109]. Finally, macrophage activation by multiple stimuli (TNF-α, peptidoglycan, etc.) diminishes the susceptibility of the phagosome and more generally, the endolysosomal network, to disruption by LLO and other stresses [112]. The molecular basis for this increased resistance of the endolysosomal network to multiple physical stressors remains to be elucidated. The process of autophagy also limits host cell invasion by targeting newly internalized bacteria that reside in damaged phagosomes [113, 114]. LLO was shown to be necessary and sufficient to activate autophagy of the phagosome. However, the impact of autophagy varies with the nature of the infected cells: autophagy decreases infection of fibroblasts, but not of bone marrow derived macrophages [115]. It is important to highlight that autophagy does not affect *L. monocytogenes* growth in the cytosol since several bacterial factors prevent cytosolic recognition of the bacterium by the autophagy pathway (Fig. 9.1) [113, 114, 116].

Escape from the Secondary Vacuole

The secretion of LLO, PI-PLC, and PC-PLC is also critical for *L. monocytogenes* escape from the secondary vacuole during cell-to-cell spreading [46, 48]. To demonstrate the role of LLO in this process, LLO-deficient *L. monocytogenes* were noncovalently coated with recombinant LLO. This coating was sufficient to promote bacterial escape from the primary vacuole, allowing the bacterium to proliferate in the cytosol and undergo F-actin-based motility; however, bacteria were unable to escape from the secondary vacuole and failed to spread to adjacent cells [46]. Also, the construction of *L. monocytogenes* wild type and *plcA/plcB* double deletion mutant strains that express *hly* under the control of an IPTG-inducible promoter, confirmed that both LLO and the PLCs are critical for *L. monocytogenes* escape from the secondary vacuole [117, 118]. Electron microscopy analysis of infected macrophages showed that in the absence of continued LLO expression, but in the presence of the PLCs, *L. monocytogenes* remained predominantly trapped in secondary vacuoles made of one membrane. This observation indicates that the PLCs are critical for the disruption of the inner membrane (originating from the donor cell), whereas LLO is mostly involved in the disorganization of the outer membrane (originating from the recipient cell) [117]. Consistent with this model, when *L. monocytogenes* is transmitted to a recipient cell in which LLO was dispensable for the disruption of the primary vacuole, LLO was also dispensable for dissolution of the secondary vesicle [117].

Mechanisms Reducing the Cytotoxicity of Listeriolysin O

In the cytosol, *L. monocytogenes* continuously produces LLO in order to spread from cell to cell [119]. *L. monocytogenes* tightly controls LLO synthesis and activity in order to disrupt vacuolar membranes without killing the infected cell. It has been well-established that *L. monocytogenes* mutant strains with increased LLO expression or activity escape from primary and secondary vacuoles, but are less virulent because they lyse the host cells, destroying their replicative niche [120, 121]. Also, *hly* replacement by genes encoding other CDCs, such as PFO, ALO (anthrolysin O, produced by *Bacillus anthracis*), or ILY (intermedilysin O, produced by *Streptococcus intermedius*), under the control of the *hly* promoter, allowed *L. monocytogenes* vacuolar escape, but resulted in increased host cell lysis and loss of virulence in animal models [122–124]. Therefore, unlike other CDCs, LLO displays unique characteristics that limit its cytotoxicity. Additional studies revealed that the low cytotoxicity of LLO results from the combination of several processes including translational repression in the cytosol, pH-dependent denaturation, and degradation by the proteasome.

pH-Dependent Denaturation of Listeriolysin O

LLO displays a relatively short lifetime in biological fluids and the cytosol, due to its propensity to aggregate at neutral pH (developed in section "Regulation of LLO Activity by pH, Oxidation, and Cholesterol"). However, it should be noted that the pH-dependent denaturation of LLO does not instantaneously nor completely abrogate its activity. For example, intravenous injection of recombinant LLO in mice has multiple biological effects and can be lethal to the animal [35, 125]. Therefore, LLO has acquired additional features to decrease its cytotoxicity.

Listeriolysin O Expression and Degradation in the Cytosol

Sequence comparison between LLO and the more cytotoxic PFO revealed the presence of an additional region in the LLO N-terminal domain. Deletion of the corresponding DNA sequence from the *L. monocytogenes* genome increased LLO toxicity and severely decreased bacterial virulence. However, this deletion did not affect the efficiency of pore-formation or vacuolar escape, indicating that the N-terminal region does not control LLO activity [120, 126]. This sequence contains eukaryotic PEST-like motifs (enriched in proline (P), glutamic acid (D), aspartic acid (E) and serine (S) or threonine (T)), known to target proteins for degradation by the proteasome. Therefore, it was initially believed that the PEST-like sequences target LLO for degradation. Additional studies disproved this hypothesis. Even though LLO is phosphorylated and ubiquitinated within the PEST-like sequences and is degraded by the proteasome, LLO proteasomal degradation is independent of the PEST-like sequences. Furthermore, inhibition of the proteasome did not significantly affect the cytotoxicity of wild type *L. monocytogenes* [62]. It was then proposed that the 5' coding sequence of the *hly* mRNA, corresponding to the N-terminal PEST-like sequences, negatively controls LLO translation in the cytosol [62]. Indeed, this sequence was shown to specifically decrease LLO synthesis during *L. monocytogenes* exponential growth in cell culture medium and in host cell cytosol [120]. When this sequence was inserted in the gene coding for PFO, PFO synthesis was also decreased [120]. Interestingly, when the PEST-like sequences were deleted, the resulting excess of cytosolic LLO was degraded by the proteasome, decreasing LLO cytotoxicity. Therefore, proteasomal degradation is an additional process that protects host cells in case of excessive production of LLO. LLO proteasomal degradation was proposed to involve the N-end rule pathway [127]. Finally, several hours post-infection, LLO was associated with numerous perinuclear aggregates of polyubiquitinated proteins in murine bone marrow derived macrophages [128]. The mechanism of formation and the fate of such LLO aggregates are unknown. It is possible that the pH-dependent denaturation of LLO facilitates their formation. Also, such aggregates may sequester LLO and favor its subsequent degradation by the proteasome or the autophagy pathways. Formation of aggregates may represent an additional mechanism to reduce LLO cytosolic toxicity [128].

Does Cytosolic Listeriolysin O Cause Organelle and/or Plasma Membrane Lesions?

Mechanisms that decrease LLO concentration and activity in the cytosol are sufficient to ensure host cell survival; however, these mechanisms may not fully inhibit the activity of LLO in the cytosol. Moderate membrane damage caused by cytosolic LLO is likely difficult to observe, since host cells can undergo rapid repair of their organelles and plasma membrane. In favor of the idea that cytosolic LLO can target and perforate host cell membranes, it has been reported that, at late stages of cellular infection, a LLO-dependent Ca^{2+} rise is responsible for NF-kappa B activation and production of IL-6 [129]. The authors concluded that LLO released by *L. monocytogenes* in the cytosol damages the plasma membrane, leading to an influx of Ca^{2+}, which activates cytokine production without killing infected cells [129]. Also, a more recent study suggested that LLO perforates the plasma membrane of bone marrow-derived macrophages at late stages of infection. Interestingly, the plasma membrane of caspase $7^{-/-}$ macrophages was more damaged, revealing a role for this cysteine protease in maintaining the integrity of infected cells [130]. Therefore, LLO secreted into the cytosol may undergo moderate perforation of host cell membranes affecting the biology of infected cells. Such a process is likely to be detected at late stages of cell infection when the number of replicating bacteria is substantial. Studying LLO activity in the cytosol emerges as a relevant research focus. In this line of investigation, several points need to be clarified: (i) it is necessary to confirm that cytosolic LLO perforates host cell membranes and determine which membranes are preferentially targeted by LLO; (ii) establish if and how such membrane damage affects the course of infection; and if so, (iii) whether membrane damage favors the host or the pathogen. Finally, future studies will focus on elucidating the plasma membrane and organelle repair mechanisms, which are still poorly understood.

Listeriolysin O is an Extracellular Pore-Forming Toxin

In addition to disrupting the primary and secondary vacuoles, LLO has been recently proposed to exert novel roles during the *L. monocytogenes* intracellular lifecycle. Indeed, a large body of evidence supports the idea that LLO secreted by extracellular *L. monocytogenes* perforates the host cell plasma membrane before bacterial entry, activating multiple signaling pathways. If cells are exposed to a moderate number of bacteria, the plasma membrane repair process maintains the integrity of the host cell. Nevertheless, the signaling activity of extracellular LLO modifies the biology of the host cell, potentially influencing the course of infection. Indeed, extracellular LLO was proposed to control *L. monocytogenes* internalization into nonphagocytic cells, vacuolar escape, organelle homeostasis, transcriptional activity, and the production of inflammatory mediators (Fig. 9.3). It is clear that we are only starting to appreciate the complexity and diversity of the activities of extracellular LLO.

Listeriolysin O Released by Extracellular L. monocytogenes *Activates Multiple Signaling Pathways*

Listeriolysin O Perforates the Host Cell Plasma Membrane

Extracellular *L. monocytogenes* perforates host cells in a LLO-dependent fashion. Perforation is indicated by a rapid, LLO-dependent influx of extracellular Ca^{2+} upon addition of *L. monocytogenes* to animal cells [75]. Cell perforation can also be directly detected by measuring the entry of small cell-impermeant fluorescent dyes into host cells, as for example ethidium homodimer or propidium iodide [89, 131]. Plasma membrane perforation is a very dangerous situation for host cells, which must immediately eliminate the toxin pore in order to survive.

Host Cells Repair their Plasma Membrane

Animal cells are frequently wounded in tissues subjected to mechanical stress such as skeletal muscles, heart, intestine, lung, and skin. Mechanical lesions can span

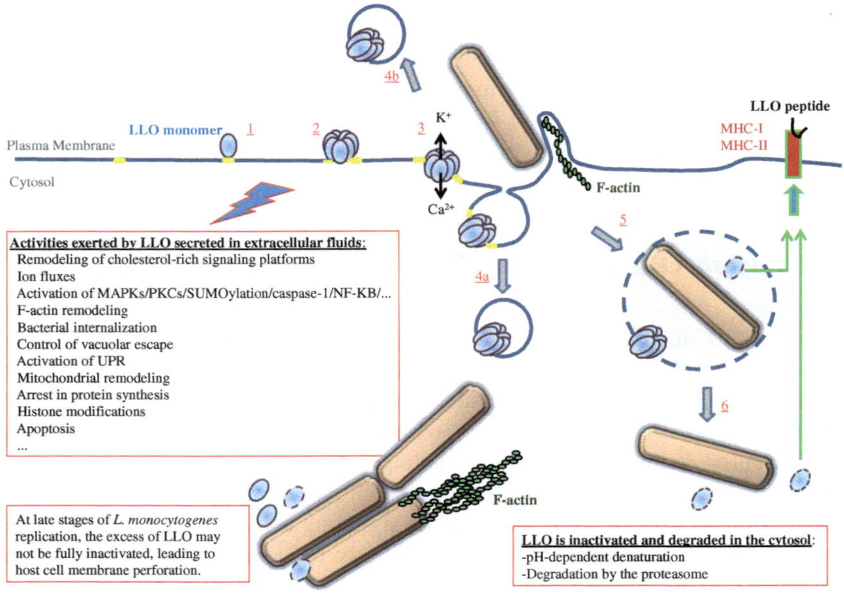

Fig. 9.3 Model for the multifaceted activity of LLO during the *L. monocytogenes* intracellular lifecycle. LLO is secreted as a water-soluble monomer that binds to cholesterol in the host plasma membrane (*1*). The toxin assembles into a prepore complex (*2*), which then vertically collapses to form the transmembrane pore, leading to rapid fluxes of ions and small molecules (*3*). It is proposed that LLO interaction with and remodeling of lipid raft signaling platforms (in *yellow*), and fluxes of ions and molecules across the LLO pore complex, activate multiple signaling pathways. Downstream events include plasma membrane repair, which may involve the internalization (*4a*) and/or extrusion (*4b*) of LLO vesicles; F-actin remodeling and activation of the endocytic uptake of the bacterium (*5*). Also, signaling initiated by LLO at the plasma membrane may influence vacuolar escape (*6*). Independently of bacterial internalization, extracellular LLO exerts multiple effects on target cells including modifications of ER and mitochondria; protein synthesis; histone modifications; and apoptosis. The signaling activities of LLO though activation of transcription factors and histone modifications affect the production of inflammatory mediators; however, it is not always clear whether LLO exerts a pro- or anti-inflammatory function. LLO produced in the primary and secondary (not shown in the figure) vacuoles release the bacterium into the cytosol. Once LLO is produced in the cytosol, several mechanisms decrease its activity and production. Nevertheless, at late stages of intracellular replication, when the number of bacteria is very high, an excess of LLO may not be efficiently inactivated leading to cell damage. Finally, digestion of LLO in the phagosome or cytosol generates peptides for presentation by Class II or I MHC molecules, activating the adaptive immune response

several micrometers and do not spontaneously reseal. The immediate consequence of membrane damage is a massive influx of extracellular Ca^{2+} into the cytosol, which is sensed as a membrane damage signal, triggering a complex multistep Ca^{2+}-dependent membrane repair program [132–134]. Pore-forming proteins of the host immune system (perforin, complement membrane attack complex) or toxins produced by pathogens can also injure the host cell plasma membrane. Despite their relatively small size ($\leq$50 nm), these transmembrane pores are very stable. If too many pores are formed, the target cell is rapidly lysed. However, active processes can restore the integrity of animal cells exposed to lower concentrations of toxins [135, 136]. Recent studies proposed that host cells injured by the CDC SLO are repaired by Ca^{2+}-dependent endocytosis of the toxin pores. The model for the repair of cells damaged by SLO shares some similarities with the repair model of large mechanical wounds. In this model, the influx of extracellular Ca^{2+} activates Ca^{2+}-sensitive fusogenic proteins, such as the members of the SNARE family of proteins, inducing rapid lysosome fusion with the plasma membrane [137–139]. The exocytosed lysosomal enzyme acid sphingomyelinase (ASM) then converts sphingomyelin into ceramides on the outer leaflet of the plasma membrane. Ceramides were shown to induce inward curvature of the plasma membrane, which likely facilitates the endocytosis of SLO pores [139, 140]. Once internalized, the toxin pores are trafficked to, and degraded within the lumen of multivesicular bodies [141]. Alternatively, other studies proposed that vesicles containing the SLO pores bud from the plasma membrane and are released in the extracellular compartment. This process is also activated upon Ca^{2+} influx and involves a family of Ca^{2+}-regulated cytosolic proteins, the annexins [135, 142]. Resealing of the plasma membrane of cells exposed to low concentrations of LLO is Ca^{2+}-dependent, but K^{+}-, F-actin-, dynamin-, and clathrin-independent, which is compatible with the two proposed models [89]. Whether these two repair mechanisms coexist in a same cell, are cell type-dependent, or depend upon the extent of membrane damage remains to be elucidated. Also, it is possible that these models only incompletely describe the complex process of plasma membrane repair. Importantly, resealing of the plasma membrane may not always suffice to maintain cell viability. Indeed, if the rise in cytosolic Ca^{2+} is too high, a state of "Ca^{2+} overload" is reached, leading to the activation of cell death pathways despite recovery of the plasma membrane integrity [143].

Extracellular Listeriolysin O is a Potent Signaling Molecule

LLO signaling is not limited to the activation of the plasma membrane repair process. LLO activates multiple signaling pathways via two major mechanisms. First, similar to other CDCs, LLO is thought to bind to and reorganize cholesterol-rich microdomains, affecting the dynamics and signaling activities of these signaling platforms [144]. Second, CDCs form a large pore complex that measures 30–50 nm, allowing ions and small molecules to diffuse across the plasma membrane, thereby eliciting multiple signaling events. For example, Ca^{2+} and K^{+} fluxes subsequent to cell perforation by LLO and other pore-forming toxins activate multiple pathways. Also, cell depolarization, Na^{+} and Cl^{-} perturbations, exposure to the oxidative extracellular environment, loss of ATP and small proteins, etc., all likely affect signaling.

The Versatility of Ca^{2+} and K^+ Signaling

Eukaryotic cells maintain a very low cytosolic concentration of Ca^{2+} (~100 nM) in comparison to the extracellular environment (~1 mM). The endoplasmic reticulum (ER) and mitochondria also store Ca^{2+}. This provides cells a means for rapid and sensitive signaling by increasing the cytosolic Ca^{2+} concentration through the activation of diverse channels and pumps on cytoplasmic and organelle membranes. Ca^{2+} is a universal second messenger that regulates a large array of cellular processes. How a cytosolic rise in Ca^{2+} controls many different pathways can be explained by the fact that cells integrate the amplitude, duration, pattern, and localization of the Ca^{2+} rises in any given cellular context. In the light of the central role Ca^{2+} plays, it is not surprising that pathogens have evolved to manipulate Ca^{2+} signaling. Upon interaction with host cells, LLO released by *L. monocytogenes* rapidly induces a rise in intracellular Ca^{2+}, which was shown to result from: (1) Ca^{2+} influx across LLO pores and activation of Ca^{2+} channels at the plasma membrane; (2) release of Ca^{2+} from intracellular stores via PLC-dependent activation of Ca^{2+} channels, (3) release of intracellular Ca^{2+} stores due to a direct insult of the ER [75, 145–148]. Many signaling pathways activated by extracellular LLO are linked to the rise in intracellular Ca^{2+}. For example, a rise in intracellular Ca^{2+} affects *L. monocytogenes* internalization and organelle homeostasis. However, if the Ca^{2+} rise is too high, the situation of "calcium overload" can damage organelles such as the mitochondria and ER, potentially activating cell death pathways.

Cells maintain a high intracellular concentration of K^+ (~140 mM) in comparison to the extracellular environment (~5 mM). The efflux of K^+ upon LLO perforation is also responsible for various cellular effects including activation of protein phosphorylation, autophagy, and transcriptional regulation [149]. Numerous Ca^{2+}-sensitive transducers are known to respond to cytosolic variations in Ca^{2+} concentration; however, it is unclear how host cells sense and respond to variation in intracellular K^+. Adding more complexity, the combinatorial effect of simultaneous ionic and molecular fluxes should be considered when studying how extracellular LLO controls host cell signaling [131].

Activation of MAPKs

LLO and many other pore-forming toxins are known to activate members of the large family of the mitogen-activated protein kinases (MAPKs) which include the extracellular signal-regulated kinase (ERK), p38, and Jun N-terminal kinase (JNK) subfamilies. These protein kinases transduce environmental and developmental signals into essential cell responses such as differentiation, apoptosis, and inflammation [150]. These kinases act through regulating transcription, but also exert more rapid effects that are independent of de novo protein synthesis. LLO is a potent activator of all three subfamilies, ERK1/2, p38, and JNK [149, 151]. Activation of ERK and p38 is independent of Ca^{2+} influx, but requires K^+ efflux [149].

Furthermore, ERK and p38 activation is important for recovering ion homeostasis following perforation by LLO [149]. Reflecting the diversity of events regulated by these transducers, it was reported that MEK-1/ERK2 is required for *L. monocytogenes* internalization into epithelial cells [151, 152].

SUMOylation

Another class of post-transcriptional modification more recently attributed to LLO is SUMOylation [93]. SUMO (Small Ubiquitin-like Modifier) is an ubiquitin-like polypeptide that can be covalently conjugated to proteins by three enzymes designated E1, E2, and E3. SUMOylation is reversible, as deSUMOylation proteases, the "deSUMOylases", cleave the SUMO group from the target protein. SUMOylation controls multiple cellular processes such as intracellular transport, transcriptional regulation, stress responses, cell cycle, and apoptosis [153, 154]. *L. monocytogenes* was shown to decrease the level of SUMOylated proteins in infected cells, in a LLO-dependent fashion [93]. This effect is due to the degradation of the unique human E2 enzyme, Ubc9, possibly via an aspartyl-protease. Demonstrating the relevance of this pathway during pathogenesis, Ubc9 degradation was also observed in mice infected by *L. monocytogenes* [93]. As a consequence of Ubc9 degradation, there is an overall decrease in protein SUMOylation in *L. monocytogenes* infected cells. LLO alone, as well as other CDCs, is sufficient to exert this activity. Since increasing the global level of SUMOylated proteins by SUMO overexpression protects cells against *L. monocytogenes* infection, it was proposed that deSUMOylation subsequent to cell exposure to LLO favors pathogenesis. Which proteins are deSUMOylated during *L. monocytogenes* infection and how SUMOylated proteins protect host cells from infection remain to be established.

Roles of Extracellular Listeriolysin O in **L. monocytogenes** *Internalization and Vacuolar Escape*

Extracellular Listeriolysin O is Sufficient to Activate *L. monocytogenes* Internalization

A particularity of *L. monocytogenes* is its ability to infect a large variety of normally nonphagocytic cells. To invade nonphagocytic cells, bacterial surface adhesins anchor the bacterium to the host cell, while a second class of virulence factors, the invasins, activates the host cell internalization machineries. Some virulence factors can exert the dual function of adhesin and invasin. For example, InlA promotes both bacterial attachment and internalization by interacting with the cell adhesion molecule E-cadherin on epithelial cells. InlB only acts as an invasin by activating several receptors including HGF-Rc or c-Met [34]. *L. monocytogenes* uptake by normally nonphagocytic cells is poorly efficient when compared to professional phagocytes. Therefore, *L. monocytogenes* internalization into nonphagocytic cells is likely a limiting factor during pathogenesis. Thus, multiple adhesins and invasins may be required to facilitate *L. monocytogenes* internalization. Also with the wide range of cells *L. monocytogenes* infects, it is not surprising that this bacterium employs several adhesins and invasins to target diverse host cell receptors. It was reported that LLO alone can increase the efficiency of bacterial internalization into nonphagocytic cells, likely by mediating an influx of the second messenger Ca^{2+} [145]. It was shown more recently that LLO can act as an invasin, i.e. LLO is sufficient to induce the internalization of *L. monocytogenes* into some nonphagocytic cell lines [89]. This study showed for the first time that a pore-forming toxin can activate bacterial internalization. This novel internalization pathway requires host cell tyrosine kinase activity, F-actin polymerization, dynamin, but is microtubule- and clathrin-independent [131]. The organization of the plasma membrane into signaling, cholesterol-rich microdomains is essential for *L. monocytogenes* internalization [155, 156]. Thus, LLO binding to cholesterol-enriched microdomains may favor the creation a favorable signaling microenvironment. Furthermore, only native LLO, but not the LLO prepore-locked variant could induce bacterial internalization, revealing that formation of the LLO pore complex is strictly required for bacterial entry. This led to the hypothesis that membrane perforation is sufficient to activate F-actin remodeling and the host cell internalization machineries. In support of this, the CDC PLY also induces *L. monocytogenes* internalization and the parasite *Trypanosoma cruzi* was shown to perforate the host cell plasma membrane to stimulate its internalization [89, 157]. More recent studies propose that a combination of Ca^{2+}-influx and K^+-efflux is required for the LLO-dependent internalization pathway. Further linking ion fluxes to bacterial internalization, cell treatment with both Ca^{2+} and K^+ ionophores, in the absence of LLO, could induce the internalization of large cargoes such as 1 μm polystyrene beads or bacteria [131]. Thus, damaging the host cell plasma membrane emerges as an invasion strategy, shared by unrelated pathogens, that involves ion fluxes across the perforated plasma membrane. This invasion mechanism is relevant to many pathogens since pore-forming proteins are commonly produced by viruses, bacteria, and eukaryotic intracellular pathogens. The relationship between membrane perforation and pathogen internalization is certainly complex. Although some CDCs are required for uptake of *Streptococcus* and *Listeria* species, SLO exerts an anti-phagocytic activity for group A *Streptococcus* [89, 158, 159]. Additional studies are necessary to determine how membrane perforation can positively or, in some instances, negatively, regulate bacterial internalization.

Extracellular Listeriolysin O Controls the Efficiency of Vacuolar Escape

It was proposed that host cell signaling activated by LLO before or during bacterial internalization influences subsequent vacuolar escape. Indeed, extracellular LLO and PI-PLC activate the translocation of PKC βII on endosomal membranes, which controls *L. monocytogenes* escape from the phagosome in murine macrophages [102]. Therefore, early signaling by extracellular LLO may importantly affect the composition and properties of the phagosome and the properties of the endosomal network that fuses with the phagosome. Further studies are necessary to better understand the role of extracellular LLO and PKC βII in endosomal trafficking and bacterial escape.

Effects of Extracellular Listeriolysin O on Organelle Homeostasis and Transcription

LLO secreted by *L. monocytogenes* in the extracellular compartment or recombinant LLO added exogenously to animal cells affect the ER and mitochondria. These effects result most likely from perforation of the plasma membrane, giving rise to an increase in intracellular Ca^{2+}. Interestingly, it is proposed that the ER and mitochondria responses to LLO perforation then regulate positively or negatively host cell invasion.

Perturbation of the Endoplasmic Reticulum

The ER ensures essential cellular functions including posttranslational modification, folding, and sorting of newly synthesized proteins, lipid metabolism, and Ca^{2+} storage. In case of stress, the accumulation of unfolded proteins in the ER lumen is sensed by three ER-associated signal transducer proteins (PERK, ATF6, and IRE1) that activate the unfolded protein response (UPR). The UPR is conserved among all mammalian species and ensures cell survival until homeostasis is restored [160]. This response is characterized by a decrease in protein translation to limit the incoming pool of proteins in the ER, and by transcription of UPR target genes [161]. However, if the stress is too severe to be resolved, the UPR leads to apoptotic cell death [160]. Viral and bacterial pathogens have been shown to perturb ER homeostasis, activating the UPR. In particular, extracellular *L. monocytogenes* activates the UPR in a LLO-dependent fashion and recombinant LLO, added exogenously to the cell culture medium, was sufficient to activate the UPR [162]. How LLO activates the UPR is not known. It can be speculated that LLO indirectly triggers this response via perturbation of Ca^{2+} homeostasis. Indeed, cell exposure to LLO leads to depletion of the ER Ca^{2+} store, which is known to stimulate the UPR [163, 164]. Apoptotic cell death via activation of the ER-resident procaspase-12 upstream from the pro-apoptotic caspase-3 was observed in a substantial proportion of cells infected with *L. monocytogenes* [162]. Cell treatment with UPR-activating chemicals before infection decreased the intracellular load of *L. monocytogenes*. Therefore, UPR activation can be considered as a host cell response that decreases *L. monocytogenes* infection and may lead to cell death, thereby destroying the bacterial replication niche [162]. Similarly, other studies on the pore-forming Crystal (Cry) produced by the invertebrate pathogen *Bacillus thuringiensis* have revealed that the UPR sensor IRE1, activated downstream from p38MAPK, protects *Caenorhabditis elegans* and mammalian cell lines against Cry toxins [165]. In addition to the UPR, other pathways likely control protein synthesis in cells exposed to pore-forming toxins, since recombinant LLO halted protein synthesis in a K^+-dependent, but UPR-independent fashion [149]. This process was proposed to maintain the target cell in a quiescent state, while it recovers from toxin attack [149]. The ER is physically and functionally connected to the mitochondrial network [166]. It is therefore not surprising that pore forming toxins such as LLO affect the homeostasis of both ER and mitochondria.

Transient Fragmentation of the Mitochondrial Network

Infection of epithelial cells by *L. monocytogenes* is paralleled with a transient fragmentation of the mitochondrial network [94]. LLO was identified as the virulence factor responsible for this effect, since recombinant LLO, but not LLO-deficient bacteria, induced mitochondrial fragmentation [94]. LLO also caused a loss of mitochondrial membrane potential and a decrease in respiration and cellular ATP. Mechanistically, these effects require an intact F-actin cytoskeleton and involve the ER rather than the canonical mitochondrial fission and fusion proteins Drp1 and Opa 1 [167]. Importantly, fragmentation of the mitochondrial network was transient and the cellular levels of ATP were recovered after a few hours [94]. The influx of extracellular Ca^{2+} was strictly required for mitochondrial fragmentation and recombinant LLO variants with lowered capacity to form pores were inactive, revealing the involvement of plasma membrane perforation by LLO in mitochondrial remodeling. Interestingly, this process was observed to be "all or nothing", which could reflects that a threshold level of LLO and subsequently of intracellular Ca^{2+} concentration is required [166]. Blocking mitochondrial remodeling could modulate the efficiency of *L. monocytogenes* invasion of host cells [94]. Mitochondrial fragmentation may maintain host cell viability by limiting an otherwise lethal Ca^{2+} overload since mitochondrial remodeling was not paralleled with apoptosis [94]. Such a protective response would advantage the bacterium by mediating host cell survival. It is also possible that *L. monocytogenes* affects the cell bioenergetics to favor infection.

Roles of Extracellular Listeriolysin O in Inflammatory and Cell Death Pathways

Transcriptional Reprogramming via Histone Modifications

The DNA of all eukaryotic cells is wrapped around histone octamers made of two copies of each of four histones H2A, H2B, H3, and H4, forming nucleosomes. Posttranslational modifications of histones by acetylation, methylation, phosphorylation, or other modifications regulate gene expression, repair, replication, and recombination [168]. Pathogens also target histones in order to control the transcriptional activity of the infected host. LLO secreted by extracellular *L. monocytogenes* or recombinant LLO, but not LLO-deficient bacteria, induces the dephosphorylation of Ser10 on histone H3 and deacetylation of histone H4 [49]. Although it was initially thought that this function was independent of the formation of LLO pore complex, a more recent study by the same authors showed that this effect indeed results from an efflux of K^+ subsequent to LLO pore formation [169]. It was proposed that *L. monocytogenes* exploits histone modification to decrease the expression of some immunity genes, thereby lowering inflammation [49].

Activation of Innate and Adaptive Immunity

LLO affects the inflammatory response of the host through several processes. First, the toxin activates the inflammatory response by releasing *L. monocytogenes* into the cytosol, where the bacterium directly activates multiple NOD-like receptors (NLRs) [170–172]. Second, LLO alone released in the extracellular environment activates caspase-1 and the transcription of inflammatory genes. More specifically, K^+ efflux subsequent to plasma membrane perforation by LLO or several other pore-forming toxins activates the NALP3 inflammasome leading to caspase-1-dependent processing of pro-IL1β [169, 173, 174]. Caspase-1 was also shown to induce the translocation of the transcription factor SREBP to the nucleus, activating transcription of genes controlling lipid metabolism and cell survival [173, 175]. *L. monocytogenes* invasion of endothelial cells activates the pro-inflammatory transcription factor NF-KB and cytokine production in a LLO-dependent fashion [176]. Indeed, plasma membrane perforation by extracellular or intracellular LLO is responsible for NF- KB activation, surface expression of adhesion molecules, and the production of pro-inflammatory cytokines in epithelial cells [129, 176]. It was also proposed that LLO and other CDCs activate Toll-like receptor 4, which then modulates the inflammatory response [177].

Additionally, LLO is an immunogenic protein that contains several T and B cells epitopes. Two CD4+ (corresponding to the amino acids 189–201 and 215–226) and one CD8+ (amino acids 91–99) immunodominant epitopes have been identified in LLO [178]. Thus LLO, which is indispensable for pathogenesis, will ultimately be used by the immuno-competent host to drive the activation of immune cells, permanently eliminating the pathogen. In addition to providing major antigens, LLO is an immunomodulatory molecule. Several highly promising anti-tumor vaccine candidates use LLO or LLO variants as adjuvants [178]. The molecular basis of the LLO adjuvant properties is still unclear, but highlights the key role LLO plays in the control of innate and adaptive immune responses. It is proposed that LLO immunomodulatory activities are independent of its pore-forming activity [179].

Activation of Cell Death Pathways by Extracellular Listeriolysin O

L. monocytogenes has long been known to cause rapid—one to three days post-infection—apoptotic lesions in the spleen, lymph nodes, brain, and liver of infected mice [180]. In the liver, lesions are associated with infected hepatocytes, suggesting that cellular infection is responsible for activating the cell death pathway [181]. Although *L. monocytogenes* can activate apoptosis in macrophages and dendritic cells, T lymphocytes is the predominant cell type involved in apoptotic lesions in vivo [180]. In the spleen, massive lesions containing apoptotic T lymphocytes were observed independently of the mouse and *L. monocytogenes* strains and route of infection [125, 180]. Several lines of evidence strongly support that LLO secreted in the extracellular environment is responsible for lymphocyte apoptosis. First, apoptotic lymphocytes were not infected with *L. monocytogenes,* but free *L. monocytogenes* were observed in the affected areas. Thus, an extracellular factor such as LLO is likely responsible for apoptosis. Second, apoptosis was LLO-dependent, required LLO secretion, and could be induced in the draining lymph nodes following injection of recombinant LLO in the mouse footpads [125, 182]. Lymphocyte apoptosis decreased with bacterial burden when LLO-neutralizing antibodies were injected in infected mice [183, 184]. Finally, in support of a direct role of LLO, sublytic concentrations of LLO induce lymphocyte apoptosis in vitro [125]. Indeed, LLO can activate caspase-dependent (rapid) and independent (slow) pathways leading to apoptosis. A rapid activation of capase-3, -6, and -9, was observed in cells treated with LLO and was followed by exposure of phosphatidylserine on the outer leaflet of the plasma membrane, mitochondrial depolarization, and DNA fragmentation [125]. Granzymes play a major role in LLO-induced T-cell apoptosis in vitro and in vivo [185]. LLO is likely endocytosed and acts intracellularly to disrupt the lytic granules, thereby releasing granzymes into the cytosol, where the enzymes activate apoptosis [185]. Importantly, it was shown that LLO facilitates the production of type I interferon, which sensitizes T lymphocytes to the apoptogenic function of LLO [186]. Apoptosis facilitates *L. monocytogenes* infection by decreasing host resistance and facilitating bacterial proliferation [187]. Thus the role of extracellular LLO in lymphocyte apoptosis is clearly characterized in vitro and in vivo. Importantly, this extracellular activity of LLO substantially affects the course of infection.

Concluding Remarks

LLO has emerged as a multifunctional toxin that regulates the intracellular lifecycle of *L. monocytogenes* in diverse cell types and affects the innate and adaptive immune responses. Many more studies are necessary to better delineate the mechanisms of action of this toxin. In particular, how LLO facilitates the disruption of the endocytic vacuoles containing *L. monocytogenes* is still unclear. LLO secreted by extracellular bacteria also targets and affects the biology of multiple cell types, from epithelial cells to T lymphocytes. Extracellular LLO may control several stages of the *L. monocytogenes* intracellular lifecycle including bacterial internalization, vacuolar escape, and efficient intracellular replication. Extracellular LLO may also affect many cell types that the bacterium does not infect. For example, LLO can modulate the production of inflammatory messengers and cell death pathways independently of host cell invasion. A major challenge in the field is now to distinguish the extracellular from intracellular activities of LLO in vivo, and to determine which, if not all, of the host cell responses to LLO observed in cell culture models play a substantial role during pathogenesis. A fascinating aspect of LLO is its dual role during infection: whereas LLO is indispensable for pathogenesis, it is a major immunogenic molecule that ultimately elicits *Listeria monocytogenes* killing by the immunocompetent host. Due to its unique set of properties, LLO has emerged as a promising tool for the development of vaccine adjuvants. However, the molecular basis of the LLO immunomodulatory activity is poorly understood. Gaining knowledge on the interplay between LLO and host cells will importantly benefit basic research on pathogen-host interaction, pore-forming toxins, immunity, as well as the design of vaccines.

Acknowledgements This work was made possible by Grant R01AI107250 (Stephanie Seveau) from the National Institutes of Health (NIAID) and its contents are solely the responsibility of the author and do not necessarily represent the official views of the NIAID.

References

1. Oevermann A, Zurbriggen A, Vandevelde M (2010) Rhombencephalitis caused by Listeria monocytogenes in humans and ruminants: a zoonosis on the rise? Interdiscip Perspect Infect Dis 2010:632513
2. Swaminathan B, Gerner-Smidt P (2007) The epidemiology of human listeriosis. Microbes Infect 9:1236–1243
3. Linnan MJ, Mascola L, Lou XD, Goulet V, May S, Salminen C, Hird DW, Yonekura ML, Hayes P, Weaver R et al (1988) Epidemic listeriosis associated with Mexican-style cheese. N Engl J Med 319:823–828
4. Murray EGD, Webb RA, HBR S (1926) A disease of rabbits characterized by a large mononuclear leucoytosis caused by a hitherto undescribed bacillus Bacterium monocytogenes. J Pathol Bacteriol 29:407–439

5. Schlech WF 3rd, Lavigne PM, Bortolussi RA, Allen AC, Haldane EV, Wort AJ, Hightower AW, Johnson SE, King SH, Nicholls ES, Broome CV (1983) Epidemic listeriosis–evidence for transmission by food. N Engl J Med 308:203–206
6. Barbuddhe SB, Chakraborty T (2009) Listeria as an enteroinvasive gastrointestinal pathogen. Curr Top Microbiol Immunol 337:173–195
7. Begley M, Gahan CG, Hill C (2002) Bile stress response in Listeria monocytogenes LO28: adaptation, cross-protection, and identification of genetic loci involved in bile resistance. Appl Environ Microbiol 68:6005–6012
8. Chaturongakul S, Raengpradub S, Wiedmann M, Boor KJ (2008) Modulation of stress and virulence in Listeria monocytogenes. Trends Microbiol 16:388–396
9. Dussurget O, Cabanes D, Dehoux P, Lecuit M, Buchrieser C, Glaser P, Cossart P (2002) Listeria monocytogenes bile salt hydrolase is a PrfA-regulated virulence factor involved in the intestinal and hepatic phases of listeriosis. Mol Microbiol 45:1095–1106
10. Grif K, Patscheider G, Dierich MP, Allerberger F (2003) Incidence of fecal carriage of Listeria monocytogenes in three healthy volunteers: a one-year prospective stool survey. Eur J Clin Microbiol Infect Dis 22:16–20
11. Ooi ST, Lorber B (2005) Gastroenteritis due to Listeria monocytogenes. Clin Infect Dis 40:1327–1332
12. Wing EJ, Gregory SH (2002) Listeria monocytogenes: clinical and experimental update. J Infect Dis 185:S18–S24
13. Alonzo F 3rd, Bobo LD, Skiest DJ, Freitag NE (2011) Evidence for subpopulations of Listeria monocytogenes with enhanced invasion of cardiac cells. J Med Microbiol 60:423–434
14. Disson O, Lecuit M (2012) Targeting of the central nervous system by Listeria monocytogenes. Virulence 3:213–221
15. Goulet V, Hebert M, Hedberg C, Laurent E, Vaillant V, De Valk H, Desenclos JC (2012) Incidence of listeriosis and related mortality among groups at risk of acquiring listeriosis. Clin Infect Dis 54:652–660
16. Poulsen KP, Czuprynski CJ (2013) Pathogenesis of listeriosis during pregnancy. Anim Health Res Rev 14:30–39
17. Ramaswamy V, Cresence VM, Rejitha JS, Lekshmi MU, Dharsana KS, Prasad SP, Vijila HM (2007) Listeria–review of epidemiology and pathogenesis. J Microbiol Immunol Infect 40:4–13
18. Robbins JR, Bakardjiev AI (2012) Pathogens and the placental fortress. Curr Opin Microbiol 15:36–43
19. Temple ME, Nahata MC (2000) Treatment of listeriosis. Ann Pharmacother 34(5):656–661
20. Le Monnier A, Abachin E, Beretti JL, Berche P, Kayal S (2011) Diagnosis of Listeria monocytogenes meningoencephalitis by real-time PCR for the hly gene. J Clin Microbiol 49:3917–3923
21. Vazquez-Boland JA, Dominguez-Bernal G, Gonzalez-Zorn B, Kreft J, Goebel W (2001) Pathogenicity islands and virulence evolution in Listeria. Microbes Infect 3:571–584
22. Vazquez-Boland JA, Kuhn M, Berche P, Chakraborty T, Dominguez-Bernal G, Goebel W, Gonzalez-Zorn B, Wehland J, Kreft J (2001) Listeria pathogenesis and molecular virulence determinants. Clin Microbiol Rev 14:584–640
23. Perry KJ, Higgins DE (2013) A differential fluorescence-based genetic screen identifies Listeria monocytogenes determinants required for intracellular replication. J Bacteriol. doi: 10.1128/JB.00210-13 (in press)
24. Glaser P, Frangeul L, Buchrieser C, Rusniok C, Amend A, Baquero F, Berche P, Bloecker H, Brandt P, Chakraborty T, Charbit A, Chetouani F, Couve E, de Daruvar A, Dehoux P, Domann E, Dominguez-Bernal G, Duchaud E, Durant L, Dussurget O, Entian KD, Fsihi H, Garcia-del Portillo F, Garrido P, Gautier L, Goebel W, Gomez-Lopez N, Hain T, Hauf J, Jackson D, Jones LM, Kaerst U, Kreft J, Kuhn M, Kunst F, Kurapkat G, Madueno E, Maitournam A, Vicente JM, Ng E, Nedjari H, Nordsiek G, Novella S, de Pablos B, Perez-Diaz JC, Purcell R, Remmel B, Rose M, Schlueter T, Simoes N, Tierrez A, Vazquez-Boland

JA, Voss H, Wehland J, Cossart P (2001) Comparative genomics of Listeria species. Science 294:849–852
25. Joseph B, Przybilla K, Stuhler C, Schauer K, Slaghuis J, Fuchs TM, Goebel W (2006) Identification of Listeria monocytogenes genes contributing to intracellular replication by expression profiling and mutant screening. J Bacteriol 188:556–568
26. Gaillard JL, Berche P, Mounier J, Richard S, Sansonetti P (1987) In vitro model of penetration and intracellular growth of Listeria monocytogenes in the human enterocyte-like cell line Caco-2. Infect Immun 55:2822–2829
27. Braun L, Ohayon H, Cossart P (1998) The InlB protein of Listeria monocytogenes is sufficient to promote entry into mammalian cells. Mol Microbiol 27:1077–1087
28. Dramsi S, Biswas I, Maguin E, Braun L, Mastroeni P, Cossart P (1995) Entry of Listeria monocytogenes into hepatocytes requires expression of inlB, a surface protein of the internalin multigene family. Mol Microbiol 16:251–261
29. Gaillard JL, Berche P, Frehel C, Gouin E, Cossart P (1991) Entry of L. monocytogenes into cells is mediated by internalin, a repeat protein reminiscent of surface antigens from gram-positive cocci. Cell 65:1127–1141
30. Jonquieres R, Bierne H, Fiedler F, Gounon P, Cossart P (1999) Interaction between the protein InlB of Listeria monocytogenes and lipoteichoic acid: a novel mechanism of protein association at the surface of gram-positive bacteria. Mol Microbiol 34:902–914
31. Jonquieres R, Pizarro-Cerda J, Cossart P (2001) Synergy between the N- and C-terminal domains of InlB for efficient invasion of non-phagocytic cells by Listeria monocytogenes. Mol Microbiol 42:955–965
32. Marino M, Banerjee M, Jonquieres R, Cossart P, Ghosh P (2002) GW domains of the Listeria monocytogenes invasion protein InlB are SH3-like and mediate binding to host ligands. EMBO J 21:5623–5634
33. Pizarro-Cerda J, Kuhbacher A, Cossart P (2012) Entry of Listeria monocytogenes in mammalian epithelial cells: an updated view. Cold Spring Harb Perspect Med 2:a010009
34. Seveau S, Pizarro-Cerda J, Cossart P (2007) Molecular mechanisms exploited by Listeria monocytogenes during host cell invasion. Microbes Infect 9:1167–1175
35. Geoffroy C, Gaillard JL, Alouf JE, Berche P (1987) Purification, characterization, and toxicity of the sulfhydryl-activated hemolysin listeriolysin O from Listeria monocytogenes. Infect Immun 55:1641–1646
36. Geoffroy C, Gaillard JL, Alouf JE, Berche P (1989) Production of thiol-dependent haemolysins by Listeria monocytogenes and related species. J Gen Microbiol 135:481–487
37. Portnoy DA, Jacks PS, Hinrichs DJ (1988) Role of hemolysin for the intracellular growth of Listeria monocytogenes. J Exp Med 167:1459–1471
38. Cossart P, Vicente MF, Mengaud J, Baquero F, Perez-Diaz JC, Berche P (1989) Listeriolysin O is essential for virulence of Listeria monocytogenes: direct evidence obtained by gene complementation. Infect Immun 57:3629–3636
39. Gaillard JL, Berche P, Sansonetti P (1986) Transposon mutagenesis as a tool to study the role of hemolysin in the virulence of Listeria monocytogenes. Infect Immun 52:50–55
40. Kathariou S, Metz P, Hof H, Goebel W (1987) Tn916-induced mutations in the hemolysin determinant affecting virulence of Listeria monocytogenes. J Bacteriol 169:1291–1297
41. Mengaud J, Chenevert J, Geoffroy C, Gaillard JL, Cossart P (1987) Identification of the structural gene encoding the SH-activated hemolysin of Listeria monocytogenes: listeriolysin O is homologous to streptolysin O and pneumolysin. Infect Immun 55:3225–3227
42. Kocks C, Gouin E, Tabouret M, Berche P, Ohayon H, Cossart P (1992) L. monocytogenes-induced actin assembly requires the actA gene product, a surface protein. Cell 68:521–531
43. Welch MD, Rosenblatt J, Skoble J, Portnoy DA, Mitchison TJ (1998) Interaction of human Arp2/3 complex and the Listeria monocytogenes ActA protein in actin filament nucleation. Science 281:105–108

44. Mounier J, Ryter A, Coquis-Rondon M, Sansonetti PJ (1990) Intracellular and cell-to-cell spread of Listeria monocytogenes involves interaction with F-actin in the enterocytelike cell line Caco-2. Infect Immun 58:1048–1058
45. Tilney LG, Portnoy DA (1989) Actin filaments and the growth, movement, and spread of the intracellular bacterial parasite, Listeria monocytogenes. J Cell Biol 109:1597–1608
46. Gedde MM, Higgins DE, Tilney LG, Portnoy DA (2000) Role of listeriolysin O in cell-to-cell spread of Listeria monocytogenes. Infect Immun 68:999–1003
47. Sun AN, Camilli A, Portnoy DA (1990) Isolation of Listeria monocytogenes small-plaque mutants defective for intracellular growth and cell-to-cell spread. Infect Immun 58:3770–3778
48. Smith GA, Marquis H, Jones S, Johnston NC, Portnoy DA, Goldfine H (1995) The two distinct phospholipases C of Listeria monocytogenes have overlapping roles in escape from a vacuole and cell-to-cell spread. Infect Immun 63:4231–4237
49. Hamon MA, Batsche E, Regnault B, Tham TN, Seveau S, Muchardt C, Cossart P (2007) Histone modifications induced by a family of bacterial toxins. Proc Natl Acad Sci USA 104:13467–13472
50. Hamon MA, Ribet D, Stavru F, Cossart P (2012) Listeriolysin O: the Swiss army knife of Listeria. Trends Microbiol 20:360–368
51. Chakraborty T, Leimeister-Wachter M, Domann E, Hartl M, Goebel W, Nichterlein T, Notermans S (1992) Coordinate regulation of virulence genes in Listeria monocytogenes requires the product of the prfA gene. J Bacteriol 174:568–574
52. de las Heras A, Cain RJ, Bielecka MK, Vazquez-Boland JA (2011) Regulation of Listeria virulence: PrfA master and commander. Curr Opin Microbiol 14:118–127
53. Leimeister-Wachter M, Haffner C, Domann E, Goebel W, Chakraborty T (1990) Identification of a gene that positively regulates expression of listeriolysin, the major virulence factor of listeria monocytogenes. Proc Natl Acad Sci USA 87:8336–8340
54. Freitag NE, Youngman P, Portnoy DA (1992) Transcriptional activation of the Listeria monocytogenes hemolysin gene in Bacillus subtilis. J Bacteriol 174:1293–1298
55. Behari J, Youngman P (1998) Regulation of hly expression in Listeria monocytogenes by carbon sources and pH occurs through separate mechanisms mediated by PrfA. Infect Immun 66:3635–3642
56. Datta AR, Kothary MH (1993) Effects of glucose, growth temperature, and pH on listeriolysin O production in Listeria monocytogenes. Appl Environ Microbiol 59:3495–3497
57. Milenbachs AA, Brown DP, Moors M, Youngman P (1997) Carbon-source regulation of virulence gene expression in Listeria monocytogenes. Mol Microbiol 23:1075–1085
58. Johansson J, Mandin P, Renzoni A, Chiaruttini C, Springer M, Cossart P (2002) An RNA thermosensor controls expression of virulence genes in Listeria monocytogenes. Cell 110:551–561
59. Leimeister-Wachter M, Domann E, Chakraborty T (1992) The expression of virulence genes in Listeria monocytogenes is thermoregulated. J Bacteriol 174:947–952
60. Bubert A, Sokolovic Z, Chun SK, Papatheodorou L, Simm A, Goebel W (1999) Differential expression of Listeria monocytogenes virulence genes in mammalian host cells. Mol Gen Genet 261:323–336
61. Shen A, Higgins DE (2005) The 5' untranslated region-mediated enhancement of intracellular listeriolysin O production is required for Listeria monocytogenes pathogenicity. Mol Microbiol 57:1460–1473
62. Schnupf P, Portnoy DA, Decatur AL (2006) Phosphorylation, ubiquitination and degradation of listeriolysin O in mammalian cells: role of the PEST-like sequence. Cell Microbiol 8:353–364
63. Mengaud J, Vicente MF, Chenevert J, Pereira JM, Geoffroy C, Gicquel-Sanzey B, Baquero F, Perez-Diaz JC, Cossart P (1988) Expression in Escherichia coli and sequence analysis of the listeriolysin O determinant of Listeria monocytogenes. Infect Immun 56:766–772

64. Arnett E, Lehrer RI, Pratikhya P, Lu W, Seveau S (2011) Defensins enable macrophages to inhibit the intracellular proliferation of Listeria monocytogenes. Cell Microbiol 13:635–651
65. Lehrer RI, Jung G, Ruchala P, Wang W, Micewicz ED, Waring AJ, Gillespie EJ, Bradley KA, Ratner AJ, Rest RF, Lu W (2009) Human alpha-defensins inhibit hemolysis mediated by cholesterol-dependent cytolysins. Infect Immun 77:4028–4040
66. Arnett E, Seveau S (2011) The multifaceted activities of mammalian defensins. Curr Pharm Des 17:4254–4269
67. Koster S, Hudel M, Chakraborty T, Yildiz O (2013) Crystallization and X-ray crystallographic analysis of the cholesterol-dependent cytolysin listeriolysin O from Listeria monocytogenes. Acta Crystallogr, Sect F: Struct Biol Cryst Commun 69:1212–1215
68. Dunstone MA, Tweten RK (2012) Packing a punch: the mechanism of pore formation by cholesterol dependent cytolysins and membrane attack complex/perforin-like proteins. Curr Opin Struct Biol 22:342–349
69. Rossjohn J, Feil SC, McKinstry WJ, Tweten RK, Parker MW (1997) Structure of a cholesterol-binding, thiol-activated cytolysin and a model of its membrane form. Cell 89:685–692
70. Heuck AP, Moe PC, Johnson BB (2010) The cholesterol-dependent cytolysin family of gram-positive bacterial toxins. Subcell Biochem 51:551–577
71. Hotze EM, Tweten RK (2012) Membrane assembly of the cholesterol-dependent cytolysin pore complex. Biochim Biophys Acta 1818:1028–1038
72. Ramachandran R, Heuck AP, Tweten RK, Johnson AE (2002) Structural insights into the membrane-anchoring mechanism of a cholesterol-dependent cytolysin. Nat Struct Biol 9:823–827
73. Solovyova AS, Nollmann M, Mitchell TJ, Byron O (2004) The solution structure and oligomerization behavior of two bacterial toxins: pneumolysin and perfringolysin O. Biophys J 87:540–552
74. Palmer M, Harris R, Freytag C, Kehoe M, Tranum-Jensen J, Bhakdi S (1998) Assembly mechanism of the oligomeric streptolysin O pore: the early membrane lesion is lined by a free edge of the lipid membrane and is extended gradually during oligomerization. EMBO J 17:1598–1605
75. Repp H, Pamukci Z, Koschinski A, Domann E, Darji A, Birringer J, Brockmeier D, Chakraborty T, Dreyer F (2002) Listeriolysin of Listeria monocytogenes forms Ca2+-permeable pores leading to intracellular Ca2+ oscillations. Cell Microbiol 4:483–491
76. Shaughnessy LM, Hoppe AD, Christensen KA, Swanson JA (2006) Membrane perforations inhibit lysosome fusion by altering pH and calcium in Listeria monocytogenes vacuoles. Cell Microbiol 8:781–792
77. Ramachandran R, Tweten RK, Johnson AE (2004) Membrane-dependent conformational changes initiate cholesterol-dependent cytolysin oligomerization and intersubunit beta-strand alignment. Nat Struct Mol Biol 11:697–705
78. Soltani CE, Hotze EM, Johnson AE, Tweten RK (2007) Specific protein-membrane contacts are required for prepore and pore assembly by a cholesterol-dependent cytolysin. J Biol Chem 282:15709–15716
79. Farrand AJ, LaChapelle S, Hotze EM, Johnson AE, Tweten RK (2010) Only two amino acids are essential for cytolytic toxin recognition of cholesterol at the membrane surface. Proc Natl Acad Sci USA 107:4341–4346
80. Bavdek A, Gekara NO, Priselac D, Gutierrez Aguirre I, Darji A, Chakraborty T, Macek P, Lakey JH, Weiss S, Anderluh G (2007) Sterol and pH interdependence in the binding, oligomerization, and pore formation of Listeriolysin O. Biochemistry 46:4425–4437
81. Coconnier MH, Lorrot M, Barbat A, Laboisse C, Servin AL (2000) Listeriolysin O-induced stimulation of mucin exocytosis in polarized intestinal mucin-secreting cells: evidence for toxin recognition of membrane-associated lipids and subsequent toxin internalization through caveolae. Cell Microbiol 2:487–504

82. Gelber SE, Aguilar JL, Lewis KL, Ratner AJ (2008) Functional and phylogenetic characterization of Vaginolysin, the human-specific cytolysin from Gardnerella vaginalis. J Bacteriol 190:3896–3903
83. Giddings KS, Zhao J, Sims PJ, Tweten RK (2004) Human CD59 is a receptor for the cholesterol-dependent cytolysin intermedilysin. Nat Struct Mol Biol 11:1173–1178
84. Jacobs T, Darji A, Frahm N, Rohde M, Wehland J, Chakraborty T, Weiss S (1998) Listeriolysin O: cholesterol inhibits cytolysis but not binding to cellular membranes. Mol Microbiol 28:1081–1089
85. Dowd KJ, Tweten RK (2012) The cholesterol-dependent cytolysin signature motif: a critical element in the allosteric pathway that couples membrane binding to pore assembly. PLoS Pathog 8:e1002787
86. Shatursky O, Heuck AP, Shepard LA, Rossjohn J, Parker MW, Johnson AE, Tweten RK (1999) The mechanism of membrane insertion for a cholesterol-dependent cytolysin: a novel paradigm for pore-forming toxins. Cell 99:293–299
87. Shepard LA, Heuck AP, Hamman BD, Rossjohn J, Parker MW, Ryan KR, Johnson AE, Tweten RK (1998) Identification of a membrane-spanning domain of the thiol-activated pore-forming toxin Clostridium perfringens perfringolysin O: an alpha-helical to beta-sheet transition identified by fluorescence spectroscopy. Biochemistry 37:14563–14574
88. Hotze EM, Wilson-Kubalek EM, Rossjohn J, Parker MW, Johnson AE, Tweten RK (2001) Arresting pore formation of a cholesterol-dependent cytolysin by disulfide trapping synchronizes the insertion of the transmembrane beta-sheet from a prepore intermediate. J Biol Chem 276:8261–8268
89. Vadia S, Arnett E, Haghighat AC, Wilson-Kubalek EM, Tweten RK, Seveau S (2011) The pore-forming toxin listeriolysin O mediates a novel entry pathway of L. monocytogenes into human hepatocytes. PLoS Pathog 7:e1002356
90. Schnupf P, Portnoy DA (2007) Listeriolysin O: a phagosome-specific lysin. Microbes Infect 9:1176–1187
91. Bavdek A, Kostanjsek R, Antonini V, Lakey JH, Dalla Serra M, Gilbert RJ, Anderluh G (2012) pH dependence of listeriolysin O aggregation and pore-forming ability. FEBS J 279:126–141
92. Schuerch DW, Wilson-Kubalek EM, Tweten RK (2005) Molecular basis of listeriolysin O pH dependence. Proc Natl Acad Sci USA 102:12537–12542
93. Ribet D, Hamon M, Gouin E, Nahori MA, Impens F, Neyret-Kahn H, Gevaert K, Vandekerckhove J, Dejean A, Cossart P (2010) Listeria monocytogenes impairs SUMOylation for efficient infection. Nature 464:1192–1195
94. Stavru F, Bouillaud F, Sartori A, Ricquier D, Cossart P (2011) Listeria monocytogenes transiently alters mitochondrial dynamics during infection. Proc Natl Acad Sci USA 108:3612–3617
95. Glomski IJ, Gedde MM, Tsang AW, Swanson JA, Portnoy DA (2002) The Listeria monocytogenes hemolysin has an acidic pH optimum to compartmentalize activity and prevent damage to infected host cells. J Cell Biol 156:1029–1038
96. Michel E, Reich KA, Favier R, Berche P, Cossart P (1990) Attenuated mutants of the intracellular bacterium Listeria monocytogenes obtained by single amino acid substitutions in listeriolysin O. Mol Microbiol 4:2167–2178
97. Singh R, Jamieson A, Cresswell P (2008) GILT is a critical host factor for Listeria monocytogenes infection. Nature 455:1244–1247
98. Burrack LS, Harper JW, Higgins DE (2009) Perturbation of vacuolar maturation promotes listeriolysin O-independent vacuolar escape during Listeria monocytogenes infection of human cells. Cell Microbiol 11:1382–1398
99. Grundling A, Gonzalez MD, Higgins DE (2003) Requirement of the Listeria monocytogenes broad-range phospholipase PC-PLC during infection of human epithelial cells. J Bacteriol 185:6295–6307

100. Marquis H, Doshi V, Portnoy DA (1995) The broad-range phospholipase C and a metalloprotease mediate listeriolysin O-independent escape of Listeria monocytogenes from a primary vacuole in human epithelial cells. Infect Immun 63:4531–4534
101. Camilli A, Tilney LG, Portnoy DA (1993) Dual roles of plcA in Listeria monocytogenes pathogenesis. Mol Microbiol 8:143–157
102. Wadsworth SJ, Goldfine H (2002) Mobilization of protein kinase C in macrophages induced by Listeria monocytogenes affects its internalization and escape from the phagosome. Infect Immun 70:4650–4660
103. Beauregard KE, Lee KD, Collier RJ, Swanson JA (1997) pH-dependent perforation of macrophage phagosomes by listeriolysin O from Listeria monocytogenes. J Exp Med 186:1159–1163
104. Henry R, Shaughnessy L, Loessner MJ, Alberti-Segui C, Higgins DE, Swanson JA (2006) Cytolysin-dependent delay of vacuole maturation in macrophages infected with Listeria monocytogenes. Cell Microbiol 8:107–119
105. Radtke AL, Anderson KL, Davis MJ, DiMagno MJ, Swanson JA, O'Riordan MX (2011) Listeria monocytogenes exploits cystic fibrosis transmembrane conductance regulator (CFTR) to escape the phagosome. Proc Natl Acad Sci USA 108:1633–1638
106. Lam GY, Fattouh R, Muise AM, Grinstein S, Higgins DE, Brumell JH (2011) Listeriolysin O suppresses phospholipase C-mediated activation of the microbicidal NADPH oxidase to promote Listeria monocytogenes infection. Cell Host Microbe 10:627–634
107. Madden JC, Ruiz N, Caparon M (2001) Cytolysin-mediated translocation (CMT): a functional equivalent of type III secretion in gram-positive bacteria. Cell 104:143–152
108. Shaughnessy LM, Swanson JA (2007) The role of the activated macrophage in clearing Listeria monocytogenes infection. Front Biosci 12:2683–2692
109. Arnett E, Vadia S, Nackerman CC, Oghumu S, Satoskar AR, McLeish KR, Uriarte SM, Seveau S (2014) The pore-forming toxin listeriolysin O is degraded by neutrophil metalloproteiase-8 and is fails to mediate Listeria monocytogenes intracellular survival in neutrophils. J Immunol. doi:10.4049/jimmunol.1301302 (in press)
110. Myers JT, Tsang AW, Swanson JA (2003) Localized reactive oxygen and nitrogen intermediates inhibit escape of Listeria monocytogenes from vacuoles in activated macrophages. J Immunol 171:5447–5453
111. del Cerro-Vadillo E, Madrazo-Toca F, Carrasco-Marin E, Fernandez-Prieto L, Beck C, Leyva-Cobian F, Saftig P, Alvarez-Dominguez C (2006) Cutting edge: a novel nonoxidative phagosomal mechanism exerted by cathepsin-D controls Listeria monocytogenes intracellular growth. J Immunol 176:1321–1325
112. Davis MJ, Gregorka B, Gestwicki JE, Swanson JA (2012) Inducible renitence limits Listeria monocytogenes escape from vacuoles in macrophages. J Immunol 189:4488–4495
113. Birmingham CL, Canadien V, Gouin E, Troy EB, Yoshimori T, Cossart P, Higgins DE, Brumell JH (2007) Listeria monocytogenes evades killing by autophagy during colonization of host cells. Autophagy 3:442–451
114. Py BF, Lipinski MM, Yuan J (2007) Autophagy limits Listeria monocytogenes intracellular growth in the early phase of primary infection. Autophagy 3:117–125
115. Meyer-Morse N, Robbins JR, Rae CS, Mochegova SN, Swanson MS, Zhao Z, Virgin HW, Portnoy D (2010) Listeriolysin O is necessary and sufficient to induce autophagy during Listeria monocytogenes infection. PLoS ONE 5:e8610
116. Dortet L, Mostowy S, Samba-Louaka A, Gouin E, Nahori MA, Wiemer EA, Dussurget O, Cossart P (2011) Recruitment of the major vault protein by InlK: a Listeria monocytogenes strategy to avoid autophagy. PLoS Pathog 7:e1002168
117. Alberti-Segui C, Goeden KR, Higgins DE (2007) Differential function of Listeria monocytogenes listeriolysin O and phospholipases C in vacuolar dissolution following cell-to-cell spread. Cell Microbiol 9:179–195
118. Dancz CE, Haraga A, Portnoy DA, Higgins DE (2002) Inducible control of virulence gene expression in Listeria monocytogenes: temporal requirement of listeriolysin O during intracellular infection. J Bacteriol 184:5935–5945

119. Villanueva MS, Sijts AJ, Pamer EG (1995) Listeriolysin is processed efficiently into an MHC class I-associated epitope in Listeria monocytogenes-infected cells. J Immunol 155:5227–5233
120. Decatur AL, Portnoy DA (2000) A PEST-like sequence in listeriolysin O essential for Listeria monocytogenes pathogenicity. Science 290:992–995
121. Glomski IJ, Decatur AL, Portnoy DA (2003) Listeria monocytogenes mutants that fail to compartmentalize listerolysin O activity are cytotoxic, avirulent, and unable to evade host extracellular defenses. Infect Immun 71:6754–6765
122. Frehel C, Lety MA, Autret N, Beretti JL, Berche P, Charbit A (2003) Capacity of ivanolysin O to replace listeriolysin O in phagosomal escape and in vivo survival of Listeria monocytogenes. Microbiology 149:611–620
123. Jones S, Portnoy DA (1994) Characterization of Listeria monocytogenes pathogenesis in a strain expressing perfringolysin O in place of listeriolysin O. Infect Immun 62:5608–5613
124. Wei Z, Schnupf P, Poussin MA, Zenewicz LA, Shen H, Goldfine H (2005) Characterization of Listeria monocytogenes expressing anthrolysin O and phosphatidylinositol-specific phospholipase C from Bacillus anthracis. Infect Immun 73:6639–6646
125. Carrero JA, Calderon B, Unanue ER (2004) Listeriolysin O from Listeria monocytogenes is a lymphocyte apoptogenic molecule. J Immunol 172:4866–4874
126. Lety MA, Frehel C, Dubail I, Beretti JL, Kayal S, Berche P, Charbit A (2001) Identification of a PEST-like motif in listeriolysin O required for phagosomal escape and for virulence in Listeria monocytogenes. Mol Microbiol 39:1124–1139
127. Schnupf P, Zhou J, Varshavsky A, Portnoy DA (2007) Listeriolysin O secreted by Listeria monocytogenes into the host cell cytosol is degraded by the N-end rule pathway. Infect Immun 75:5135–5147
128. Viala JP, Mochegova SN, Meyer-Morse N, Portnoy DA (2008) A bacterial pore-forming toxin forms aggregates in cells that resemble those associated with neurodegenerative diseases. Cell Microbiol 10:985–993
129. Tsuchiya K, Kawamura I, Takahashi A, Nomura T, Kohda C, Mitsuyama M (2005) Listeriolysin O-induced membrane permeation mediates persistent interleukin-6 production in Caco-2 cells during Listeria monocytogenes infection in vitro. Infect Immun 73:3869–3877
130. Cassidy SK, Hagar JA, Kanneganti TD, Franchi L, Nunez G, O'Riordan MX (2012) Membrane damage during Listeria monocytogenes infection triggers a caspase-7 dependent cytoprotective response. PLoS Pathog 8:e1002628
131. Vadia S, Seveau S (2014) Fluxes of Ca2+ and K+ are required for the LLO-dependent internalization pathway of Listeria monocytogenes. Infect Immun (in press)
132. Abreu-Blanco MT, Verboon JM, Parkhurst SM (2011) Single cell wound repair: dealing with life's little traumas. Bioarchitecture 1:114–121
133. McNeil PL (2002) Repairing a torn cell surface: make way, lysosomes to the rescue. J Cell Sci 115:873–879
134. Steinhardt RA (2005) The mechanisms of cell membrane repair: a tutorial guide to key experiments. Ann N Y Acad Sci 1066:152–165
135. Draeger A, Monastyrskaya K, Babiychuk EB (2011) Plasma membrane repair and cellular damage control: the annexin survival kit. Biochem Pharmacol 81:703–712
136. Idone V, Tam C, Goss JW, Toomre D, Pypaert M, Andrews NW (2008) Repair of injured plasma membrane by rapid Ca2+-dependent endocytosis. J Cell Biol 180:905–914
137. Rodriguez A, Webster P, Ortego J, Andrews NW (1997) Lysosomes behave as Ca2+-regulated exocytic vesicles in fibroblasts and epithelial cells. J Cell Biol 137:93–104
138. Steinhardt RA, Bi G, Alderton JM (1994) Cell membrane resealing by a vesicular mechanism similar to neurotransmitter release. Science 263:390–393
139. Tam C, Idone V, Devlin C, Fernandes MC, Flannery A, He X, Schuchman E, Tabas I, Andrews NW (2010) Exocytosis of acid sphingomyelinase by wounded cells promotes endocytosis and plasma membrane repair. J Cell Biol 189:1027–1038

140. Zha X, Pierini LM, Leopold PL, Skiba PJ, Tabas I, Maxfield FR (1998) Sphingomyelinase treatment induces ATP-independent endocytosis. J Cell Biol 140:39–47
141. Corrotte M, Fernandes MC, Tam C, Andrews NW (2012) Toxin pores endocytosed during plasma membrane repair traffic into the lumen of MVBs for degradation. Traffic 13:483–494
142. Potez S, Luginbuhl M, Monastyrskaya K, Hostettler A, Draeger A, Babiychuk EB (2011) Tailored protection against plasmalemmal injury by annexins with different Ca2+ sensitivities. J Biol Chem 286:17982–17991
143. Dong Z, Saikumar P, Weinberg JM, Venkatachalam MA (2006) Calcium in cell injury and death. Annu Rev Pathol 1:405–434
144. Gekara NO, Weiss S (2004) Lipid rafts clustering and signalling by listeriolysin O. Biochem Soc Trans 32:712–714
145. Dramsi S, Cossart P (2003) Listeriolysin O-mediated calcium influx potentiates entry of Listeria monocytogenes into the human Hep-2 epithelial cell line. Infect Immun 71:3614–3618
146. Gekara NO, Westphal K, Ma B, Rohde M, Groebe L, Weiss S (2007) The multiple mechanisms of Ca2+ signalling by listeriolysin O, the cholesterol-dependent cytolysin of Listeria monocytogenes. Cell Microbiol 9:2008–2021
147. Rose F, Zeller SA, Chakraborty T, Domann E, Machleidt T, Kronke M, Seeger W, Grimminger F, Sibelius U (2001) Human endothelial cell activation and mediator release in response to Listeria monocytogenes virulence factors. Infect Immun 69:897–905
148. Wadsworth SJ, Goldfine H (1999) Listeria monocytogenes phospholipase C-dependent calcium signaling modulates bacterial entry into J774 macrophage-like cells. Infect Immun 67:1770–1778
149. Gonzalez MR, Bischofberger M, Freche B, Ho S, Parton RG, van der Goot FG (2011) Pore-forming toxins induce multiple cellular responses promoting survival. Cell Microbiol 13:1026–1043
150. Arthur JS, Ley SC (2013) Mitogen-activated protein kinases in innate immunity. Nat Rev Immunol 13:679–692
151. Tang P, Sutherland CL, Gold MR, Finlay BB (1998) Listeria monocytogenes invasion of epithelial cells requires the MEK-1/ERK-2 mitogen-activated protein kinase pathway. Infect Immun 66:1106–1112
152. Weiglein I, Goebel W, Troppmair J, Rapp UR, Demuth A, Kuhn M (1997) Listeria monocytogenes infection of HeLa cells results in listeriolysin O-mediated transient activation of the Raf-MEK-MAP kinase pathway. FEMS Microbiol Lett 148:189–195
153. Flotho A, Melchior F (2013) Sumoylation: a regulatory protein modification in health and disease. Annu Rev Biochem 82:357–385
154. Ribet D, Cossart P (2010) SUMOylation and bacterial pathogens. Virulence 1:532–534
155. Seveau S, Bierne H, Giroux S, Prevost MC, Cossart P (2004) Role of lipid rafts in E-cadherin– and HGF-R/Met–mediated entry of Listeria monocytogenes into host cells. J Cell Biol 166:743–753
156. Seveau S, Tham TN, Payrastre B, Hoppe AD, Swanson JA, Cossart P (2007) A FRET analysis to unravel the role of cholesterol in Rac1 and PI 3-kinase activation in the InlB/Met signalling pathway. Cell Microbiol 9:790–803
157. Fernandes MC, Cortez M, Flannery AR, Tam C, Mortara RA, Andrews NW (2011) Trypanosoma cruzi subverts the sphingomyelinase-mediated plasma membrane repair pathway for cell invasion. J Exp Med 208:909–921
158. Logsdon LK, Hakansson AP, Cortes G, Wessels MR (2011) Streptolysin o inhibits clathrin-dependent internalization of group a streptococcus. MBio 2:e00332-e00310
159. Sukeno A, Nagamune H, Whiley RA, Jafar SI, Aduse-Opoku J, Ohkura K, Maeda T, Hirota K, Miyake Y, Kourai H (2005) Intermedilysin is essential for the invasion of hepatoma HepG2 cells by Streptococcus intermedius. Microbiol Immunol 49:681–694
160. Sovolyova N, Healy S, Samali A, Logue SE (2014) Stressed to death—mechanisms of ER stress-induced cell death. Biol Chem 395:1–13

161. Carrara M, Prischi F, Ali MM (2013) UPR signal activation by luminal sensor domains. Int J Mol Sci 14:6454–6466
162. Pillich H, Loose M, Zimmer KP, Chakraborty T (2012) Activation of the unfolded protein response by Listeria monocytogenes. Cell Microbiol 14:949–964
163. Gekara NO, Groebe L, Viegas N, Weiss S (2008) Listeria monocytogenes desensitizes immune cells to subsequent Ca2+ signaling via listeriolysin O-induced depletion of intracellular Ca2+ stores. Infect Immun 76:857–862
164. Schroder M, Kaufman RJ (2005) The mammalian unfolded protein response. Annu Rev Biochem 74:739–789
165. Bischof LJ, Kao CY, Los FC, Gonzalez MR, Shen Z, Briggs SP, van der Goot FG, Aroian RV (2008) Activation of the unfolded protein response is required for defenses against bacterial pore-forming toxin in vivo. PLoS Pathog 4:e1000176
166. de Brito OM, Scorrano L (2010) An intimate liaison: spatial organization of the endoplasmic reticulum-mitochondria relationship. EMBO J 29:2715–2723
167. Stavru F, Palmer AE, Wang C, Youle RJ, Cossart P (2013) Atypical mitochondrial fission upon bacterial infection. Proc Natl Acad Sci USA 110:16003–16008
168. Bannister AJ, Kouzarides T (2011) Regulation of chromatin by histone modifications. Cell Res 21:381–395
169. Hamon MA, Cossart P (2011) K+ efflux is required for histone H3 dephosphorylation by Listeria monocytogenes listeriolysin O and other pore-forming toxins. Infect Immun 79:2839–2846
170. Eitel J, Suttorp N, Opitz B (2010) Innate immune recognition and inflammasome activation in listeria monocytogenes infection. Front Microbiol 1:149
171. O'Riordan M, Yi CH, Gonzales R, Lee KD, Portnoy DA (2002) Innate recognition of bacteria by a macrophage cytosolic surveillance pathway. Proc Natl Acad Sci USA 99:13861–13866
172. Opitz B, Puschel A, Beermann W, Hocke AC, Forster S, Schmeck B, van Laak V, Chakraborty T, Suttorp N, Hippenstiel S (2006) Listeria monocytogenes activated p38 MAPK and induced IL-8 secretion in a nucleotide-binding oligomerization domain 1-dependent manner in endothelial cells. J Immunol 176:484–490
173. Gurcel L, Abrami L, Girardin S, Tschopp J, van der Goot FG (2006) Caspase-1 activation of lipid metabolic pathways in response to bacterial pore-forming toxins promotes cell survival. Cell 126:1135–1145
174. Meixenberger K, Pache F, Eitel J, Schmeck B, Hippenstiel S, Slevogt H, N'Guessan P, Witzenrath M, Netea MG, Chakraborty T, Suttorp N, Opitz B (2010) Listeria monocytogenes-infected human peripheral blood mononuclear cells produce IL-1beta, depending on listeriolysin O and NLRP3. J Immunol 184:922–930
175. Im SS, Osborne TF (2012) Protection from bacterial-toxin-induced apoptosis in macrophages requires the lipogenic transcription factor sterol regulatory element binding protein 1a. Mol Cell Biol 32:2196–2202
176. Kayal S, Lilienbaum A, Poyart C, Memet S, Israel A, Berche P (1999) Listeriolysin O-dependent activation of endothelial cells during infection with Listeria monocytogenes: activation of NF-kappa B and upregulation of adhesion molecules and chemokines. Mol Microbiol 31:1709–1722
177. Park JM, Ng VH, Maeda S, Rest RF, Karin M (2004) Anthrolysin O and other gram-positive cytolysins are toll-like receptor 4 agonists. J Exp Med 200:1647–1655
178. Sun R, Liu Y (2013) Listeriolysin O as a strong immunogenic molecule for the development of new anti-tumor vaccines. Hum Vaccin Immunother 9:1058–1068
179. Wallecha A, Wood L, Pan ZK, Maciag PC, Shahabi V, Paterson Y (2013) Listeria monocytogenes-derived listeriolysin O has pathogen-associated molecular pattern-like properties independent of its hemolytic ability. Clin Vaccine Immunol 20:77–84
180. Carrero JA, Unanue ER (2012) Mechanisms and immunological effects of apoptosis caused by Listeria monocytogenes. Adv Immunol 113:157–174

181. Rogers HW, Callery MP, Deck B, Unanue ER (1996) Listeria monocytogenes induces apoptosis of infected hepatocytes. J Immunol 156:679–684
182. Carrero JA, Calderon B, Vivanco-Cid H, Unanue ER (2009) Recombinant Listeria monocytogenes expressing a cell wall-associated listeriolysin O is weakly virulent but immunogenic. Infect Immun 77:4371–4382
183. Edelson BT, Cossart P, Unanue ER (1999) Cutting edge: paradigm revisited: antibody provides resistance to Listeria infection. J Immunol 163:4087–4090
184. Edelson BT, Unanue ER (2001) Intracellular antibody neutralizes Listeria growth. Immunity 14:503–512
185. Carrero JA, Vivanco-Cid H, Unanue ER (2008) Granzymes drive a rapid listeriolysin O-induced T cell apoptosis. J Immunol 181:1365–1374
186. Carrero JA, Calderon B, Unanue ER (2004) Type I interferon sensitizes lymphocytes to apoptosis and reduces resistance to Listeria infection. J Exp Med 200:535–540
187. Carrero JA, Unanue ER (2007) Impact of lymphocyte apoptosis on the innate immune stages of infection. Immunol Res 38:333–341
188. Cossart P (2011) Illuminating the landscape of host-pathogen interactions with the bacterium Listeria monocytogenes. Proc Natl Acad Sci USA 108:19484–19491
189. Rajabian T, Gavicherla B, Heisig M, Muller-Altrock S, Goebel W, Gray-Owen SD, Ireton K (2009) The bacterial virulence factor InlC perturbs apical cell junctions and promotes cell-to-cell spread of Listeria. Nat Cell Biol 11:1212–1218

Chapter 10
Perforin: A Key Pore-Forming Protein for Immune Control of Viruses and Cancer

Jerome Thiery and Judy Lieberman

Abstract Perforin (PFN) is the key pore-forming molecule in the cytotoxic granules of immune killer cells. Expressed only in killer cells, PFN is the rate-limiting molecule for cytotoxic function, delivering the death-inducing granule serine proteases (granzymes) into target cells marked for immune elimination. In this chapter we describe our current understanding of how PFN accomplishes this task. We discuss where PFN is expressed and how its expression is regulated, the biogenesis and storage of PFN in killer cells and how they are protected from potential damage, how it is released, how it delivers Granzymes into target cells and the consequences of PFN deficiency.

Keywords Cytotoxic granules · Cytotoxicity · Cytotoxic T cells · Familial hemophagocytic lymphohistiocytosis · Granzymes · MACPF · Natural killer cells · Perforin · Pore-forming protein

Abbreviations

ADCC	Antibody-dependent cell-mediated cytotoxicity
APC	Antigen presenting cells
ASM	Acid sphingomyelinase
CDCs	Cholesterol-dependent cytolysins
CI-MPR	Cation-independent mannose-6-phosphate receptor
c-SMAC	The central region of the immune synapse
CTL	Cytotoxic T lymphocytes
EOMES	Eomesodermin
ER	Endoplasmic reticulum

J. Thiery
INSERM U753, University Paris Sud and Gustave Roussy Cancer Campus, Villejuif, France
e-mail: jerome.thiery@gustaveroussy.fr

J. Lieberman (✉)
Program in Cellular and Molecular Medicine, Boston Children's Hospital and Harvard Medical School, Boston, MA, USA
e-mail: judy.lieberman@childrens.harvard.edu

G. Anderluh and R. Gilbert (eds.), *MACPF/CDC Proteins - Agents of Defence, Attack and Invasion*, Subcellular Biochemistry 80, DOI: 10.1007/978-94-017-8881-6_10,

FHL	Familial hemophagocytic lymphohistiocytosis
GNLY	Granulysin
Gzm	Granzymes
GvHD	Graft-versus-host disease
HGH	Hemophagocytic lymphohistiocytosis
IL2	Interleukin-2
IS	Immune synapse
LCMV	Lymphochoriomeningitis virus
LCR	Locus control region
MAC	Membrane attack complex
MACPF	Membrane attack complex/perforin
MTOC	Microtubule organizing center
NK	Natural killer
PFN	Perforin
SLO	Streptolysin O
SNARE	Soluble N-ethylmaleimide-sensitive factor accessory protein receptor
TCR	T cell receptor

Introduction

Immune cytotoxic (killer) cells protect us from intracellular infection by triggering programmed cell death to eliminate infected cells; they also help protect us, although less effectively, from transformed cancer cells. The main killer cells are natural killer (NK) cells of the innate immune system and $CD8^+$ cytotoxic T lymphocytes (CTL) of adaptive immunity, although some $CD4^+$ T lymphocytes, especially T_H1 and T_{reg} cells, also deploy the specialized cell death-inducing machinery. All killer lymphocytes contain specialized secretory lysosomes, called cytotoxic granules, filled with death-inducing serine proteases, called granzymes (Gzm, "granule enzyme") (Fig. 10.1). When the killer cell recognizes a cell targeted for elimination, the cytotoxic granules move to the immune synapse (IS) formed with the target cell and fuse their membranes with the killer cell plasma membrane, dumping their contents into the IS. This process is called granule exocytosis (Figs. 10.2, 10.3). The cytotoxic granules also contain perforin (PFN), a pore-forming protein encoded by the *PRF1* gene, which has an N-terminal membrane attack complex/perforin (MACPF) domain (reviewed in [101]) that is similar to the pore forming domain of the cholesterol-dependent cytolysins (CDCs) from $Gram^+$ bacteria, such as perfringolysin O [77], and to complement components C6–C9 of the membrane attack complex (MAC) [49, 86, 98] (Fig. 10.4). PFN, which is expressed only in killer lymphocytes, delivers the death-inducing Gzms into the target cell cytoplasm. The Gzm proteases activate cell death by cleaving specific target proteins in the cytoplasm and concentrate within mitochondria and

(a)

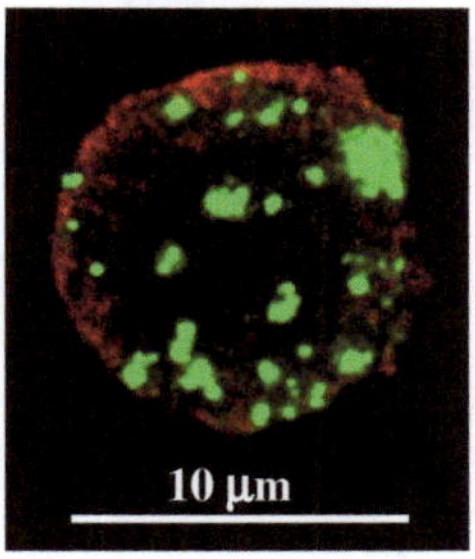

(b)

CTL and NK cell cytotoxic granule contents

Dense core	Function
Perforin	Pore formation; internalization of granzymes
Granzymes	Serine proteases
Granulysin	Microbicidal activity
Calreticulin	Calcium storage and perforin inhibitor
Cathepsin C	Pro-granzyme processing
Serglycin	Proteoglycan matrix
periphery (resident lysosomal proteins)	
Lamp-1	
Lamp-2	Lysosomal membrane proteins
CD63	
H+ ATPase	Lysosomal acidification
Rab27a	
Munc 13-4	Granule exocytosis
VAMP7/VAMP8	
Cathepsin B	Protect CTL against perforin

Fig. 10.1 Key components of cytotoxic granules. **a** Perforin (*green*) is localized within the cytotoxic granules of a Natural Killer (NK) cell. The plasma membrane is stained *red*. **b** The cytolytic effector molecules, perforin, granzymes, and granulysin, are bound in the granule core to the serglycin proteoglycan. Calreticulin in the granule scavenges Ca^{2+} to prevent perforin membrane insertion. Cytotoxic granules also contain molecules found in all lysosomes, such as Lamp1 (CD107a), CD63 and cathepsins, as well as membrane-associated proteins specific to secretory lysosomes, such as vesicle-associated soluble N-ethylmaleimide-sensitive factor accessory complex component (VAMP)7 or VAMP8, Munc13-4, and Rab27a, which are essential for granule exocytosis. Cathepsins B and C play a special role in cytotoxic granules—cathepsin C removes 2 N-terminal amino acids from the pro-granzymes to produce the active enzyme; lysosomal membrane-associated cathepsin B may help protect the killer cell from membrane damage in the immune synapse by cleaving and inactivating perforin. Other cathepsins may substitute for these cathepsins when they are absent or mutated

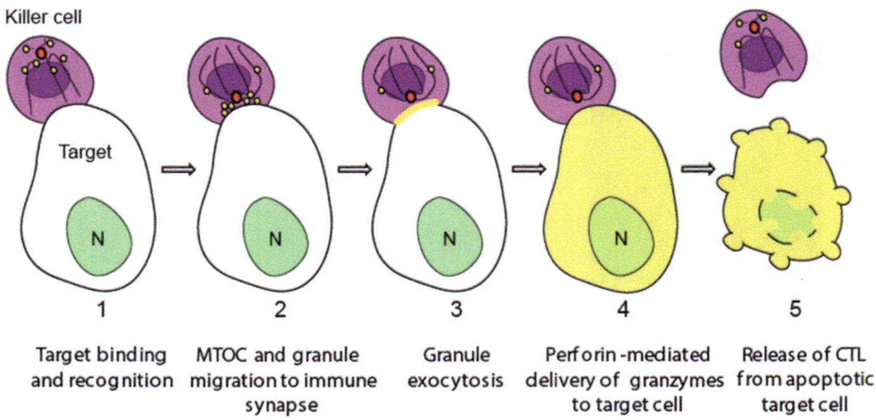

Fig. 10.2 Steps in granule-mediated cytotoxicity. After the killer cell recognizes a target cell (*1*), an immune synapse is formed at the interface and the microtubule organizing center moves to the synapse, reorganizing the microtubule network (*2*). Cytotoxic granules move along microtubules to dock at the killer cell membrane. Granule membranes fuse with the killer cell plasma membrane, releasing their contents (*yellow*) into the immune synapse (*3*). Perforin delivers the granzymes into the cytosol of target cells (*4*) where they initiate apoptotic death (*5*). The granzymes concentrate in the nucleus of target cells. The killer cell then detaches from the dying cell and is free to seek out additional targets

nuclei, where many key Gzm protein substrates reside. In the encounter with a target cell, the killer cell remains unharmed; thus killer cells are *serial* killers that can detach from one target to seek and destroy others. Target cells destroyed by cytotoxic granules die a highly regulated programmed cell death (apoptosis), rather than necrosis. Programmed cell death minimizes inflammation and damage to nearby tissue since target cells and their debris are rapidly removed by immune phagocytes, especially macrophages. Although killer lymphocytes can also activate programmed cell death by recognizing and activating death receptors on target cells, granule-mediated cell death is responsible for immune control of viral and intracellular bacterial infection and cancer.

PFN is the only molecule known to deliver the Gzms into target cells in vivo. Mice genetically deficient in *Prf1* are profoundly immunodeficient, being unable to protect themselves from viruses and prone to develop lymphoma. Humans genetically deficient in functional PFN are also impaired in their ability to handle intracellular infection and can develop an often-fatal inflammatory syndrome due to unresolved infection that can only be treated by bone marrow transplantation (see Chap. 11). Individuals bearing hypomorphic mutations that do not completely eliminate PFN function can be asymptomatic until adulthood and are susceptible to developing lymphoma. Recently a weakly paralogous protein PFN-2 that contains a MACPF domain and is expressed from the *MPEG1* gene mainly in macrophages has also been identified and is hypothesized to also form membrane pores [54]. PFN-2 may play a role in macrophage defense against bacteria.

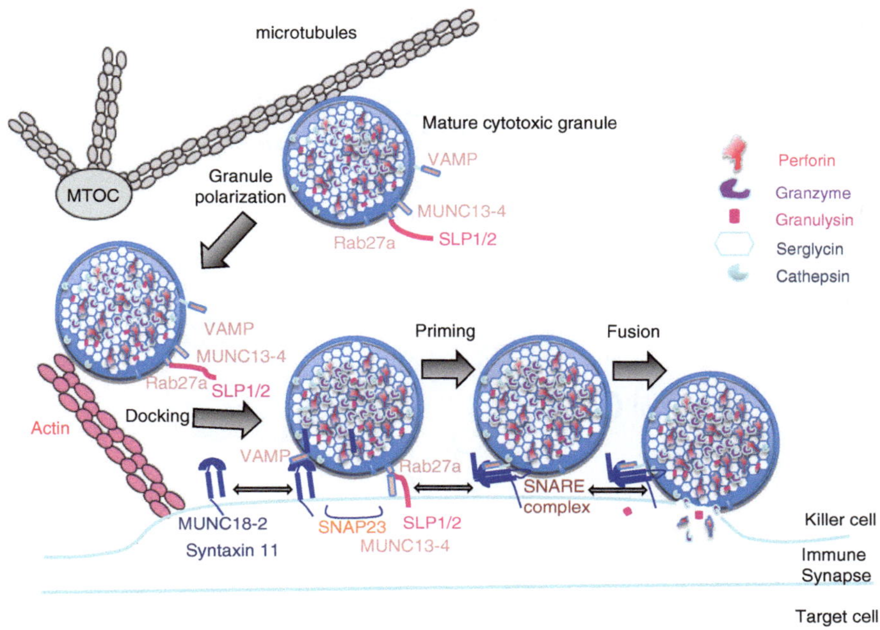

Fig. 10.3 Model of granule exocytosis. In response to antigen recognition, the mature cytotoxic granule moves along microtubules, to dock at the cell membrane at the immune synapse probably with assistance from the actin-myosin cytoskeleton. A cytotoxic granule vesicle-associated soluble N-ethylmaleimide-sensitive factor accessory (SNARE) complex component (VAMP) protein binds to Munc18-2, which is associated with plasma membrane syntaxin 11. Cytotoxic granule proteins Rab27a and Munc13-4, in association with a synaptotagmin SLP1 or SLP2, help anchor the granule to the membrane. A SNARE complex forms between plasma membrane SNAP23 and syntaxin 11 and granule membrane VAMP to initiate fusion of the granule membrane to the plasma membrane. Following membrane fusion, the cytotoxic granule contents are released into the immune synapse. After fusion, granule membrane-associated cathepsin B (*not shown*) is displayed on the killer cell membrane and protects it from perforin membrane damage. Figure adapted from [20]

In this chapter we describe our current understanding of how PFN functions in killer cells. PFN multimerizes to form pores in cholesterol-containing membranes in a Ca^{2+}-dependent manner. The structure of PFN, based on the recent crystallization of monomeric PFN and single-particle cryo-EM reconstruction of PFN pores, is described in Chaps. 4 and 6. Here we discuss where PFN is expressed and how its expression is regulated, the biogenesis and storage of PFN in killer cells and how they are protected from it, how it is released, how it delivers Gzms into target cells and the consequences of PFN deficiency. Although the simplest model for PFN-mediated Gzm delivery into target cells would be through PFN pores in the target cell plasma membrane, this model does not fit the data. Plasma membrane delivery leads to cell membrane damage and rapid target cell death by necrosis, rather than the characteristic slower and non-inflammatory immune-mediated death

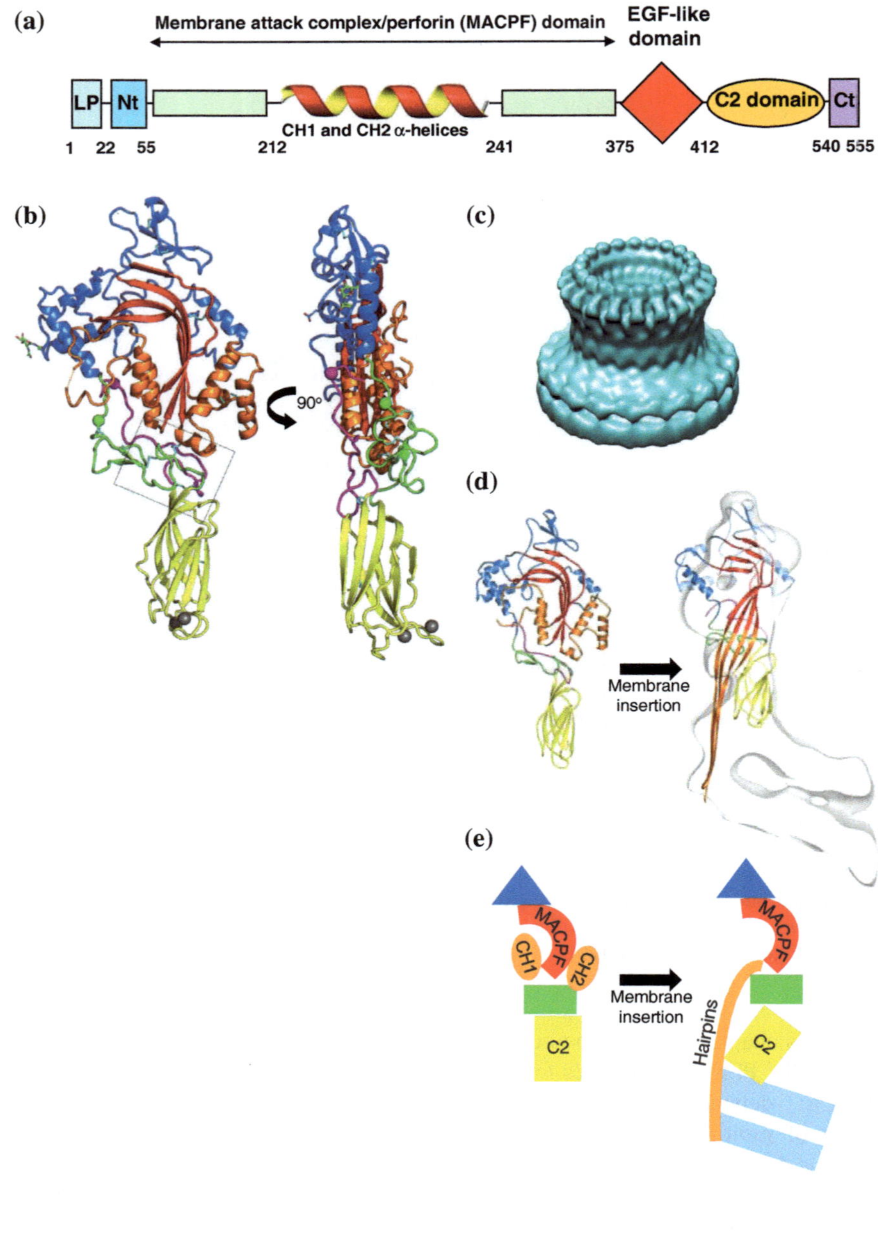
(a)
Membrane attack complex/perforin (MACPF) domain
EGF-like domain
LP
Nt
CH1 and CH2 α-helices
C2 domain
Ct
1
22
55
212
241
375
412
540
555
(b)
90°
(c)
(d)
Membrane insertion
(e)
MACPF
CH1
CH2
C2
Membrane insertion
Hairpins
MACPF
C2

◀ **Fig. 10.4** Perforin structure and insertion in the plasma membrane. **a** The PFN monomer contains an N-terminal domain (Nt) with a leader peptide (LP), followed by a membrane attack complex/perforin (MACPF) domain (composed of complement homology domains joined by two alpha-helical domains). A calcium-binding C2 domain, responsible for membrane binding, followed by a C-terminal peptide (Ct) are linked to the MACPF domain by an epidermal growth factor (EGF) domain. **b** Crystal structure of the PFN monomer, color coded with the MACPF domain in *red*, the CH1 and CH2 helices in *orange*, the EGF domain in *green*, and the C2 domain in *yellow*. The *grey balls* indicate calcium binding to the C2 domain in the crystal structure. **c** Cryoelectron microscopy reconstruction of the large PFN pore. **d**, **e** Model of PFN conformational change induced by membrane binding of the C2 domain to form a multimerized pore, based on the structure of the monomer and the reconstruction of PFN pore densities. Domains in **e** are color coded as in **b**. *Light Blue* in **e** indicates the membrane. **b**–**e** are adapted from [48]

by apoptosis. Here we discuss the experimental basis for an alternate model for Gzm delivery (Fig. 10.5). In this model, PFN causes transient cell membrane damage that mobilizes the target cell to repair the damaged membrane rapidly and remove PFN and Gzms on the membrane by endocytosis. Gzms are then released from endosomes when PFN forms large pores in the endosomal membrane.

Perforin is Only Expressed by Killer Lymphocytes

NK cells and $CD8^+$ T cells are the major classes of PFN-expressing killer lymphocytes. Non-lymphoid cells, B lymphocytes and noncytolytic T cells do not express PFN. Cytotoxicity and *PFN1* expression are tightly regulated and correlated. When T cells are released from the thymus and before they encounter antigen, they are "naïve"—they do not express *Prf1* and are not cytotoxic. About 5–7 days after naïve CD8 T cells encounter antigen recognized by their T cell receptor (TCR), they differentiate into effector cytotoxic cells that express PFN and some of the Gzms [36]. At the same time, they down-regulate adhesive and chemokine receptor molecules that retain them in lymph nodes and acquire receptors that allow them to traffic to tissue sites of infection and tumor invasion. Professional antigen presenting cells (APC), macrophages and dendritic cells, are the most effective cells for activating cytotoxic function. Induction of cytotoxic genes requires not only antigen-receptor activation, but also costimulation, and is greatly enhanced by APCs that have been stimulated by danger- and pathogen-associated pattern recognition receptors or by exogenous inflammatory and anti-viral cytokines, including the Type I interferons, IL-1 and IFN-γ. The combination of these signals leads to production of IL-2 and other related γ_c-dependent cytokines (IL-4, IL-7, IL-9, IL-15 and IL-21), which are needed to activate PFN gene expression. This insures that cytotoxicity is only triggered by *bona fide* antigens on infected cells or cancers and not by "self" antigens that the TCR might recognize. Activated killer $CD8^+$ T cells also begin to express the Fcγ receptor CD16, also present on cytotoxic NK cells, which enables them to recognize and lyse target

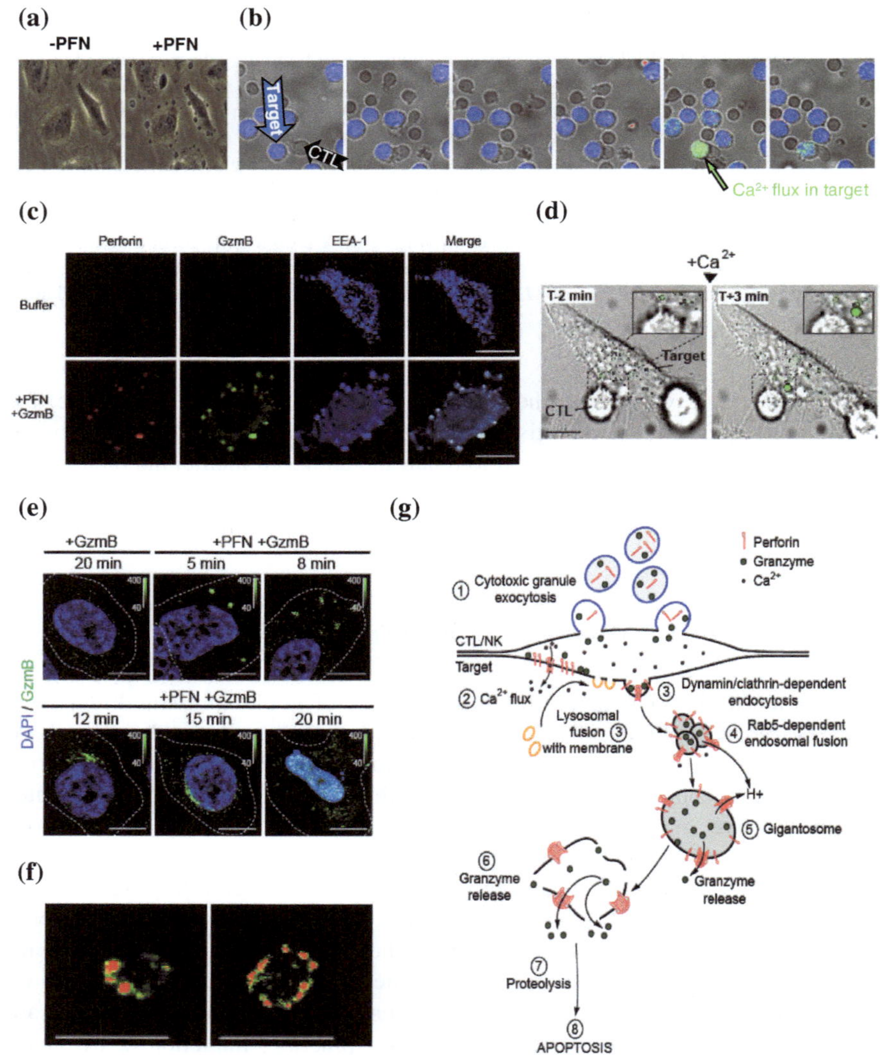
(a)
-PFN
+PFN
(b)
Target
CTL
Ca2+ flux in target
(c)
Perforin
GzmB
EEA-1
Merge
Buffer
+PFN
+GzmB
(d)
+Ca2+
T-2 min
T+3 min
Target
CTL
(e)
+GzmB
+PFN +GzmB
20 min
5 min
8 min
12 min
15 min
20 min
DAPI / GzmB
(f)
(g)
Perforin
Granzyme
Ca2+
1 Cytotoxic granule exocytosis
CTL/NK
Target
2 Ca2+ flux
3 Dynamin/clathrin-dependent endocytosis
Lysosomal fusion 3 with membrane
4 Rab5-dependent endosomal fusion
H+
5 Gigantosome
6 Granzyme release
Granzyme release
7 Proteolysis
8 APOPTOSIS

◀ **Fig. 10.5** Model of perforin delivery of granzymes into the target cell. **a** PFN treatment of HeLa cells causes dramatic membrane perturbation and blebbing. **b** Killer cell degranulation causes a transient calcium influx in target cells that persists for a few minutes. In this experiment from [45] PHA-activated human cytotoxic T lymphocytes (CTL) were incubated with Fura-2-loaded, anti-cluster of differentiation 3 (CD3)-coated U937 cells and images were obtained every 30 s. The Fura-2 indicator dye is *blue* when calcium is low and *green* when it is elevated. **c** PFN and granzyme B are endocytosed into giant EEA-1—staining endosomes (gigantosomes). When HeLa cells are treated with PFN and granzyme B, within 5 min, granzyme B (*green*) concentrates in gigantosomes and is released beginning after about 10 min. Later the released granzyme concentrates in the target cell nucleus (*blue*). **c** is reprinted from [94]. **d** Large EEA-1^+ endosomes (*green*) form in a target cell after CTL degranulation. EGFP-EEA-1 transfected HeLa target cells were incubated with specific CTL in the absence of calcium to allow cell conjugation. After 2 min, $CaCl_2$ was added to induce CTL degranulation. Enlarged endosomes form in the target cell within minutes following CTL degranulation. **e** The addition of PFN to granzyme B causes fluorescent granzyme B (*green*) uptake into gigantosomes. After ~10 min, granzymes are released into cytosol and concentrate after 20 min in the target cell nucleus, stained with DAPI (*blue*). **f** High magnification confocal section of representative gigantosomes stained 7 min after HeLa cell treatment with human PFN showing highly localized PFN staining in clumps on the endosomal membrane. **g** Model for PFN delivery of granzymes. After cytotoxic granule exocytosis into the immunological synapse (*1*), PFN multimerizes in the target-cell membrane to form small pores through which Ca^{2+} enters (*2*), triggering a plasma membrane repair response (*3*) in which lysosomes fuse with the damaged plasma membrane and PFN and granzymes are rapidly internalized by dynamin and clathrin-dependent endocytosis. PFN and granzyme-containing endosomes then fuse together by rapid Rab5-dependent homotypic fusion in response to the transient Ca^{2+} flux (*4*) to form gigantosomes. Within gigantosomes, PFN continues to multimerize to form new and possibly bigger pores, preventing acidification and causing some granzyme release (*5*), before inducing endosomal rupture and complete granzyme release into the target-cell cytoplasm (*6*), where they initiate proteolysis (*7*) leading to programmed cell death (*8*). **d–g** are reprinted from [93]. Scale bars 10 μm (**c–e**) or 5 μm (**f**)

cells that have been coated with antibodies in a process called antibody-dependent cell-mediated cytotoxicity (ADCC). In situations of persistent and extensive antigenic stimulation, however, such as occurs in tumors and chronic viral infection, many effector T cells with surface protein expression of $CD8^+$ no longer express PFN and are not cytotoxic [97, 108, 112]. Effector $CD8^+$ T cells that lack cytotoxicity are termed "exhausted".

PFN is also present in NKT cells (a group of T cells co-expressing a TCR and NK cell surface receptors [63]). Some murine $CD4^+$ T cells, especially T_H1 cells generated by viral infection, also express *Prf1* mRNA, but about 20 times less than activated $CD8^+$ T cells [12, 13]. Effector-like $\gamma\delta$ T cells and human $CD4^+CD25^+$ regulatory T cells [33] also express PFN and can be cytotoxic.

Most effector cells in an immediate immune response die within a few weeks, but some survive and develop into memory cells. Memory cells down-regulate cytotoxic effector proteins, but the kinetics of down-regulation varies with the molecule and with the particularities of the immunostimulatory environment [19, 78]. PFN, as the limiting molecule for cytotoxicity, is down-regulated more rapidly than other effector molecules. Memory $CD8^+$ T cells rapidly reacquire cytotoxic capability within hours of restimulation. The molecular basis for this rapid response is not well understood, although recent studies suggest that in memory $CD8^+$ T cells the chromatin of cytolytic effector gene promoters and of eomesodermin, the master transcription factor that regulates $CD8^+$ T cell effector genes, bear epigenetic marks that poise them for transcription compared to naïve T cells [1, 2, 24, 62, 113]. These cells might also store *Prf1* and *Gzm* mRNAs that can be rapidly translated upon activation.

It takes a week to ten days for naïve $CD8^+$ T cells to proliferate and differentiate into a large population of antigen-specific CTLs. In the meantime, NK cells mediate the immediate response to intracellular infection in individuals that have not been vaccinated or previously exposed. NK activating receptors recognize cell surface changes on tumors, stressed cells and infected cells, such as down-regulation of MHC/HLA molecules or cell surface expression of nonclassical MHC molecules, such as MICA and MICB, that are induced by stress. Freshly minted NK cells (at least in mice, but not studied in humans) do not immediately express Gzms and PFN [26]. They constitutively express mRNAs for *Gzma* and *Gzmb* and *Prf1*, but only have detectable GzmA protein. Hence, they have limited cytotoxicity. However, PFN and Gzms and cytotoxicity are up-regulated rapidly when NK cell activating receptors are engaged. Less differentiated NK cells that highly express the neural cell adhesion molecule NCAM or CD56 and lack CD16 are weakly or not cytotoxic, while more differentiated $CD56^{dim}CD16^+$ NK cells are potent killer cells [100]. Once an NK cell has acquired cytolytic activity, the *PRF1* gene is thought to be constitutively transcribed [79]. In the circulation $CD56^{dim}$ NK cells have about a log more PFN than $CD56^{bright}$ NK cells.

Perforin Gene Regulation

In mammals and marsupials, PFN is encoded by a single-copy gene, but multiple variants have been described in fish and amphibians, most likely as a result of genome duplication [9]. *PRF1* is closely related in sequence to the terminal complement genes in fish and other species. In mammals another more weakly related pore-forming protein, found in macrophages (MPEG-1, 12 % amino acid identity), may have been a *PRF1* ancestor, since it can be traced evolutionarily to sponges [54]. Sequence alignments show a high degree of conservation of *Prfl* genes from different species. This high conservation may be related to its non-redundant function as a target cell membrane damaging agent, essential for cytotoxicity. Despite a 30 % divergence in sequence, mouse and human PFN can functionally substitute for each other.

Both human *PRF1* and mouse *Prfl* genes are located on chromosome 10 and share a common simple structure comprised of three small exons that span 6 kb [70] as well as 5′- and 3′-untranslated regions. The core promoter and sites of transcription initiation have been mapped, as well as cis-acting functional sequences in the proximal region of the *PRF1* gene. The *PRF1* promoter is GC-rich, does not possess a TATA-box and has one major and several minor sites of transcription initiation [50]. The promoter is moderately active and not specific for killer cells. However, basal transcriptional activity is repressed in non-cytotoxic cells by a sequence upstream by −240 bp and an unidentified transcriptional repressor [115]. The two key transcription factors, T-bet (*TBX21*) and eomesodermin (*EOMES*), that belong to the T-box family are the key master regulators of cytotoxic gene expression and survival of committed $CD8^+$ memory cells [31, 67, 90, 107]. After naïve $CD8^+$ T cell activation, T-bet is induced before eomesodermin [18]. Notch signaling and the Runx3 transcription factor upregulate *Eomes*, but also directly upregulate expression of *Prf1* [17, 18]. Mice deficient in both *Tbx21* and *Eomes* genes are unable to control tumors and intracellular infection [5, 40, 117]. They develop a wasting syndrome caused by anomalous differentiation to IL-17-secreting cells, suggesting that these two transcription factors not only positively regulate cytotoxic gene expression and other genes required for CTL survival and function, but also suppress differentiation to alternate lineages. An uncharacterized Ets family transcription factor also supports *PRF1* promoter activity in killer cells [110].

Two enhancers at −15 and −1 kb have also been described. Interleukin-2 (IL2) signalling induces the STAT-5 transcription factor, which stimulates these enhancers, leading to expression of both human *PRF1* and mouse *Prf1* genes [114]. The −1 kb enhancer links *PRF1* gene expression to signalling from other cytokines through STAT3 or STAT4 [109, 111]. Additional transcription factors can also bind to the −15 and −1 kb enhancers including NF-κB, NFAT, Ikaros and AP-1 [116]. However, the presence of these two enhancers is not sufficient for optimal and physiological expression of PFN. PFN expression is controlled by an extended 150 kb cis-regulatory "territory" that includes a locus control region (LCR) regulating the developmental and activation-specific expression of PFN only in T cells and NK cells [69].

Perforin Structure, Mechanism of Membrane Insertion and Pore Formation

Mature PFN is a 533 amino acid protein consisting of three domains: an N-terminal MACPF domain, an intermediate EGF-like domain and a C-terminal Ca^{2+}-binding C2 domain responsible for the initial Ca^{2+}-dependent binding of PFN to membrane surfaces (Fig. 10.4). At high concentrations PFN multimerizes in a cholesterol- and Ca^{2+}-dependent manner in the plasma membrane of cells to form 5–15 nm pores [48, 59, 71, 81, 98]. Recent crosslinking and biophysical studies suggest that PFN may form at least two types of pores in membranes—small unstable pores composed of about seven monomers, and much larger stable pores [72, 93]. Cryoelectron microscopy reconstructions suggest that the large pores are composed of ~19–24 subunits and have a lumen large enough for Gzm monomers or GzmA dimers to pass through easily. Membrane pore formation by PFN can be separated into three stages, each involving structural transitions: membrane binding, multimerization and formation of a transmembrane pore. PFN mainly forms pores in cholesterol-containing membranes. However, it can also less efficiently form pores in cholesterol poor membranes that are found in bacteria, fungi and parasites [72]. It is largely inactive against pathogens, which do not synthesize cholesterol.

The PFN C2 domain, which initiates docking with the lipid membrane in a Ca^{2+}-dependent manner, contains key aspartic acid residues that coordinate binding of up to four Ca^{2+} atoms [105]. The CDCs contain a Thr-Leu motif that binds to cholesterol [25] and causes two nearby hydrophobic loops to insert into the lipid bilayer and anchor the CDC monomer perpendicularly to the membrane surface [74]. However, PFN lacks this motif.

The N-terminal MACPF region of the mature 67 kDa protein (residues 44–410 of the human protein) is homologous to domains in complement proteins C6, C7, C8α, C8β and C9 that form the complement MAC. The crystal structure of monomeric mouse PFN was recently solved [48]. The MACPF domain is similar in structure to that of bacterial pore-forming cholesterol-dependent cytolysins, although they have been hypothesized to insert into membranes in opposite orientations (although this is uncertain [30]). The PFN (and CDC) MACPF domain are composed of a bent and twisted four-stranded β-sheet flanked by clusters of α-helices. During pore formation the α-helices unwind to insert into lipid membranes as amphipathic β-strands, which assemble into a transmembrane β-barrel during multimerization [75, 83]. PFN multimerization involves the association of oppositely charged residues on two "flat faces" of the MACPF domain (R213 and E343) [7]. Between the N-terminal MACPF domain and the C2 domain, a central EGF-like domain lies. Little is known about the function of this domain. However, in the PFN monomer, the EGF-like domain interacts with the α-helices of the MACPF domain and may help fix their orientation to the membrane [48].

Perforin Synthesis and Storage

PFN is translated with a leader sequence that directs it to the endoplasmic reticulum (ER), but because it could potentially cause self-destruction of killer cells, the biosynthesis and storage of PFN is carefully controlled. PFN only forms pores at neutral pH, which means that the neutral, Ca^{2+}-rich ER milieu would be ideal for activating it. The inactivity of PFN in the ER was initially ascribed to its synthesis as an inactive precursor, which is only activated by removal of a carboxy-terminal glycosylated peptide in cytotoxic granules [99]. Removal of this C-terminal peptide was proposed to be necessary for Ca^{2+} to bind to the C2 domain to enable membrane binding. However, this idea is not supported by the solved structure of monomeric PFN in which the C-terminal peptide and the C2 domain are not close enough to interact [48]. More recent data show that PFN is active within the ER, but that glycosylation at 2 sites, in the MACPF domain and in the C-terminal peptide, leads to its rapid trafficking to the Ca^{2+}-poor Golgi and thence to safe storage in acidic cytotoxic granules [11]. A conserved C-terminal tryptophan residue, working through an unknown mechanism, facilitates the rapid transport from the ER to the Golgi [11]. Mutation of the terminal tryptophan leads to enhanced death of the killer cell. In the ER and cytotoxic granules, PFN interacts with calreticulin, an ER chaperone and Ca^{2+}-binding protein, which may also inhibit PFN membrane insertion until after the granules are released [21, 27].

Cytotoxic granules are acidic, electron-dense, specialized secretory lysosomes [68] (Fig. 10.1). These granules are mobilized like secretory vesicles in other secretory cells, such as neurosecretory vesicles near the synapses of neurons and melanin-containing vesicles of melanocytes. The acidic environment of the granules not only inhibits PFN binding to Ca^{2+} [105], but also contributes to PFN stability. Indeed, PFN levels in CTLs treated with concanamycin, an inhibitor of the vacuolar H^{+}-ATPase, which disrupts granule acidification, are dramatically decreased [44]. In addition to PFN, cytotoxic granules contain the Gzms and a cationic pore-forming molecule, granulysin (GNLY), which is homologous to the saposins and selectively active at disrupting negatively charged bacterial and possibly fungal and parasite cell membranes. GNLY is expressed in humans and nonhuman primates and orthologues are found in some other species (pigs, cows and horses), but not in rodents. Like PFN, GNLY expression is restricted to cytotoxic cells. The positively charged cytotoxic effector molecules are bound in the granule to an acidic proteoglycan, called serglycin, after its many Ser-Gly repeats [53, 57], which has been proposed as another mechanism for reducing PFN multimerization within the granules [29]. However, although the killer cells of serglycin-deficient mice store less GzmB, they have normal amounts of PFN and GzmA and have unimpaired cytotoxicity [34].

In addition to these specialized molecules, the cytotoxic granules also contain lysosomal enzymes, the cathepsins, and internal lysosomal membrane proteins, such as CD107 (Lamp1). The outside of the granule membrane binds SNARE (soluble N-ethylmaleimide-sensitive factor accessory protein receptor) proteins, synaptotagmins and Rab GTPases, that regulate vesicular trafficking and cytotoxic granule release. Some of these molecules, including Rab27a and Munc13-4, which are important for granule exocytosis, are only incorporated into cytotoxic granules as they mature by fusion of cytotoxic granules with specialized exocytic vesicles, formed in secretory cells by fusion of late endosomes and recycling endosomes (Fig. 10.3). Some of the granule-associated molecules associate with lysosomes in all cells, while some have a specialized function in killer cells.

During target cell killing, PFN is released from the cytotoxic granules into the IS formed between the target and killer cell (Fig. 10.2, see below). In these conditions, PFN is free to act on the target, but also on the effector cell plasma membrane. How the killer cell membrane is protected from PFN is still not completely clear. The granule membrane protein cathepsin B is incorporated in the killer cell plasma membrane when cytotoxic granules fuse with the killer cell plasma membrane during granule exocytosis. Cathepsin B can cleave PFN and potentially inactivate any PFN redirected toward the killer cell [4]. However, CTLs from mice deficient in cathepsin-B survive target cell encounters in vitro and in vivo [6]. A possible explanation for these seemingly contradictory results could be that other membrane-bound granule cathepsins also inactivate PFN when cathepsin B is absent.

Perforin and Cytotoxic Protein Release into the Immune Synapse

When CTL and NK cells form an IS with a target cell, engagement of activating receptors, including the T cell receptor, killer cell activating receptors, and Fc receptors, activates the killer cell to destroy the target cell (Figs. 10.3, 10.4). Their activation for cytolysis is enhanced by binding of CD8 or CD4, costimulatory receptors and adhesion molecules like LFA-1, which cluster in well-defined concentric rings within the IS. Killer cell activation causes a Ca^{2+} flux that induces lytic granules to cluster around the microtubule organizing center (MTOC) and then align along the immunological synapse [10, 20, 22, 23, 46, 51, 64] (Figs. 10.2, 10.3). Granules move to the immune synapse via both the microtubule network and actin cytoskeleton. Cytotoxic granules then dock to the killer cell plasma membrane in the central region of the immune synapse (c-SMAC). Cytotoxic granule docking is orchestrated by binding of Rab27a on the cytosolic side of the mature granule membrane with synaptotagmin-like proteins, SLP1 or SLP2, which are anchored in the cell membrane. Docked granules are then primed for fusion by the interaction of Munc13-4 on their surface with syntaxin 11 on the killer cell membrane. This triggers the formation of a SNARE complex, the molecular machine for granule membrane fusion, between cytotoxic granule VAMP7 or VAMP8 with syntaxin 11 and SNAP23 on the cell membrane. Granule membrane fusion also requires participation of Munc18-2 to trigger the conformational activation of the SNARE complex. Acid sphingomyelinase may be required for the contraction of docked cytotoxic granules and expulsion of their contents [37]. Once in the presumably pH neutral IS, PFN is probably released from serglycin [53], and can bind Ca^{2+} to become activated for membrane insertion.

How Perforin Delivers Granzymes into Target Cells

PFN was originally hypothesized to form large pores in the target cell plasma membrane that allow Gzms to passively diffuse into the target cell. The electron microscopy images of large PFN pores in cell membranes on which this idea was based were all generated by applying very high concentrations of PFN to cells. At these concentrations, PFN causes necrotic cell death by irreversibly damaging the target cell membrane. However, during killer cell-induced death, the simple plasma membrane pore model does not fit the data (Fig. 10.5). In this cell membrane pore model, Gzms enter the cytosol directly but during killer cell lysis or when cells are treated with the sublytic concentrations of PFN that deliver Gzms into cells to induce apoptosis, Gzms are not detected at first in the cytosol, but instead are initially endocytosed into clathrin-coated vesicles and transported to endosomes [45, 57, 93, 94].

Gzms can bind to the cation-independent mannose-6-phosphate receptor (CI-MPR) or heparin receptors on target cells [60], but cells lacking these receptors are killed just as efficiently [47, 96]. Because the cell membrane is negatively charged and the Gzms are very basic (pI $\sim$10), the Gzms bind to cell surfaces independently of any receptor via electrostatic interactions [8, 84]. Receptor independence insures that all cells can be targeted and viruses or cancer cells cannot evade immune surveillance by down-regulating a cellular receptor. In the absence of PFN, bound Gzms are inefficiently endocytosed, but because they do not escape from endocytic compartments, there is no cell death without PFN [14, 28, 85].

We found that PFN indeed forms plasma membrane pores in the target cell, but these pores are small and transient and only allow small dyes to begin to enter the cytosol before the damaged area is walled off. However, Ca^{2+} flows into the target cell through these pores and remains elevated for a few minutes. Because intracellular Ca^{2+} is low in cells with an intact cell membrane, the cell senses a Ca^{2+} influx as a sign that the plasma membrane has been breached. The elevated Ca^{2+} triggers a rapid cellular membrane damage response (also known as cellular "wound-healing" [76]) in which intracellular vesicles move to the plasma membrane and fuse with it to patch holes, removing and internalizing into endosomes any damaged membrane [39, 55, 92, 93]. Some of the damaged membrane may also be removed by blebbing. When the membrane repair response is inhibited then because the cell membrane remains leaky, target cells die by necrosis instead of by the slower, regulated, noninflammatory and energy-dependent apoptosis. Normally, however, membrane wound healing immediately activates endocytosis leading to the rapid internalization of the membrane-bound granzymes, granulysin and PFN. Elevated cytosolic Ca^{2+} activates endosomal fusion, and the resulting granzyme- and PFN-containing endosomes fuse to form giant endosomes $\sim$10 times larger than normal that have been termed gigantosomes. In the endosomal membrane, PFN forms larger and more stable pores through which Gzms begin to leak out into the cytosol. The leakiness of the endosome prevents its acidification, allowing PFN to remain active, and about 10–15 min after cell death has been triggered the gigantosomes become unstable and rupture, releasing any remaining cargo to the cytosol where they activate programmed cell death. Although endosomal uptake, gigantosome formation, rupture and Gzm release have been visualized in cells treated with sublytic PFN and Gzms and during T cell and NK cell-mediated killing, this model is controversial and some researchers still think that the original plasma membrane pore model is correct [52].

Bacterial and viral endosomolysins can substitute for PFN in vitro (and are widely used as laboratory reagents for intracellular delivery [14]) and potentially might play a similar role in vivo in limited circumstances. A similar repair of injured plasma membrane by rapid Ca^{2+}-dependent endocytosis has also been demonstrated with the pore forming protein streptolysin O (SLO) [39]. The release of lysosomal acid sphingomyelinase (ASM) by SLO-wounded cells promotes endocytosis and membrane lesion removal [91]. When ASM is released from the injured cell, it converts sphingomyelin in the outer leaflet of the plasma membrane to ceramide [38]. PFN also induces the formation of large endosome-like invaginations in artificial liposomes [73].

The idea that PFN activity at the plasma membrane leads to pores that are either too small or too rapidly removed to deliver Gzms directly to the cytosol, but that PFN pores formed in endosomal membranes are functional for Gzm delivery to the cytosol is consistent with a recent study of PFN pore conductance and cryo-electron microcroscopy in planar lipid bilayers and unilamellar vesicles of different lipid compositions and sizes [72]. In this study two different membrane-bound PFN conformations were observed, which were interpreted as pre-pore and pore states of the protein. Small, highly unstable pores preceded the development of stably open and larger pores that retain a size distribution [72]. In another study, PFN monomers formed at the plasma membrane included arc-like structures, representing incomplete PFN pores [58]. These studies suggest that the rapid membrane repair response interferes with the formation of larger pores on the plasma membrane, restricting Gzm entry, but that PFN multimerizes into larger stable pores on the gigantosome membrane, where the membrane repair response doesn't operate.

Perforin Deficiency and Disease

Prf1 knockout mice are severely impaired in their defence against viruses and immune surveillance of tumors [3, 35, 42, 43, 87, 106]. Moreover, in response to lymphochoriomeningitis virus (LCMV), *Prf1*$^{-/-}$ mice develop a hemophagocytic lymphohistiocytosis (HLH) syndrome that resembles the syndrome associated with *PRF1* insufficiency in humans (also called familial hemophagocytic lymphohistiocytosis (FHL) [41, 88]). *Prf1*$^{-/-}$ mice are also less able to clear infection with the intracellular bacterium *Listeria monocytogenes* [42]. They are more susceptible to chemically or oncogenically induced tumors, including 3-methylcholanthrene-induced fibrosarcoma [82] and oncogene-driven mammary adenocarcinoma [89]. As they age, *Prf1*$^{-/-}$ mice also spontaneously develop highly aggressive disseminated B-cell lymphomas [87]. In a large cohort ($n > 800$) of aging *Prf1*$^{-/-}$ mice, additional tumors developed including thymic lymphomas, sarcomas, and lung adenocarcinoma. The severe immune consequences of *Prf1* deficiency can be contrasted by the mild phenotypes associated with Gzm deficiency, likely because the Gzms have redundant functions. Mice deficient in any one of the 10 Gzms, or even of the 5 Gzms in the Gzm B cluster, only have subtle differences compared to wild-type animals. Requirements for a single Gzm have been shown in some cases by specific immune challenges. For example, Gzm A-deficient mice are more susceptible to the poxvirus ectromelia [61] and Gzm B-deficient mice have reduced graft-versus-host disease (GvHD) [32]. No human clinical syndromes have been identified with gene mutations of the 5 human Gzm genes.

In humans, *PRF1* bi-allelic mutations are extremely rare, but have been identified in 30–60 % of young children suffering from a rare autosomal recessive disorder called FHL (FHL2 or Type 2 FHL) [88] that is often fatal if not treated with bone marrow transplantation. Mutations that interfere with cytotoxic granule exocytosis (*UNC13D* mutations encoding for Munc13-4 (FHL3); *STX11* mutations encoding for syntaxin-11 (FHL4) and *STXBP2* mutations encoding for Munc18-2 (FHL5)) cause the same syndrome. The genetic defect in some cases has not been identified (FHL1). Killer cell cytotoxicity is profoundly impaired in these patients. FHL is an immune homeostasis disorder characterized by uncontrolled activation and proliferation of $CD4^+$ and $CD8^+$ T cells and macrophage activation [101], which is likely secondary to chronic unresolved viral infection, especially of herpesviruses. Macrophage activation is driven by excessive production of IFN-γ by activated $CD8^+$ T cells [65, 66]. In these patients, activated macrophages in the spleen and bone marrow phagocytose erythrocytes, leukocytes, and platelets, resulting in severe pancytopenia and anemia. Moreover, the uncontrolled secretion of inflammatory cytokines (IL-1, IL-6, and TNF-α) by these activated macrophages results in severe fever and other symptoms.

Sequencing of *PRF1* mutations in FHL patients has identified nonsense, frameshift and missense mutations that disrupt PFN synthesis, folding or activity [102, 104]. Some patients with hypomorphic bi-allelic mutations that affect PFN folding or stability have milder disease that is not diagnosed until adulthood. These patients are prone to develop leukemia and lymphoma [16]. The A91 V allele (the most common *PRF1* genetic variant, found in 3–17 % of Caucasians) leads primarily to misfolding of PFN rather than complete loss of function [95, 105]. The killer cells of patients with A91 V mono or bi-allelic mutations have reduced cytolytic activity [103]. A91 V mono or bi-allelic mutations predispose to acute lymphoblastic leukemia [80], anaplastic large cell lymphoma [15] and BCR-ABL^+ acute lymphoblastic leukemia [56].

References

1. Araki Y, Fann M, Wersto R, Weng NP (2008) Histone acetylation facilitates rapid and robust memory CD8 T cell response through differential expression of effector molecules (eomesodermin and its targets: perforin and granzyme B). J Immunol 180:8102–8108
2. Araki Y, Wang Z, Zang C, Wood WH 3rd, Schones D, Cui K, Roh TY, Lhotsky B, Wersto RP, Peng W, Becker KG, Zhao K, Weng NP (2009) Genome-wide analysis of histone methylation reveals chromatin state-based regulation of gene transcription and function of memory CD8+ T cells. Immunity 30:912–925
3. Badovinac VP, Hamilton SE, Harty JT (2003) Viral infection results in massive CD8+ T cell expansion and mortality in vaccinated perforin-deficient mice. Immunity 18:463–474
4. Balaji KN, Schaschke N, Machleidt W, Catalfamo M, Henkart PA (2002) Surface cathepsin B protects cytotoxic lymphocytes from self-destruction after degranulation. J Exp Med 196:493–503

5. Banerjee A, Gordon SM, Intlekofer AM, Paley MA, Mooney EC, Lindsten T, Wherry EJ, Reiner SL (2010) Cutting edge: the transcription factor eomesodermin enables CD8+ T cells to compete for the memory cell niche. J Immunol 185:4988–4992
6. Baran K, Ciccone A, Peters C, Yagita H, Bird PI, Villadangos JA, Trapani JA (2006) Cytotoxic T lymphocytes from cathepsin B-deficient mice survive normally in vitro and in vivo after encountering and killing target cells. J Biol Chem 281:30485–30491
7. Baran K, Dunstone M, Chia J, Ciccone A, Browne KA, Clarke CJ, Lukoyanova N, Saibil H, Whisstock JC, Voskoboinik I, Trapani JA (2009) The molecular basis for perforin oligomerization and transmembrane pore assembly. Immunity 30:684–695
8. Bird CH, Sun J, Ung K, Karambalis D, Whisstock JC, Trapani JA, Bird PI (2005) Cationic sites on granzyme B contribute to cytotoxicity by promoting its uptake into target cells. Mol Cell Biol 25:7854–7867
9. Bischofberger M, Gonzalez MR, van der Goot FG (2009) Membrane injury by pore-forming proteins. Curr Opin Cell Biol 21:589–595
10. Bossi G, Booth S, Clark R, Davis EG, Liesner R, Richards K, Starcevic M, Stinchcombe J, Trambas C, Dell'Angelica EC, Griffiths GM (2005) Normal lytic granule secretion by cytotoxic T lymphocytes deficient in BLOC-1, -2 and -3 and myosins Va, VIIa and XV. Traffic 6:243–251
11. Brennan AJ, Chia J, Browne KA, Ciccone A, Ellis S, Lopez JA, Susanto O, Verschoor S, Yagita H, Whisstock JC, Trapani JA, Voskoboinik I (2011) Protection from endogenous perforin: glycans and the C terminus regulate exocytic trafficking in cytotoxic lymphocytes. Immunity 34:879–892
12. Brown DM, Dilzer AM, Meents DL, Swain SL (2006) CD4 T cell-mediated protection from lethal influenza: perforin and antibody-mediated mechanisms give a one-two punch. J Immunol 177:2888–2898
13. Brown DM, Kamperschroer C, Dilzer AM, Roberts DM, Swain SL (2009) IL-2 and antigen dose differentially regulate perforin- and FasL-mediated cytolytic activity in antigen specific CD4 + T cells. Cell Immunol 257:69–79
14. Browne KA, Blink E, Sutton VR, Froelich CJ, Jans DA, Trapani JA (1999) Cytosolic delivery of granzyme B by bacterial toxins: evidence that endosomal disruption, in addition to transmembrane pore formation, is an important function of perforin. Mol Cell Biol 19:8604–8615
15. Cannella S, Santoro A, Bruno G, Pillon M, Mussolin L, Mangili G, Rosolen A, Arico M (2007) Germline mutations of the perforin gene are a frequent occurrence in childhood anaplastic large cell lymphoma. Cancer 109:2566–2571
16. Chia J, Yeo KP, Whisstock JC, Dunstone MA, Trapani JA, Voskoboinik I (2009) Temperature sensitivity of human perforin mutants unmasks subtotal loss of cytotoxicity, delayed FHL, and a predisposition to cancer. Proc Natl Acad Sci USA 106:9809–9814
17. Cho OH, Shin HM, Miele L, Golde TE, Fauq A, Minter LM, Osborne BA (2009) Notch regulates cytolytic effector function in CD8+ T cells. J Immunol 182:3380–3389
18. Cruz-Guilloty F, Pipkin ME, Djuretic IM, Levanon D, Lotem J, Lichtenheld MG, Groner Y, Rao A (2009) Runx3 and T-box proteins cooperate to establish the transcriptional program of effector CTLs. J Exp Med 206:51–59
19. Cui W, Kaech SM (2010) Generation of effector CD8+ T cells and their conversion to memory T cells. Immunol Rev 236:151–166
20. de Saint Basile G, Menasche G, Fischer A (2010) Molecular mechanisms of biogenesis and exocytosis of cytotoxic granules. Nat Rev Immunol 10:568–579
21. Dupuis M, Schaerer E, Krause KH, Tschopp J (1993) The calcium-binding protein calreticulin is a major constituent of lytic granules in cytolytic T lymphocytes. J Exp Med 177:1–7
22. Dustin ML (2009) The cellular context of T cell signaling. Immunity 30:482–492
23. Dustin ML, Long EO (2010) Cytotoxic immunological synapses. Immunol Rev 235:24–34
24. Fann M, Godlove JM, Catalfamo M, Wood WH 3rd, Chrest FJ, Chun N, Granger L, Wersto R, Madara K, Becker K, Henkart PA, Weng NP (2006) Histone acetylation is associated

with differential gene expression in the rapid and robust memory CD8(+) T-cell response. Blood 108:3363–3370
25. Farrand AJ, LaChapelle S, Hotze EM, Johnson AE, Tweten RK (2010) Only two amino acids are essential for cytolytic toxin recognition of cholesterol at the membrane surface. Proc Natl Acad Sci USA 107:4341–4346
26. Fehniger TA, Cai SF, Cao X, Bredemeyer AJ, Presti RM, French AR, Ley TJ (2007) Acquisition of murine NK cell cytotoxicity requires the translation of a pre-existing pool of granzyme B and perforin mRNAs. Immunity 26:798–811
27. Fraser SA, Karimi R, Michalak M, Hudig D (2000) Perforin lytic activity is controlled by calreticulin. J Immunol 164:4150–4155
28. Froelich CJ, Orth K, Turbov J, Seth P, Gottlieb R, Babior B, Shah GM, Bleackley RC, Dixit VM, Hanna W (1996) New paradigm for lymphocyte granule-mediated cytotoxicity. Target cells bind and internalize granzyme B, but an endosomolytic agent is necessary for cytosolic delivery and subsequent apoptosis. J Biol Chem 271:29073–29079
29. Galvin JP, Spaeny-Dekking LH, Wang B, Seth P, Hack CE, Froelich CJ (1999) Apoptosis induced by granzyme B-glycosaminoglycan complexes: implications for granule-mediated apoptosis in vivo. J Immunol 162:5345–5350
30. Gilbert RJ, Mikelj M, Dalla Serra M, Froelich CJ, Anderluh G (2013) Effects of MACPF/CDC proteins on lipid membranes. Cell Mol Life Sci 70:2083–2098
31. Glimcher LH, Townsend MJ, Sullivan BM, Lord GM (2004) Recent developments in the transcriptional regulation of cytolytic effector cells. Nat Rev Immunol 4:900–911
32. Graubert TA, Russell JH, Ley TJ (1996) The role of granzyme B in murine models of acute graft-versus-host disease and graft rejection. Blood 87:1232–1237
33. Grossman WJ, Verbsky JW, Barchet W, Colonna M, Atkinson JP, Ley TJ (2004) Human T regulatory cells can use the perforin pathway to cause autologous target cell death. Immunity 21:589–601
34. Grujic M, Braga T, Lukinius A, Eloranta ML, Knight SD, Pejler G, Abrink M (2005) Serglycin-deficient cytotoxic T lymphocytes display defective secretory granule maturation and granzyme B storage. J Biol Chem 280:33411–33418
35. Gupta M, Greer P, Mahanty S, Shieh WJ, Zaki SR, Ahmed R, Rollin PE (2005) CD8-mediated protection against Ebola virus infection is perforin dependent. J Immunol 174:4198–4202
36. Hamann D, Baars PA, Rep MH, Hooibrink B, Kerkhof-Garde SR, Klein MR, van Lier RA (1997) Phenotypic and functional separation of memory and effector human CD8+ T cells. J Exp Med 186:1407–1418
37. Herz J, Pardo J, Kashkar H, Schramm M, Kuzmenkina E, Bos E, Wiegmann K, Wallich R, Peters PJ, Herzig S, Schmelzer E, Kronke M, Simon MM, Utermohlen O (2009) Acid sphingomyelinase is a key regulator of cytotoxic granule secretion by primary T lymphocytes. Nat Immunol 10:761–768
38. Holopainen JM, Angelova MI, Kinnunen PK (2000) Vectorial budding of vesicles by asymmetrical enzymatic formation of ceramide in giant liposomes. Biophys J 78:830–838
39. Idone V, Tam C, Goss JW, Toomre D, Pypaert M, Andrews NW (2008) Repair of injured plasma membrane by rapid Ca2+-dependent endocytosis. J Cell Biol 180:905–914
40. Intlekofer AM, Banerjee A, Takemoto N, Gordon SM, Dejong CS, Shin H, Hunter CA, Wherry EJ, Lindsten T, Reiner SL (2008) Anomalous type 17 response to viral infection by CD8+ T cells lacking T-bet and eomesodermin. Science 321:408–411
41. Jordan MB, Hildeman D, Kappler J, Marrack P (2004) An animal model of hemophagocytic lymphohistiocytosis (HLH): CD8+ T cells and interferon gamma are essential for the disorder. Blood 104:735–743
42. Kagi D, Ledermann B, Burki K, Hengartner H, Zinkernagel RM (1994) CD8+ T cell-mediated protection against an intracellular bacterium by perforin-dependent cytotoxicity. Eur J Immunol 24:3068–3072

43. Kagi D, Ledermann B, Burki K, Seiler P, Odermatt B, Olsen KJ, Podack ER, Zinkernagel RM, Hengartner H (1994) Cytotoxicity mediated by T cells and natural killer cells is greatly impaired in perforin-deficient mice. Nature 369:31–37
44. Kataoka T, Togashi K, Takayama H, Takaku K, Nagai K (1997) Inactivation and proteolytic degradation of perforin within lytic granules upon neutralization of acidic pH. Immunology 91:493–500
45. Keefe D, Shi L, Feske S, Massol R, Navarro F, Kirchhausen T, Lieberman J (2005) Perforin triggers a plasma membrane-repair response that facilitates CTL induction of apoptosis. Immunity 23:249–262
46. Krzewski K, Coligan JE (2012) Human NK cell lytic granules and regulation of their exocytosis. Front Immunol 3:335
47. Kurschus FC, Bruno R, Fellows E, Falk CS, Jenne DE (2005) Membrane receptors are not required to deliver granzyme B during killer cell attack. Blood 105:2049–2058
48. Law RH, Lukoyanova N, Voskoboinik I, Caradoc-Davies TT, Baran K, Dunstone MA, D'Angelo ME, Orlova EV, Coulibaly F, Verschoor S, Browne KA, Ciccone A, Kuiper MJ, Bird PI, Trapani JA, Saibil HR, Whisstock JC (2010) The structural basis for membrane binding and pore formation by lymphocyte perforin. Nature 468:447–451
49. Lichtenheld MG, Olsen KJ, Lu P, Lowrey DM, Hameed A, Hengartner H, Podack ER (1988) Structure and function of human perforin. Nature 335:448–451
50. Lichtenheld MG, Podack ER (1992) Structure and function of the murine perforin promoter and upstream region. Reciprocal gene activation or silencing in perforin positive and negative cells. J Immunol 149:2619–2626
51. Liu D, Bryceson YT, Meckel T, Vasiliver-Shamis G, Dustin ML, Long EO (2009) Integrin-dependent organization and bidirectional vesicular traffic at cytotoxic immune synapses. Immunity 31:99–109
52. Lopez JA, Susanto O, Jenkins MR, Lukoyanova N, Sutton VR, Law RH, Johnston A, Bird CH, Bird PI, Whisstock JC, Trapani JA, Saibil HR, Voskoboinik I (2013) Perforin forms transient pores on the target cell plasma membrane to facilitate rapid access of granzymes during killer cell attack. Blood 121:2659–2668
53. Masson D, Peters PJ, Geuze HJ, Borst J, Tschopp J (1990) Interaction of chondroitin sulfate with perforin and granzymes of cytolytic T-cells is dependent on pH. Biochemistry 29:11229–11235
54. McCormack R, de Armas LR, Shiratsuchi M, Ramos JE, Podack ER (2013) Inhibition of intracellular bacterial replication in fibroblasts is dependent on the perforin-like protein (perforin-2) encoded by macrophage-expressed gene 1. J Innate Immun 5:185–194
55. McNeil PL, Kirchhausen T (2005) An emergency response team for membrane repair. Nat Rev Mol Cell Biol 6:499–505
56. Mehta PA, Davies SM, Kumar A, Devidas M, Lee S, Zamzow T, Elliott J, Villanueva J, Pullen J, Zewge Y, Filipovich A (2006) Perforin polymorphism A91 V and susceptibility to B-precursor childhood acute lymphoblastic leukemia: a report from the Children's Oncology Group. Leukemia 20:1539–1541
57. Metkar SS, Wang B, Aguilar-Santelises M, Raja SM, Uhlin-Hansen L, Podack E, Trapani JA, Froelich CJ (2002) Cytotoxic cell granule-mediated apoptosis: perforin delivers granzyme B-serglycin complexes into target cells without plasma membrane pore formation. Immunity 16:417–428
58. Metkar SS, Wang B, Catalan E, Anderluh G, Gilbert RJ, Pardo J, Froelich CJ (2011) Perforin rapidly induces plasma membrane phospholipid flip-flop. PLoS ONE 6:e24286
59. Millard PJ, Henkart MP, Reynolds CW, Henkart PA (1984) Purification and properties of cytoplasmic granules from cytotoxic rat LGL tumors. J Immunol 132:3197–3204
60. Motyka B, Korbutt G, Pinkoski MJ, Heibein JA, Caputo A, Hobman M, Barry M, Shostak I, Sawchuk T, Holmes CF, Gauldie J, Bleackley RC (2000) Mannose 6-phosphate/insulin-like growth factor II receptor is a death receptor for granzyme B during cytotoxic T cell-induced apoptosis. Cell 103:491–500

61. Mullbacher A, Ebnet K, Blanden RV, Hla RT, Stehle T, Museteanu C, Simon MM (1996) Granzyme A is critical for recovery of mice from infection with the natural cytopathic viral pathogen, ectromelia. Proc Natl Acad Sci USA 93:5783–5787
62. Northrop JK, Wells AD, Shen H (2008) Cutting edge: chromatin remodeling as a molecular basis for the enhanced functionality of memory CD8 T cells. J Immunol 181:865–868
63. Ohkawa T, Seki S, Dobashi H, Koike Y, Habu Y, Ami K, Hiraide H, Sekine I (2001) Systematic characterization of human CD8+ T cells with natural killer cell markers in comparison with natural killer cells and normal CD8+ T cells. Immunology 103:281–290
64. Orange JS (2007) The lytic NK cell immunological synapse and sequential steps in its formation. Adv Exp Med Biol 601:225–233
65. Pachlopnik Schmid J, Cote M, Menager MM, Burgess A, Nehme N, Menasche G, Fischer A, de Saint Basile G (2010) Inherited defects in lymphocyte cytotoxic activity. Immunol Rev 235:10–23
66. Pachlopnik Schmid J, Ho CH, Chretien F, Lefebvre JM, Pivert G, Kosco-Vilbois M, Ferlin W, Geissmann F, Fischer A, de Saint Basile G (2009) Neutralization of IFNgamma defeats haemophagocytosis in LCMV-infected perforin- and Rab27a-deficient mice. EMBO Mol Med 1:112–124
67. Pearce EL, Mullen AC, Martins GA, Krawczyk CM, Hutchins AS, Zediak VP, Banica M, DiCioccio CB, Gross DA, Mao CA, Shen H, Cereb N, Yang SY, Lindsten T, Rossant J, Hunter CA, Reiner SL (2003) Control of effector CD8+ T cell function by the transcription factor Eomesodermin. Science 302:1041–1043
68. Peters PJ, Borst J, Oorschot V, Fukuda M, Krahenbuhl O, Tschopp J, Slot JW, Geuze HJ (1991) Cytotoxic T lymphocyte granules are secretory lysosomes, containing both perforin and granzymes. J Exp Med 173:1099–1109
69. Pipkin ME, Ljutic B, Cruz-Guilloty F, Nouzova M, Rao A, Zuniga-Pflucker JC, Lichtenheld MG (2007) Chromosome transfer activates and delineates a locus control region for perforin. Immunity 26:29–41
70. Pipkin ME, Rao A, Lichtenheld MG (2010) The transcriptional control of the perforin locus. Immunol Rev 235:55–72
71. Podack ER, Young JD, Cohn ZA (1985) Isolation and biochemical and functional characterization of perforin 1 from cytolytic T-cell granules. Proc Natl Acad Sci USA 82:8629–8633
72. Praper T, Sonnen A, Viero G, Kladnik A, Froelich CJ, Anderluh G, Dalla Serra M, Gilbert RJ (2011) Human perforin employs different avenues to damage membranes. J Biol Chem 286:2946–2955
73. Praper T, Sonnen AF, Kladnik A, Andrighetti AO, Viero G, Morris KJ, Volpi E, Lunelli L, Dalla Serra M, Froelich CJ, Gilbert RJ, Anderluh G (2011) Perforin activity at membranes leads to invaginations and vesicle formation. Proc Natl Acad Sci USA 108:21016–21021
74. Ramachandran R, Heuck AP, Tweten RK, Johnson AE (2002) Structural insights into the membrane-anchoring mechanism of a cholesterol-dependent cytolysin. Nat Struct Biol 9:823–827
75. Ramachandran R, Tweten RK, Johnson AE (2004) Membrane-dependent conformational changes initiate cholesterol-dependent cytolysin oligomerization and intersubunit beta-strand alignment. Nat Struct Mol Biol 11:697–705
76. Reddy A, Caler EV, Andrews NW (2001) Plasma membrane repair is mediated by Ca(2+)-regulated exocytosis of lysosomes. Cell 106:157–169
77. Rosado CJ, Buckle AM, Law RH, Butcher RE, Kan WT, Bird CH, Ung K, Browne KA, Baran K, Bashtannyk-Puhalovich TA, Faux NG, Wong W, Porter CJ, Pike RN, Ellisdon AM, Pearce MC, Bottomley SP, Emsley J, Smith AI, Rossjohn J, Hartland EL, Voskoboinik I, Trapani JA, Bird PI, Dunstone MA, Whisstock JC (2007) A common fold mediates vertebrate defense and bacterial attack. Science 317:1548–1551
78. Rutishauser RL, Kaech SM (2010) Generating diversity: transcriptional regulation of effector and memory CD8 T-cell differentiation. Immunol Rev 235:219–233

79. Salcedo TW, Azzoni L, Wolf SF, Perussia B (1993) Modulation of perforin and granzyme messenger RNA expression in human natural killer cells. J Immunol 151:2511–2520
80. Santoro A, Cannella S, Trizzino A, Lo Nigro L, Corsello G, Arico M (2005) A single amino acid change A91 V in perforin: a novel, frequent predisposing factor to childhood acute lymphoblastic leukemia? Haematologica 90:697–698
81. Sauer H, Pratsch L, Tschopp J, Bhakdi S, Peters R (1991) Functional size of complement and perforin pores compared by confocal laser scanning microscopy and fluorescence microphotolysis. Biochim Biophys Acta 1063:137–146
82. Schreiber TH, Podack ER (2009) A critical analysis of the tumour immunosurveillance controversy for 3-MCA-induced sarcomas. Br J Cancer 101:381–386
83. Shepard LA, Heuck AP, Hamman BD, Rossjohn J, Parker MW, Ryan KR, Johnson AE, Tweten RK (1998) Identification of a membrane-spanning domain of the thiol-activated pore-forming toxin Clostridium perfringens perfringolysin O: an alpha-helical to beta-sheet transition identified by fluorescence spectroscopy. Biochemistry 37:14563–14574
84. Shi L, Keefe D, Durand E, Feng H, Zhang D, Lieberman J (2005) Granzyme B binds to target cells mostly by charge and must be added at the same time as perforin to trigger apoptosis. J Immunol 174:5456–5461
85. Shi L, Mai S, Israels S, Browne K, Trapani JA, Greenberg AH (1997) Granzyme B (GraB) autonomously crosses the cell membrane and perforin initiates apoptosis and GraB nuclear localization. J Exp Med 185:855–866
86. Shinkai Y, Takio K, Okumura K (1988) Homology of perforin to the ninth component of complement (C9). Nature 334:525–527
87. Smyth MJ, Thia KY, Street SE, MacGregor D, Godfrey DI, Trapani JA (2000) Perforin-mediated cytotoxicity is critical for surveillance of spontaneous lymphoma. J Exp Med 192:755–760
88. Stepp SE, Dufourcq-Lagelouse R, Le Deist F, Bhawan S, Certain S, Mathew PA, Henter JI, Bennett M, Fischer A, de Saint Basile G, Kumar V (1999) Perforin gene defects in familial hemophagocytic lymphohistiocytosis. Science 286:1957–1959
89. Street SE, Zerafa N, Iezzi M, Westwood JA, Stagg J, Musiani P, Smyth MJ (2007) Host perforin reduces tumor number but does not increase survival in oncogene-driven mammary adenocarcinoma. Cancer Res 67:5454–5460
90. Szabo SJ, Kim ST, Costa GL, Zhang X, Fathman CG, Glimcher LH (2000) A novel transcription factor, T-bet, directs Th1 lineage commitment. Cell 100:655–669
91. Tam C, Idone V, Devlin C, Fernandes MC, Flannery A, He X, Schuchman E, Tabas I, Andrews NW (2010) Exocytosis of acid sphingomyelinase by wounded cells promotes endocytosis and plasma membrane repair. J Cell Biol 189:1027–1038
92. Terasaki M, Miyake K, McNeil PL (1997) Large plasma membrane disruptions are rapidly resealed by Ca2+-dependent vesicle-vesicle fusion events. J Cell Biol 139:63–74
93. Thiery J, Keefe D, Boulant S, Boucrot E, Walch M, Martinvalet D, Goping IS, Bleackley RC, Kirchhausen T, Lieberman J (2011) Perforin pores in the endosomal membrane trigger the release of endocytosed granzyme B into the cytosol of target cells. Nat Immunol 12:770–777
94. Thiery J, Keefe D, Saffarian S, Martinvalet D, Walch M, Boucrot E, Kirchhausen T, Lieberman J (2010) Perforin activates clathrin- and dynamin-dependent endocytosis, which is required for plasma membrane repair and delivery of granzyme B for granzyme-mediated apoptosis. Blood 115:1582–1593
95. Trambas C, Gallo F, Pende D, Marcenaro S, Moretta L, De Fusco C, Santoro A, Notarangelo L, Arico M, Griffiths GM (2005) A single amino acid change, A91 V, leads to conformational changes that can impair processing to the active form of perforin. Blood 106:932–937
96. Trapani JA, Sutton VR, Thia KY, Li YQ, Froelich CJ, Jans DA, Sandrin MS, Browne KA (2003) A clathrin/dynamin- and mannose-6-phosphate receptor-independent pathway for granzyme B-induced cell death. J Cell Biol 160:223–233

97. Trimble LA, Lieberman J (1998) Circulating CD8 T lymphocytes in human immunodeficiency virus-infected individuals have impaired function and downmodulate CD3 zeta, the signaling chain of the T-cell receptor complex. Blood 91:585–594
98. Tschopp J, Masson D, Stanley KK (1986) Structural/functional similarity between proteins involved in complement- and cytotoxic T-lymphocyte-mediated cytolysis. Nature 322:831–834
99. Uellner R, Zvelebil MJ, Hopkins J, Jones J, MacDougall LK, Morgan BP, Podack E, Waterfield MD, Griffiths GM (1997) Perforin is activated by a proteolytic cleavage during biosynthesis which reveals a phospholipid-binding C2 domain. EMBO J 16:7287–7296
100. Vivier E, Sorrell JM, Ackerly M, Robertson MJ, Rasmussen RA, Levine H, Anderson P (1993) Developmental regulation of a mucinlike glycoprotein selectively expressed on natural killer cells. J Exp Med 178:2023–2033
101. Voskoboinik I, Dunstone MA, Baran K, Whisstock JC, Trapani JA (2010) Perforin: structure, function, and role in human immunopathology. Immunol Rev 235:35–54
102. Voskoboinik I, Smyth MJ, Trapani JA (2006) Perforin-mediated target-cell death and immune homeostasis. Nat Rev Immunol 6:940–952
103. Voskoboinik I, Sutton VR, Ciccone A, House CM, Chia J, Darcy PK, Yagita H, Trapani JA (2007) Perforin activity and immune homeostasis: the common A91 V polymorphism in perforin results in both presynaptic and postsynaptic defects in function. Blood 110:1184–1190
104. Voskoboinik I, Thia MC, De Bono A, Browne K, Cretney E, Jackson JT, Darcy PK, Jane SM, Smyth MJ, Trapani JA (2004) The functional basis for hemophagocytic lymphohistiocytosis in a patient with co-inherited missense mutations in the perforin (PFN1) gene. J Exp Med 200:811–816
105. Voskoboinik I, Thia MC, Fletcher J, Ciccone A, Browne K, Smyth MJ, Trapani JA (2005) Calcium-dependent plasma membrane binding and cell lysis by perforin are mediated through its C2 domain: a critical role for aspartate residues 429, 435, 483, and 485 but not 491. J Biol Chem 280:8426–8434
106. Walsh CM, Matloubian M, Liu CC, Ueda R, Kurahara CG, Christensen JL, Huang MT, Young JD, Ahmed R, Clark WR (1994) Immune function in mice lacking the perforin gene. Proc Natl Acad Sci USA 91:10854–10858
107. Werneck MB, Lugo-Villarino G, Hwang ES, Cantor H, Glimcher LH (2008) T-bet plays a key role in NK-mediated control of melanoma metastatic disease. J Immunol 180:8004–8010
108. Wherry EJ (2011) T cell exhaustion. Nat Immunol 12:492–499
109. Yamamoto K, Shibata F, Miyasaka N, Miura O (2002) The human perforin gene is a direct target of STAT4 activated by IL-12 in NK cells. Biochem Biophys Res Commun 297:1245–1252
110. Youn BS, Kim KK, Kwon BS (1996) A critical role of Sp1- and Ets-related transcription factors in maintaining CTL-specific expression of the mouse perforin gene. J Immunol 157:3499–3509
111. Yu CR, Ortaldo JR, Curiel RE, Young HA, Anderson SK, Gosselin P (1999) Role of a STAT binding site in the regulation of the human perforin promoter. J Immunol 162:2785–2790
112. Zajac AJ, Blattman JN, Murali-Krishna K, Sourdive DJ, Suresh M, Altman JD, Ahmed R (1998) Viral immune evasion due to persistence of activated T cells without effector function. J Exp Med 188:2205–2213
113. Zediak VP, Wherry EJ, Berger SL (2011) The contribution of epigenetic memory to immunologic memory. Curr Opin Genet Dev 21:154–159
114. Zhang J, Scordi I, Smyth MJ, Lichtenheld MG (1999) Interleukin 2 receptor signaling regulates the perforin gene through signal transducer and activator of transcription (Stat)5 activation of two enhancers. J Exp Med 190:1297–1308

115. Zhang Y, Lichtenheld MG (1997) Non-killer cell-specific transcription factors silence the perforin promoter. J Immunol 158:1734–1741
116. Zhou J, Zhang J, Lichtenheld MG, Meadows GG (2002) A role for NF-kappa B activation in perforin expression of NK cells upon IL-2 receptor signaling. J Immunol 169:1319–1325
117. Zhu Y, Ju S, Chen E, Dai S, Li C, Morel P, Liu L, Zhang X, Lu B (2010) T-bet and eomesodermin are required for T cell-mediated antitumor immune responses. J Immunol 185:3174–3183

Chapter 11
Perforin and Human Diseases

Omar Naneh, Tadej Avčin and Apolonija Bedina Zavec

Abstract Natural killer (NK) cells and cytotoxic T lymphocytes (CTL) use a highly toxic pore-forming protein perforin (PFN) to destroy cells infected with intracellular pathogens and cells with pre-cancerous transformations. However, mutations of PFN and defects in its expression can cause an abnormal function of the immune system and difficulties in elimination of altered cells. As discussed in this chapter, deficiency of PFN due to the mutations of its gene, *PFN1*, can be associated with malignancies and severe immune disorders such as familial hemophagocytic lymphohistiocytosis (FHL) and macrophage activation syndrome. On the other hand, overactivity of PFN can turn the immune system against autologous cells resulting in other diseases such as systemic lupus erythematosus, polymyositis, rheumatoid arthritis and cutaneous inflammation. PFN also has a crucial role in the cellular rejection of solid organ allografts and destruction of pancreatic β-cells resulting in type 1 diabetes. These facts highlight the importance of understanding the biochemical characteristics of PFN.

Keywords Perforin · Familial hemophagocytic lymphohistiocytosis · Systemic lupus erythematosus · Polymyositis · Rheumatoid arthritis · Cutaneous inflammation

Abbreviations

AD	Atopic dermatitis
CD4	Cluster of differentiation 4
CD8	Cluster of differentiation 8

O. Naneh · A. Bedina Zavec (✉)
Laboratory for Molecular Biology and Nanobiotechnology, National Institute of Chemistry, Hajdrihova 19, 1000 Ljubljana, Slovenia
e-mail: polona.bedina@ki.si

T. Avčin
Department of Allergology Rheumatology and Clinical Immunology Children's Hospital, University Medical Center Ljubljana, Bohoričeva 20, 1525 Ljubljana, Slovenia

G. Anderluh and R. Gilbert (eds.), *MACPF/CDC Proteins - Agents of Defence, Attack and Invasion*, Subcellular Biochemistry 80, DOI: 10.1007/978-94-017-8881-6_11,

CDC	Cholesterol dependent cytolysin
CIA	Collagen-induced arthritis
CL	Cytotoxic lymphocytes
CTL	Cytotoxic T lymphocytes
DM	Dermatomyositis
DMD	Duchenne muscular dystrophy
EGF	Epidermal growth factor regulatory domain
FasL	Fas ligand, a type-II transmembrane protein that belongs to the TNF family
FHL	Familial hemophagocytic lymphohistiocytosis
FHL-2	Familial hemophagocytic lymphohistiocytosis type 2
HGMD	Human gene mutation database
LCMV	Lymphocytic choriomeningitis virus
LCMV-GP	Glycoprotein of lymphocytic choriomeningitis virus
MACPF	Membrane attack complex-PFN
MHC	Major histocompatibility complex
NK	Natural killer
PFN	Perforin
PM	Polymyositis
SLE	Systemic lupus erythematosus
SLEDAI	SLE disease activity index
Tc1	Type 1 CD8 T cells
Tc2	Type 2 CD8 T cells
TNF	Tumor necrosis factor
WT	Wild type

Introduction

Cytotoxic lymphocytes (CLs), such as cytotoxic T lymphocytes (CTL) and natural killer (NK) cells, eliminate their targets with the help of the cytotoxic pore-forming protein, perforin (PFN). Upon recognition of altered cells, CLs induce formation of transient conjugates, immunological synapses, with infected or malignant cells. PFN and granzymes, pro-apoptotic serine proteases, are then vectorially delivered towards the cleft between membranes. Once released, PFN binds on the cellular membranes of targets and enables granzymes to enter the cytosol, thus inducing cell death.

PFN is a ~70 kDa pore-forming glycoprotein and a typical representative of the MACPF (Membrane Attack Complex-PFN) protein family. According to the published structure of mouse PFN, which is 68 % identical to the human homologue, PFN is a thin and key-shaped molecule [57]. It is comprised of an N-terminal MACPF/cholesterol dependent cytolysin (CDC) domain with amphipathic helices for membrane protrusion, epidermal growth factor regulatory domain (EGF) of unknown function and C-terminal C2 domain, which mediates initial, calcium-dependent membrane binding.

The PFN pore, which enables entry of granzymes into the cytosol of target cells, consists of 19–24 monomers that form a transmembrane β barrel pore with a lumen of 130–200 Å. The formation of active pores strongly depends on pH and the presence of calcium [75, 100]. Physiological conditions, such as pH 7.4 and millimolar concentrations of calcium, induce its permeabilizing activity [11, 36, 106]. Despite interactions with membranes in an acidic environment, low pH prevents the formation of active pores [60, 75]. Therefore, it is not surprising that PFN and granzymes are stored together within killer cells, in lysosome-like secretory granules, cytotoxic granules, with an acidic and calcium-free lumen.

As we will describe in this chapter, PFN inactivity or altered expression can cause severe pathological consequences. Although the first reports of PFN were published 30 years ago [73] and despite its irreplaceable role in the immune system, knowledge about PFN's pore-forming characteristics remains incomplete [32]. A deeper understanding of PFN's biochemical characteristics should highlight its clinical consequences and may identify new targets of more specific therapeutic approaches.

Perforin Deficiency

Analyses of the biochemical role of PFN have shown great importance of this protein for the elimination of cells infected with intracellular pathogens such as viruses and bacteria. PFN-deficient mice infected with lymphocytic choriomeningitis virus (LCMV) have an induced immune response but are not able to clear viral infection [50, 102]. Additionally, *Prf1*$^{-/-}$ mice are not able to survive infection with Ebola virus [34]; besides, based on recent mice model research, PFN could have a significant role in suppression of influenza virus infection [14]. *Prf1* knockout mice, infected with H1N1 influenza virus A, showed a slower time to recovery from infection and greater weight loss in comparison to their wild type (WT) counterparts. It was also shown that there was impaired viral control during the infection of PFN-deficient mice with ectromelia virus, a mouse homolog of the human smallpox virus [27]. Furthermore, it seems that PFN is significant for the elimination of intracellular bacteria such as *Listeria monocyogenes* [49] and *Francisella tularensis* [77]. Following vaccination with an attenuated strain of *F. tularensis* subsp. *novicida*, the observed mortality rate after challenging with a virulent strain is significantly higher in PFN-deficient mice than a WT control. On the other hand, PFN also has important role in the elimination of malignant tissue. *Prf1*$^{-/-}$ mice develop spontaneous lymphomas and other malignancies more readily than control, WT mice [50, 80, 87], and besides are more prone to developing a sarcoma induced by chemical carcinogen 3-methylcholanthrene [96].

Well after publishing of a nucleotide sequence for the PFN gene, *PRF1* [59], and mapping of it to the chromosomal region 10q22 [30], Stepp et al. reported the first association between PFN and a human immune disorder. Mutations at the 10q21-22 locus were identified as the cause of a rare, rapidly fatal immune disorder condition termed familial hemophagocytic lymphohistiocytosis type 2 (FHL-2) [86]. This suggests that deficiency in functional PFN may result in an impaired ability to handle intracellular infection and malignancies and that this is connected to disorders of the immune system.

Alongside with the HGMD public database (2009 update) [85], where there are currently a total of 84 reported independent missense/nonsense mutations, 13 deletions and 4 insertions, the number of known unique *PRF1* defects, which can cause PFN deficiency, is now known to be more than 120. Reported mutations are unevenly dispersed through the gene (Fig. 11.1a) and protein (Fig. 11.1b), from about 6 to 9 % of all amino acids of different domains were mutated in EGF and C2 domains, respectively. Due to its size, the MACPF domain is more often mutated than other protein regions and more than 60 % of all mutations are linked with this domain.

Based on an in silico analysis, mutations of PFN mainly destabilize the protein structure, which could result in incorrect folding of the molecule [3]. In support of these results, Chia et al. showed greater temperature instability of some missense PFN mutants [18]. Additionally, results indicated that some mutations can markedly reduce PFN activity. For example, cytotoxic activity was reduced by a D49N substitution due to the introduction of a new N-linked glycan, in turn giving incorrect protein folding [17]. However, for the majority of mutations it is still unknown how they affect the biochemical activity of PFN.

In most cases (>90 %) mutations of *PRF1* are linked with the previously mentioned FHL-2. This disease entity is a specific subtype of familial hemophagocytic lymphohistiocytosis (FHL), identified with an exaggerated and rapidly progressive inflammation caused by uncontrolled and inefficient immune response. Common features of FHL are an increased blood concentration of inflammatory cytokines [38] as an outcome of uncontrolled activation of macrophages and T cells [9]. Phenotypically FHL expresses with nonspecific clinical manifestations such as prolonged fever, hepatosplenomegaly, lymphadenopathy, consumptive coagulopathy and encephalopathy resulting in confusion and coma. Characteristic laboratory manifestations of FHL include pancytopenia, markedly elevated lactate dehydrogenase and ferritin, hypofibrinogenemia, elevated levels of triglycerides and prolonged blood coagulation tests [41, 47]. The disease can be triggered by infections, most commonly with Epstein-Barr virus, cytomegalovirus and parvovirus [39], possibly due to the relationship of viral infections to PFN deficiency described above. The frequency of FHL strongly depends on ethnic background and has been estimated from 1 in 300,000 in Japan [45], 1 in 100,000 in Texas [69] and 1 in 50,000 in Sweden [40] to 7.5 in 10,000 in Turkey, most probably due to the high rate of consanguineous marriages [35].

FHL is connected with dysfunction of PFN in about half of all cases (condition termed FHL-2). In other circumstances FHL is mostly linked to defects in proteins required for PFN processing or delivery to the target cells. Thus, deficiency of

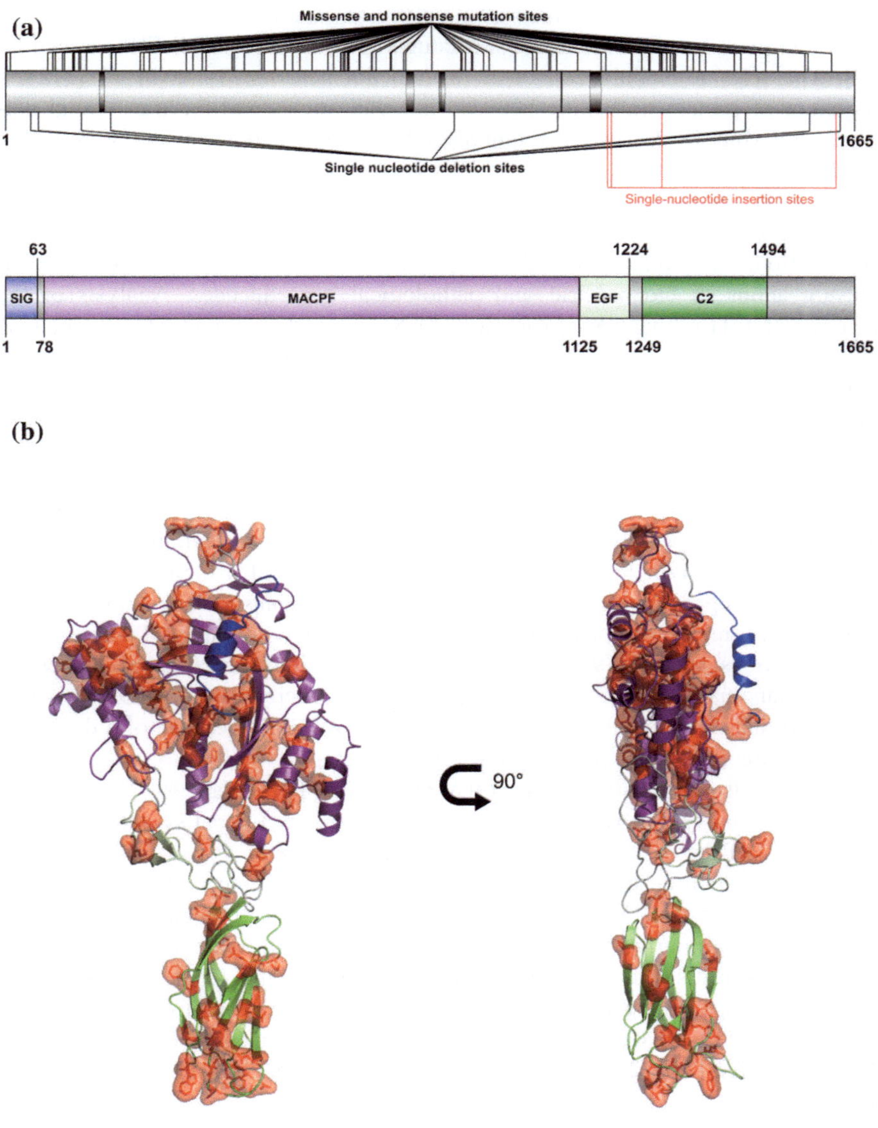

Fig. 11.1 Positions of Known Mutations within Perforin Gene and Protein Structure. **a** Domain organization and all known mutation positions in *PRF1*. Single nucleotide mutation sites are denoted with *black* or *red lines* and *dark grey areas* represent gross deletions. **b** Predicted structure of human PFN with exposed amino acids (*red*) that are known to be affected by nonsense and missense mutations. Model was built with I-TASSER server [76, 109] on known crystal structure of mouse PFN (PDB ID: 3NSJ [57]). *Blue*, signal peptide; *magenta*, MACPF domain; *cyan*, EGF domain; *green*, C2 domain

active protein Munc13-4, which is involved in maturation and exocytosis of cytotoxic granules in immune cells, accounts for one third of all FHL cases [28, 47]. The rest are mainly connected with deficiencies in Syntaxin 11 [112], a regulator of protein trafficking between late endosomes and the trans-Golgi network, and Munc18-2, which is required for the granule exocytosis in NK cells [24, 111]. As a result, a hallmark of all FHLs is decreased or totally absent killer cell activity [25, 79, 90].

The most common *PRF1* mutation is C272T that causes the A91V missense substitution, but available data do not provide a basis for making a general conclusion about its biochemical and pathological consequences. Analysis of the activity of CTLs and NKs expressing PFN-A91V shows significantly reduced cytotoxic activity [93, 99]. However, the allele frequency of the PFN mutation which causes the A91V substitution in Caucasians ranges from 3 to 17 % [16, 66] therefore it is hard to link this mutation directly with disease, since disease is not so common. Originally A91V was considered a neutral polymorphism [110] but further studies showed connection with diseases in patients with a second defective PFN allele. Clementi et al. reported that the A91V substitution in the first PFN allele and F421C, R232H or W374X substitutions in the second allele resulted in various types of lymphoma [22, 23]. On the other hand, a homozygous A91V mutation was present in children with acute lymphoblastic leukemia [66, 78]. Surprisingly, regardless of the presence of the other dominant WT allele, some PFN mutations, such as A91V, R28C and N252S, are in some cases also associated with melanoma [94]. Due to the impaired cytotoxicity of PFN-A91V, it was proposed that this mutant reduces the activity of WT PFN because of a dominant negative effect during pore formation [99].

There is also an additional report that outlines a PFN mutant A91V in primary central nervous system vasculitis, in the absence of any typical symptom of FHL [68]. *PRF1* of the described patient was present in a heterozygous state with one allele with a nonsense mutation causing E46X and the second with missense mutations causing A91V and R119W forms of PFN. The frequency of the A91V mutation was similarly found to be higher in patients with systemic onset juvenile idiopathic arthritis and macrophage activation syndrome [97]. Although FHL mostly develops during the first years of life, there are special cases describing adult onset of this disease, which are in some circumstances connected with the A91V mutation [15, 22, 29, 64]. In addition to A91V, N252S substitution in PFN was similarly found to be significantly higher in patients with autoimmune lymphoproliferative syndrome [21] and in type 1 diabetes [71].

Overall, mutations of PFN can impair its cytotoxic activity that could lead to development of various types of diseases. But due to difficulties encountered in the expression of active recombinant versions of the protein, there is still limited knowledge about the biochemical consequences of PFN deficiency.

Overactivity of Perforin

Although PFN has an essential role in immune homeostasis and surveillance against cancer and viruses, PFN overactivity can also be dangerous when PFN-dependent cytotoxicity turns against autologous cells. This characteristic has already been described as a "double-edged sword" in PFN function [13].

Systemic Lupus Erythematosus

Systemic lupus erythematosus (SLE) is a chronic autoimmune disease with multiorgan involvement (skin, kidneys, joints, central nervous system) caused by the production of pathogenic autoantibodies to nuclear antigens. Genetic predisposition plays an important role in susceptibility to SLE including genes involved in antigen/immune complex clearance, lymphoid signalling, and apoptosis [101]. Besides genetic influence, additional factors such as hormonal influence, environmental exposures, and virus infections have important contributions in disease development [46].

It has been shown that activated CD8 T lymphocytes have crucial roles in SLE pathogenesis [10]. Patients with disease flares were characterized by higher proportions of PFN- and/or granzyme B-positive lymphocytes and the frequency of these cells in peripheral blood correlated with clinical disease activity as assessed by the SLE disease activity index (SLEDAI). SLE patients also have significantly higher levels of PFN in CD4 T lymphocytes than healthy controls as shown by Western blots [55]. The flow cytometric studies confirmed significantly increased PFN expression on both CD4 and CD8 cells of SLE patients [1]. As regards NK cells, a clear decrease in NK cells was found in the blood of SLE patients, particularly in active disease, but the percentage of NK cells expressing PFN and granzyme B was higher [37].

Polymyositis

Polymyositis (PM) and dermatomyositis (DM) are chronic systemic autoimmune inflammatory diseases associated with high morbidity and functional disability. PM and DM are characterized by inflammation and weakness of proximal muscles, associated also with a skin rash in DM. PM and DM affect skeletal muscle with perimysial and endomysial infiltration of mononuclear cells. It was proposed that the infiltrating cells contribute to the pathogenesis of PM and DM by releasing cytokines such as tumor necrosis factor (TNF), and cytotoxic PFN and granzymes. The PFN distribution was studied in PM and DM patients [33] and found to be random in the cytoplasm of the muscle-infiltrating inflammatory T cells in DM patients reflecting nonspecific activation of T cells. In PM patients, PFN was located vectorially towards the target muscle fiber in 43 % of CD8 T cells. Such localization reflects the specific recognition of an antigen on the muscle fiber surface with the T cell receptor pointing to a PFN- and secretion-dependent mechanism of muscle fiber injury. In addition, the murine model of autoimmune myositis was established by induction with a recombinant skeletal muscle fast-type C protein. It was shown that C-protein in the skeletal muscle is the major myositogenic antigen and induces inflammatory lesions mimicking those of human PM [53]. Results from a murine model of PM have shown that PFN-mediated cytotoxicity by CD8 T lymphocytes primarily damages muscle fibers due to C protein-induced myositis [89]. Thus, pre-treatment with anti-CD8 monoclonal antibodies conferred resistance to C protein-induced myositis and besides, anti-TNF α monoclonal antibodies were inefficient in treating patients with PM and DM. Those results confirm previous findings that C protein-induced myositis does not require TNFα for its development, but that PFN-mediated CD8 cell cytotoxicity is definitively responsible for muscle injury.

Rheumatoid Arthritis

PFN-dependent cytotoxicity has an important role also in the initiation of tissue damage in arthritis. Collagen-induced arthritis (CIA) is the mostly studied animal model of rheumatoid arthritis, inducible in susceptible strains of mice by immunization with bovine collagen type II [95]. The development of CIA depends on T cell activation, which triggers an inflammatory cascade involving different types of leukocytes and synovial cells that produce cytokines and other mediators responsible for cartilage and bone destruction [12, 44, 52]. CD8 T cells play an important role in initiation of CIA and in providing resistance to it as was shown in CD8-deficient mice which had a lower incidence of CIA and were more susceptible to re-induction of CIA after recovery [91]. Therefore, it was proposed that PFN contributes to the pathology of arthritis in at least two ways (i) promotion of autoimmunity and (ii) destruction of target tissues. Using PFN-deficient mice, it was shown that a lack of PFN led to reduced incidence of CIA, a reduced severity, and delayed onset of disease, but some individual mice still developed severe disease [7]. It was concluded that PFN-dependent cytotoxicity has an important role in the initiation of tissue damage of arthritis; however, PFN-independent cytotoxic death pathways such as the Fas/FasL pathway might also contribute to CIA. In PFN-deficient mice, a lack of PFN led to impaired killing of autoreactive T cells although a significantly increased T cell proliferation was observed. It was shown before that PFN deficiency resulted in enhanced CD8 T cell expansion, probably because of decreased killing of antigen-presenting cells and therefore prolonged stimulation by antigen [5].

Cutaneous Inflammation

Cytotoxic CD4 and CD8 T cells containing PFN and granzyme B also play an important role in provoking cutaneous inflammation. It was found that expression of PFN and granzyme B is significantly increased in chronic inflammatory skin diseases including atopic dermatitis (AD) and psoriasis [104]. A significant enhancement of PFN and granzyme B expression was observed in lesional AD and psoriasis skin as compared with normal skin and non-lesional AD skin. However, significantly increased numbers of T cells containing cytotoxic proteins are present in the epidermis of AD patients but not in that of psoriasis patients. An increased number of T cells containing PFN and granzyme B was found in the dermis of psoriatic skin lesions, suggesting that cell-mediated cytotoxic mechanisms play an integral part also in their pathogenesis. Moreover, an increased number of T cells expressing PFN in the blood of patients with psoriasis was also shown [8]. On the other hand, peripheral NK cells are significantly reduced in AD and psoriasis, and NK cytotoxicity was affected only in AD, but not in psoriasis [61].

Cerebral Malaria

Malaria results from infection with protozoan parasites of the genus *Plasmodium* and represents a global health problem in the tropics. The exact molecular mechanism of malaria pathogenesis is still not clear, but probably parasite antigens are transferred by infected erythrocytes, taken up by endothelial cells and presented there on MHC class I antigens. CD8 T cells recognize the antigens and kill the endothelial cells. CD8 and CD4 T cells are crucial for cerebral malaria pathogenesis, which was shown by neutralizing antibodies as well as via an experimental mice model with deleted function of those cells [103]. It was shown that anti-CD8 or anti-CD4 antibodies prevent cerebral malaria even in end-stage disease in mice [42]. Further, it was shown that PFN deficient mice were completely resistant to cerebral malaria and had a reduced level of brain endothelial apoptosis [74], while deficiency of Fas ligand had no protective benefit [70]. Thus it is clear that PFN has an essential role in the pathogenesis of cerebral malaria. The result of endothelial cell death is breakdown of the blood brain barrier, cerebral edema, and further penetration of viable parasites beyond the vascular space and into the brain. Death is probably a result of overwhelming cerebral inflammation. The absence of a dendritic cell subset also resulted in complete resistance to experimental cerebral malaria [72]. C-type lectin receptors in dendritic cells have a role in cerebral malaria pathogenesis because mice deficient for them are also highly protected from the development of cerebral malaria [63]. In such mouse brains, CD8 T cell sequestration was markedly reduced, which was accompanied by reduced brain inflammation. The essential role of PFN as a mediator of endothelial cell death in cerebral malaria makes it a potential new target for therapeutic intervention [98]. Some attempts of PFN inhibition were already done in vitro by binding small molecular weight compounds [58, 62, 83, 84].

Viral Myocarditis

Viral myocarditis is a frequent complication following enteroviral infection. In PFN deficient mice it was shown that PFN has no role in viral clearance or the length of viremia but it was critical for the development of cardiac pathology [31]. A lethal dose of virus caused 90 % death in mice due to myocardical lesions, while all PFN deficient mice survived. A non-lethal dose of virus caused extensive myocardial lesions in mice, while only a mild myocarditis was observed in PFN deficient mice. Reduction of viral myocarditis in PFN deficient mice was confirmed by another group, which concluded that blockade of PFN gene expression at an early stage of enteroviral infection may limit viral myocarditis [54]. It was also reported that anti-PFN neutralizing antibody reduces myocardial injury in viral myocarditis in mice [20]. This was the first study reporting a therapeutic effect by blocking PFN function; however, these results are still not repeated and confirmed by others.

In humans, PFN was immunohistologically detected in all biopsies from patients with idiopathic dilated cardiomyopathy [4]. It was concluded that cardiac fibrosis and dilatation might be caused by CTL damage to the myocardium. Furthermore, as early as 1990 PFN-induced lesions were observed in the myocardium of patients with acute myocarditis [107].

Duchenne Muscular Dystrophy

Duchenne muscular dystrophy (DMD) involves severe muscle wasting and the premature death of afflicted individuals attributable to the absence of dystrophin, a cytoskeletal protein. CD8 T cells present in dystrophic muscle can directly trigger muscle cell death whereas CD4 T cells can be an important source of inflammatory cytokines. The dystrophic mdx mouse is the most popular animal model related to DMD. It has a point mutation within its dystrophin gene; therefore the mouse has no functional dystrophin in its muscles and experiences muscle necrosis that resembles the DMD pathology. It was shown that T cells can cause muscle loss by PFN-mediated cytotoxicity [82]. Genetic ablation of PFN in mdx mice considerably reduced fiber necrosis in mdx mice and antibody-mediated depletion of CD8 or CD4 T cells resulted in reduction in muscle histopathology [81], a significant reduction of apoptotic myonuclei concentration and a reduction in necrosis. Additionally, double-mutant mice, deficient in dystrophin and PFN showed nearly complete absence of myonuclear apoptosis. Therefore, it was concluded that CTLs contribute significantly to apoptosis and necrosis in mdx dystrophy, and that PFN-mediated killing is primarily responsible for myonuclear apoptosis.

Double Function of Perforin in Human Disease

Type 1 Diabetes

Type 1 diabetes is caused by insulin deficiency resulting from autoimmune destruction of pancreatic β-cells, and PFN-dependent cytotoxicity has been reported as a crucial mechanism for pancreatic β-cell elimination by CTLs. In model mice expressing the glycoprotein of lymphocytic choriomeningitis virus (LCMV-GP) in beta cells of the endocrine pancreas, LCMV infection leads to a T cell response resulting in a rapid development of diabetes. In PFN-deficient LCMV-GP transgenic mice, LCMV infection did not induce diabetes despite the activation of LCMV-GP-specific T cells. It was found that PFN-dependent cytotoxicity was not required for the initiation of insulitis but is crucial for the destruction of beta cells [48]. Later, PFN-deficient mice were backcrossed with the non-obese diabetes mouse strain [51]. In the non-obese diabetic mouse model of type I diabetes, PFN deficiency resulted in a highly significant reduction and delayed onset of disease. Incidence of spontaneous diabetes over a one year period was reduced from 77 % in control mice to 16 % in PFN-deficient mice. However, insulitis with infiltration of CD4 and CD8 T cells occurred similarly in both groups of mice. In PFN-deficient mice, lower incidence of disease and delayed onset was observed even when diabetes was induced by cyclophosphamide injection; these results indicate a direct role of PFN in β-cell death in vivo, while the non-obese diabetic mice bred onto a $Prf1^{-/+}$ background were partially resistant to the disease [2]. The autoreactive CD8 T cells play a major role in the destruction of pancreatic β-cells and islet inflammation. In PFN deficient mice, lymphocytic infiltration of the pancreas is also significantly reduced, perhaps because this process requires β-cell destruction [92]. In vitro tests have shown that reactive CD8 T cells can kill pancreatic β-cells by PFN or by Fas mediated cell death; but that PFN killing mechanism is highly favoured and Fas mediated cell death is observed only in the absence of PFN. However, in some transgenic mouse models diabetes is independent of PFN e.g. the mouse model expressing two rat insulin promoter (RIP)-driven transgenes leading to β-cell expression of mouse TNF [6, 43].

In contrast to the above-mentioned reports of PFN-mediated cytotoxicity, it was found that mutations in *PFN1* are susceptibility factors for type 1 diabetes development [71], with the frequency of N252S mutation in *PFN1* being significantly higher in patients than in control subjects. This fact is difficult to explain, because autoimmune damage in type 1 diabetes is mainly due to β-cell destruction by autoreactive CTL and the inflammatory response [105], especially because it was shown that PFN-mediated cytolysis has an essential role in fratricide mediated by T cell activation [88]. In addition, there are many studies in mice showing that PFN deficiency results in a highly significant reduction and delayed onset of type 1 diabetes as discussed in the previous paragraph. However, all type 1 diabetes patients with an N252S mutation in *PFN1* were heterozygotes and a mechanism to explain the association has not yet been proposed [92].

Asthma

Asthma is a chronic airway inflammation, one of the most prevalent airway diseases in the world, and is a special case as regards PFN. CD8 T cells can differentiate into two effector phenotypes, Tc1 and Tc2, secreting different cytokine patterns. Both subsets are cytotoxic via the PFN and Fas pathways. The role of CD8 T cells in asthma remains controversial: type 2 CD8 T cells (Tc2), can exacerbate inflammation by secreting type 2 cytokines such as IL-5 and IL-13 [67]. It was shown that CD8-deficient mice develop significantly lower airway hyper-responsiveness compared with WT mice. In contrast, conventional type 1 CD8 T cells (Tc1) reduce inflammation and airway hyperresponsiveness [65]. Activated CTLs can kill antigen-presenting dendritic cells, thereby limiting their ability to induce further T cell activation. It was shown that treatment with allergen-specific CTLs suppresses allergic airway inflammation and mucus production via a PFN-dependent mechanism [26]. Allergen-specific CTLs must express PFN to reduce eosinophilic inflammation and mucus production in the airway, whereas cytokine interferon γ is not critical.

Rejection of Organ Allografts

CLs are involved in the elimination of infected, transformed, or foreign cells; therefore, they have essential roles in the rejection of transplanted organs. The recipient immune system rejects organ allograft mainly by T cell effector mechanisms. CTL destruction of allogeneic vascular endothelium is important in the pathogenesis of both acute and chronic allograft rejection. In mouse endothelial cells it was shown that CTLs destroyed vascular endothelium primarily by the PFN/granzyme pathway and that the Fas/FasL pathway had only a minor contribution to apoptosis [56]. Both PFN and granzyme B molecules appear to be important for T cell-mediated destruction of allogeneic cells and both appear to be important in chronic rejection as their absence results in reduced allograft vasculopathy in mouse models of heart transplant rejection [19]. CD8 T cell deficiency prevents the acute rejection of heart allografts in a mouse model [108].

CD8 and CD4 T cells are present in acutely rejecting allografts and CD8 T cells even trigger the rejection process. Therefore, PFN and granzyme B were established as the main mediators of cellular rejection of solid organ allografts.

Conclusions

Many diseases are related to increased PFN activity or deficiency of functional PFN; therefore, the ability to affect or tune its functioning would be very beneficial from a clinical point of view. In *PFN1* mutations, addition of normal PFN or intensification of PFN activity would be desirable. In diseases with PFN overactivity, the molecular mechanisms of the diseases are often not known exactly and the PFN overactivity is usually not the cause of changes, but the consequence of some other processes such as development of autoimmunity, viral infection or a congenital defect; however, the inhibition of its functioning would be very helpful for the prevention of serious consequences that are caused by PFN overactivity. Due to the essential function of PFN in the human body, and because of the extreme complexity of its functioning at both molecular and physiological levels, the impact on PFN functioning should be very carefully designed, and specifically and temporally regulated. The development of tools for increasing and decreasing PFN's activity is therefore both extremely desirable and very demanding.

References

1. Abo-Elenein A, Shaaban D, Gheith O (2008) Flowcytometric study of expression of perforin and CD134 in patients with systemic lupus erythematosus. Egypt J Immunol 15:135–143
2. Amrani A, Verdaguer J, Anderson B, Utsugi T, Bou S, Santamaria P (1999) Perforin-independent beta-cell destruction by diabetogenic CD8(+) T lymphocytes in transgenic nonobese diabetic mice. J Clin Invest 103:1201–1209
3. An O, Gursoy A, Gurgey A, Keskin O (2013) Structural and functional analysis of perforin mutations in association with clinical data of familial hemophagocytic lymphohistiocytosis type 2 (FHL2) patients. Protein Sci 22:823–839
4. Badorff C, Noutsias M, Kühl U, Schultheiss H-P (1997) Cell-mediated cytotoxicity in hearts with dilated cardiomyopathy: correlation with interstitial fibrosis and foci of activated t lymphocytes. J Am Coll Cardiol 29:429–434
5. Badovinac VP (2000) Regulation of antigen-specific cd8+ t cell homeostasis by perforin and interferon-gamma. Science 290:1354–1357
6. Balasa B, Van Gunst K, Jung N, Balakrishna D, Santamaria P, Hanafusa T, Itoh N, Sarvetnick N (2000) Islet-specific expression of IL-10 promotes diabetes in nonobese diabetic mice independent of Fas, perforin, TNF receptor-1, and TNF receptor-2 molecules. J Immunol 165:2841–2849
7. Bauer K, Knipper A, Tu-Rapp H, Koczan D, Kreutzer H-J, Nizze H, Mix E, Thiesen H-J, Holmdahl R, Ibrahim SM (2005) Perforin deficiency attenuates collagen-induced arthritis. Arthritis Res Ther 7:R877–R884

8. Behrendt C, Gollnick H, Bonnekoh B (2000) Up-regulated perforin expression of CD8+ blood lymphocytes in generalized non-anaphylactic drug eruptions and exacerbated psoriasis. Eur J Dermatol 10:365–369
9. Billiau AD, Roskams T, Van Damme-Lombaerts R, Matthys P, Wouters C (2005) Macrophage activation syndrome: characteristic findings on liver biopsy illustrating the key role of activated, IFN-gamma-producing lymphocytes and IL-6- and TNF-alpha-producing macrophages. Blood 105:1648–1651
10. Blanco P, Pitard V, Viallard J-F, Taupin J-L, Pellegrin J-L, Moreau J-F (2005) Increase in activated CD8+ T lymphocytes expressing perforin and granzyme B correlates with disease activity in patients with systemic lupus erythematosus. Arthritis Rheum 52:201–211
11. Blumenthal R, Millard PJ, Henkart MP, Reynolds CW, Henkart Pa (1984) Liposomes as targets for granule cytolysin from cytotoxic large granular lymphocyte tumors. Proc Natl Acad Sci USA 81:5551–5555
12. Brahn E, Peacock DJ, Banquerigo ML, Liu DY (1992) Effects of tumor necrosis factor alpha (TNF-alpha) on collagen arthritis. Lymphokine Cytokine Res 11:253–256
13. Brennan AJ, Chia J, Trapani JA, Voskoboinik I (2010) Perforin deficiency and susceptibility to cancer. Cell Death Differ 17:607–615
14. Brown DM, Lee S, Garcia-Hernandez MDLL, Swain SL (2012) Multifunctional CD4 cells expressing gamma interferon and perforin mediate protection against lethal influenza virus infection. J Virol 86:6792–6803
15. Busiello R, Adriani M, Locatelli F, Galgani M, Fimiani G, Clementi R, Ursini MV, Racioppi L, Pignata C (2004) Atypical features of familial hemophagocytic lymphohistiocytosis. Blood 103:4610–4612
16. Busiello R, Fimiani G, Miano MG, Aricò M, Santoro A, Ursini MV, Pignata C (2006) A91V perforin variation in healthy subjects and FHLH patients. Int J Immunogenet 33:123–125
17. Chia J, Thia K, Brennan AJ, Little M, Williams B, Lopez JA, Trapani JA, Voskoboinik I (2012) Fatal immune dysregulation due to a gain of glycosylation mutation in lymphocyte perforin. Blood 119:1713–1716
18. Chia J, Yeo KP, Whisstock JC, Dunstone MA, Trapani JA, Voskoboinik I (2009) Temperature sensitivity of human perforin mutants unmasks subtotal loss of cytotoxicity, delayed FHL, and a predisposition to cancer. Proc Natl Acad Sci USA 106:9809–9814
19. Choy JC (2010) Granzymes and perforin in solid organ transplant rejection. Cell Death Differ 17:567–576
20. Chun-yan G, Bo H, Hong C, Hong-lei J, Xiu-zhen H (2009) Anti-perforin neutralizing antibody reduces myocardial injury in viral myocarditis. Cardiol Young 19:601–607
21. Clementi R, Chiocchetti A, Cappellano G, Cerutti E, Ferretti M, Orilieri E, Dianzani I, Ferrarini M, Bregni M, Danesino C, Bozzi V, Putti MC, Cerutti F, Cometa A, Locatelli F, Maccario R, Ramenghi U, Dianzani U (2006) Variations of the perforin gene in patients with autoimmunity/lymphoproliferation and defective Fas function. Blood 108:3079–3084
22. Clementi R, Emmi L, Maccario R, Liotta F, Moretta L, Danesino C, Aricó M (2002) Adult onset and atypical presentation of hemophagocytic lymphohistiocytosis in siblings carrying PRF1 mutations. Blood 100:2266–2267
23. Clementi R, Locatelli F, Dupré L, Garaventa A, Emmi L, Bregni M, Cefalo G, Moretta A, Danesino C, Comis M, Pession A, Ramenghi U, Maccario R, Aricò M, Roncarolo MG (2005) A proportion of patients with lymphoma may harbor mutations of the perforin gene. Blood 105:4424–4428
24. Côte M, Ménager MM, Burgess A, Mahlaoui N, Picard C, Schaffner C, Al-Manjomi F, Al-Harbi M, Alangari A, Le Deist F, Gennery AR, Prince N, Cariou A, Nitschke P, Blank U, El-Ghazali G, Ménasché G, Latour S, Fischer A, de Saint Basile G (2009) Munc18-2 deficiency causes familial hemophagocytic lymphohistiocytosis type 5 and impairs cytotoxic granule exocytosis in patient NK cells. J Clin Invest 119:3765–3773
25. Egeler RM, Shapiro R, Loechelt B, Filipovich A (1996) Characteristic immune abnormalities in hemophagocytic lymphohistiocytosis. J Pediatr Hematol Oncol 18:340–345

26. Enomoto N, Hyde E, Ma JZ-I, Yang J, Forbes-Blom E, Delahunt B, Le Gros G, Ronchese F (2012) Allergen-specific CTL require perforin expression to suppress allergic airway inflammation. J Immunol 188:1734–1741
27. Fang M, Siciliano NA, Hersperger AR, Roscoe F, Hu A, Ma X, Shamsedeen AR, Eisenlohr LC, Sigal LJ (2012) Perforin-dependent CD4+ T-cell cytotoxicity contributes to control a murine poxvirus infection. Proc Natl Acad Sci USA 109:9983–9988
28. Feldmann J, Callebaut I, Raposo G, Certain S, Bacq D, Dumont C, Lambert N, Ouachée-Chardin M, Chedeville G, Tamary H, Minard-Colin V, Vilmer E, Blanche S, Le Deist F, Fischer A, de Saint Basile G (2003) Munc13-4 is essential for cytolytic granules fusion and is mutated in a form of familial hemophagocytic lymphohistiocytosis (FHL3). Cell 115:461–473
29. Feldmann J, Le Deist F, Ouachée-Chardin M, Certain S, Alexander S, Quartier P, Haddad E, Wulffraat N, Casanova JL, Blanche S, Fischer A, de Saint Basile G (2002) Functional consequences of perforin gene mutations in 22 patients with familial haemophagocytic lymphohistiocytosis. Br J Haematol 117:965–972
30. Fink TM, Zimmer M, Weitz S, Tschopp J, Jenne DE, Lichter P (1992) Human perforin (PRF1) maps to 10q22, a region that is syntenic with mouse chromosome 10. Genomics 13:1300–1302
31. Gebhard JR, Perry CM, Harkins S, Lane T, Mena I, Asensio VC, Campbell IL, Whitton JL (1998) Coxsackievirus B3-induced myocarditis: perforin exacerbates disease, but plays no detectable role in virus clearance. Am J Pathol 153:417–428
32. Gilbert RJC, Mikelj M, Dalla Serra M, Froelich CJ, Anderluh G (2013) Effects of MACPF/CDC proteins on lipid membranes. Cell Mol Life Sci 70:2083–2098
33. Goebels N, Michaelis D, Engelhardt M, Huber S, Bender A, Pongratz D, Johnson MA, Wekerle H, Tschopp J, Jenne D, Hohlfeld R (1996) Differential expression of perforin in muscle-infiltrating T cells in polymyositis and dermatomyositis. J Clin Invest 97:2905–2910
34. Gupta M, Greer P, Mahanty S, Shieh W-J, Zaki SR, Ahmed R, Rollin PE (2005) CD8-mediated protection against Ebola virus infection is perforin dependent. J Immunol 174:4198–4202
35. Gürgey A, Göğüş S, Özyürek E, Aslan D, Gümrük F, Çetin M, Yüce A, Ceyhan M, Seçmeer G, Yetgin S, Hiçsönmez G (2003) Primary hemophagocytic lymphohistiocytosis in Turkish children. Pediatr Hematol Oncol 20:367–371
36. Henkart PA, Millard PJ, Reynolds CW, Henkart MP (1984) Cytolytic activity of purified cytoplasmic granules from cytotoxic rat large granular lymphocyte tumors. J Exp Med 160:75–93
37. Henriques A, Teixeira L, Inês L, Carvalheiro T, Gonçalves A, Martinho A, Pais ML, da Silva JAP, Paiva A (2013) NK cells dysfunction in systemic lupus erythematosus: relation to disease activity. Clin Rheumatol 32:805–813
38. Henter J-I, Elinder G, Soder O, Hansson M, Andersson B, Andersson U (1991) Hypercytokinemia in familial hemophagocytic lymphohistiocytosis. Blood 78:2918–2922
39. Henter J-I, Ehrnst A, Andersson J, Elinder G (1993) Familial hemophagocytic lymphohistiocytosis and viral infections. Acta Paediatr 82:369–372
40. Henter J-I, Elinder G, Söder O, Ost A (1991) Incidence in Sweden and clinical features of familial hemophagocytic lymphohistiocytosis. Acta Paediatr Scand 80:428–435
41. Henter J-I, Horne A, Aricó M, Egeler RM, Filipovich AH, Imashuku S, Ladisch S, McClain K, Webb D, Winiarski J, Janka G (2007) HLH-2004: Diagnostic and therapeutic guidelines for hemophagocytic lymphohistiocytosis. Pediatr Blood Cancer 48:124–131
42. Hermsen C, van de Wiel T, Mommers E, Sauerwein R, Eling W (1997) Depletion of CD4+ or CD8+ T-cells prevents *Plasmodium berghei* induced cerebral malaria in end-stage disease. Parasitology 114:7–12
43. Herrera PL, Harlan DM, Vassalli P (2000) A mouse CD8 T cell-mediated acute autoimmune diabetes independent of the perforin and Fas cytotoxic pathways: possible role of membrane TNF. Proc Natl Acad Sci USA 97:279–284

44. Hom JT, Bendele AM, Carlson DG (1988) In vivo administration with IL-1 accelerates the development of collagen-induced arthritis in mice. J Immunol 141:834–841
45. Ishii E, Ohga S, Tanimura M, Imashuku S, Sako M, Mizutani S, Miyazaki S (1998) Clinical and epidemiologic studies of familial hemophagocytic lymphohistiocytosis in Japan. Japan LCH Study Group. Med Pediatr Oncol 30:276–283
46. James JA, Harley JB, Scofield RH (2001) Role of viruses in systemic lupus erythematosus and Sjögren syndrome. Curr Opin Rheumatol 13:370–376
47. Janka GE (2012) Familial and acquired hemophagocytic lymphohistiocytosis. Annu Rev Med 63:233–246
48. Kägi D (1996) Development of insulitis without diabetes in transgenic mice lacking perforin-dependent cytotoxicity. J Exp Med 183:2143–2152
49. Kägi D, Ledermann B, Bürki K, Hengartner H, Zinkernagel RM (1994) CD8+ T cell-mediated protection against an intracellular bacterium by perforin-dependent cytotoxicity. Eur J Immunol 24:3068–3072
50. Kägi D, Ledermann B, Bürki K, Seiler P, Odermatt B, Olsen KJ, Podack ER, Zinkernagel RM, Hengartner H (1994) Cytotoxicity mediated by T cells and natural killer cells is greatly impaired in perforin-deficient mice. Nature 369:31–37
51. Kägi D, Odermatt B, Seiler P, Zinkernagel RM, Mak TW, Hengartner H (1997) Reduced incidence and delayed onset of diabetes in perforin-deficient nonobese diabetic mice. J Exp Med 186:989–997
52. Killar LM, Dunn CJ (1989) Interleukin-1 potentiates the development of collagen-induced arthritis in mice. Clin Sci 76:535–538
53. Kohyama K, Matsumoto Y (1999) C-protein in the skeletal muscle induces severe autoimmune polymyositis in Lewis rats. J Neuroimmunol 98:130–135
54. Koike H, Kanda T, Sumino H, Yokoyama T, Arai M, Motooka M, Suzuki T, Tamura J, Kobayashi I (2001) Reduction of viral myocarditis in mice lacking perforin. Res Commun Mol Pathol Pharmacol 110:229–237
55. Kozłowska A, Hrycaj P, Łącki JK, Jagodziński PP (2010) Perforin level in CD4+ T cells from patients with systemic lupus erythematosus. Rheumatol Int 30:1627–1633
56. Krupnick AS, Kreisel D, Popma SH, Balsara KR, Szeto WY, Krasinskas AM, Riha M, Wells AD, Turka LA, Rosengard BR (2002) Mechanism of T cell-mediated endothelial apoptosis. Transplantation 74:871–876
57. Law RHP, Lukoyanova N, Voskoboinik I, Caradoc-Davies TT, Baran K, Dunstone MA, D'Angelo ME, Orlova EV, Coulibaly F, Verschoor S, Browne KA, Ciccone A, Kuiper MJ, Bird PI, Trapani JA, Saibil HR, Whisstock JC (2010) The structural basis for membrane binding and pore formation by lymphocyte perforin. Nature 468:447–451
58. Lena G, Trapani JA, Sutton VR, Ciccone A, Browne KA, Smyth MJ, Denny WA, Spicer JA (2008) Dihydrofuro[3,4-c]pyridinones as inhibitors of the cytolytic effects of the pore-forming glycoprotein perforin. J Med Chem 51:7614–7624
59. Lichtenheld MG, Olsen KJ, Lu P, Lowrey DM, Hameed A, Hengartner H, Podack ER (1988) Structure and function of human perforin. Nature 335:448–451
60. Lopez JA, Susanto O, Jenkins MR, Lukoyanova N, Sutton VR, Law RHP, Johnston A, Bird CH, Bird PI, Whisstock JC, Trapani JA, Saibil HR, Voskoboinik I (2013) Perforin forms transient pores on the target cell plasma membrane to facilitate rapid access of granzymes during killer cell attack. Blood 121:2659–2668
61. Luci C, Gaudy-Marqueste C, Rouzaire P, Audonnet S, Cognet C, Hennino A, Nicolas J-F, Grob J-J, Tomasello E (2012) Peripheral natural killer cells exhibit qualitative and quantitative changes in patients with psoriasis and atopic dermatitis. Br J Dermatol 166:789–796
62. Lyons DM, Huttunen KM, Browne KA, Ciccone A, Trapani JA, Denny WA, Spicer JA (2011) Inhibition of the cellular function of perforin by 1-amino-2,4-dicyanopyrido[1,2-a]benzimidazoles. Bioorg Med Chem 19:4091–4100

63. Maglinao M, Klopfleisch R, Seeberger PH, Lepenies B (2013) The C-type lectin receptor DCIR is crucial for the development of experimental cerebral malaria. J Immunol 191:2551–2559
64. Mancebo E, Allende LM, Guzmán M, Paz-Artal E, Gil J, Urrea-Moreno R, Fernández-Cruz E, Gayà A, Calvo J, Arbós A, Durán MA, Canet R, Balanzat J, Udina MA, Vercher FJ (2006) Familial hemophagocytic lymphohistiocytosis in an adult patient homozygous for A91V in the perforin gene, with tuberculosis infection. Haematologica 91:1257–1260
65. Marsland BJ, Harris NL, Camberis M, Kopf M, Hook SM, Le Gros G (2004) Bystander suppression of allergic airway inflammation by lung resident memory CD8+ T cells. Proc Natl Acad Sci USA 101:6116–6121
66. Mehta PA, Davies SM, Kumar A, Devidas M, Lee S, Zamzow T, Elliott J, Villanueva J, Pullen J, Zewge Y, Filipovich A (2006) Perforin polymorphism A91V and susceptibility to B-precursor childhood acute lymphoblastic leukemia: a report from the Children's Oncology Group. Leukemia 20:1539–1541
67. Miyahara N, Swanson BJ, Takeda K, Taube C, Miyahara S, Kodama T, Dakhama A, Ott VL, Gelfand EW (2004) Effector CD8+ T cells mediate inflammation and airway hyper-responsiveness. Nat Med 10:865–869
68. Moshous D, Feyen O, Lankisch P, Schwarz K, Schaper J, Schneider M, Dilloo D, Laws H-J, Schwahn BC, Niehues T (2007) Primary necrotizing lymphocytic central nervous system vasculitis due to perforin deficiency in a four-year-old girl. Arthritis Rheum 56:995–999
69. Niece JA, Rogers ZR, Ahmad N, Langevin A, McClain KL (2010) Hemophagocytic lymphohistiocytosis in Texas: observations on ethnicity and race. Pediatr Blood Cancer 54:424–428
70. Nitcheu J, Bonduelle O, Combadiere C, Tefit M, Seilhean D, Mazier D, Combadiere B (2003) Perforin-dependent brain-infiltrating cytotoxic CD8+ T lymphocytes mediate experimental cerebral malaria pathogenesis. J Immunol 170:2221–2228
71. Orilieri E, Cappellano G, Clementi R, Cometa A, Ferretti M, Cerutti E, Cadario F, Martinetti M, Larizza D, Calcaterra V, D'Annunzio G, Lorini R, Cerutti F, Bruno G, Chiocchetti A, Dianzani U (2008) Variations of the perforin gene in patients with type 1 diabetes. Diabetes 57:1078–1083
72. Piva L, Tetlak P, Claser C, Karjalainen K, Renia L, Ruedl C (2012) Cutting edge: Clec9A+ dendritic cells mediate the development of experimental cerebral malaria. J Immunol 189:1128–1132
73. Podack ER, Dennert G (1983) Assembly of two types of tubules with putative cytolytic function by cloned natural killer cells. Nature 302:442–445
74. Potter S, Chan-Ling T, Ball HJ, Mansour H, Mitchell A, Maluish L, Hunt NH (2006) Perforin mediated apoptosis of cerebral microvascular endothelial cells during experimental cerebral malaria. Int J Parasitol 36:485–496
75. Praper T, Beseničar MP, Istinič H, Podlesek Z, Metkar SS, Froelich CJ, Anderluh G (2010) Human perforin permeabilizing activity, but not binding to lipid membranes, is affected by pH. Mol Immunol 47:2492–2504
76. Roy A, Kucukural A, Zhang Y (2010) I-TASSER: a unified platform for automated protein structure and function prediction. Nat Protoc 5:725–738
77. Sanapala S, Yu J-J, Murthy AK, Li W, Guentzel MN, Chambers JP, Klose KE, Arulanandam BP (2012) Perforin- and granzyme-mediated cytotoxic effector functions are essential for protection against *Francisella tularensis* following vaccination by the defined *F. tularensis* subsp. *novicida* Δ*fopC* vaccine strain. Infect Immun 80:2177–2185
78. Santoro A, Cannella S, Trizzino A, Lo Nigro L, Corsello G, Arico M (2005) A single amino acid change A91V in perforin: a novel, frequent predisposing factor to childhood acute lymphoblastic leukemia? Haematologica 90:697–698
79. Schneider EM, Lorenz I, Müller-Rosenberger M, Steinbach G, Kron M, Janka-Schaub GE (2002) Hemophagocytic lymphohistiocytosis is associated with deficiencies of cellular cytolysis but normal expression of transcripts relevant to killer-cell-induced apoptosis. Blood 100:2891–2898

80. Smyth MJ, Thia KY, Street SE, MacGregor D, Godfrey DI, Trapani JA (2000) Perforin-mediated cytotoxicity is critical for surveillance of spontaneous lymphoma. J Exp Med 192:755–760
81. Spencer MJ, Montecino-Rodriguez E, Dorshkind K, Tidball JG (2001) Helper (CD4(+)) and cytotoxic (CD8(+)) T cells promote the pathology of dystrophin-deficient muscle. Clin Immunol 98:235–243
82. Spencer MJ, Walsh CM, Dorshkind KA, Rodriguez EM, Tidball JG (1997) Myonuclear apoptosis in dystrophic mdx muscle occurs by perforin-mediated cytotoxicity. J Clin Invest 99:2745–2751
83. Spicer JA, Huttunen KM, Miller CK, Denny WA, Ciccone A, Browne KA, Trapani JA (2012) Inhibition of the pore-forming protein perforin by a series of aryl-substituted isobenzofuran-1(3H)-ones. Bioorg Med Chem 20:1319–1336
84. Spicer JA, Lena G, Lyons DM, Huttunen KM, Miller C, O'Connor PD, Bull M, Helsby N, Jamieson S, Denny WA, Ciccone A, Browne K, Lopez J, Rudd-Schmidt J, Voskoboinik I, Trapani JA (2013) Exploration of a series of 5-arylidene-2-thioxoimidazolidin-4-ones as inhibitors of the cytolytic protein perforin. J Med Chem. doi:10.1021/jm401604x
85. Stenson PD, Ball EV, Mort M, Phillips AD, Shiel JA, Thomas NST, Abeysinghe S, Krawczak M, Cooper DN (2003) Human gene mutation database (HGMD): 2003 update. Hum Mutat 21:577–581
86. Stepp SE, Dufourcq-Lagelouse R, Le Deist F, Bhawan S, Certain S, Mathew PA, Henter JI, Bennett M, Fischer A, de Saint Basile G, Kumar V (1999) Perforin gene defects in familial hemophagocytic lymphohistiocytosis. Science 286:1957–1959
87. Street SE, Cretney E, Smyth MJ (2001) Perforin and interferon-gamma activities independently control tumor initiation, growth, and metastasis. Blood 97:192–197
88. Su MW-C, Pyarajan S, Chang J-H, Yu C-L, Jin Y-J, Stierhof Y-D, Walden P, Burakoff SJ (2004) Fratricide of CD8+ cytotoxic T lymphocytes is dependent on cellular activation and perforin-mediated killing. Eur J Immunol 34:2459–2470
89. Sugihara T, Okiyama N, Suzuki M, Kohyama K, Matsumoto Y, Miyasaka N, Kohsaka H (2010) Definitive engagement of cytotoxic CD8 T cells in C protein-induced myositis, a murine model of polymyositis. Arthritis Rheum 62:3088–3092
90. Sullivan KE, Delaat CA, Douglas SD, Filipovich AH (1998) Defective natural killer cell function in patients with hemophagocytic lymphohistiocytosis and in first degree relatives. Pediatr Res 44:465–468
91. Tada Y, Ho A, Koh DR, Mak TW (1996) Collagen-induced arthritis in CD4- or CD8-deficient mice: CD8+ T cells play a role in initiation and regulate recovery phase of collagen-induced arthritis. J Immunol 156:4520–4526
92. Thomas HE, Trapani JA, Kay TWH (2010) The role of perforin and granzymes in diabetes. Cell Death Differ 17:577–585
93. Trambas C, Gallo F, Pende D, Marcenaro S, Moretta L, De Fusco C, Santoro A, Notarangelo L, Arico M, Griffiths GM (2005) A single amino acid change, A91V, leads to conformational changes that can impair processing to the active form of perforin. Blood 106:932–937
94. Trapani JA, Thia KYT, Andrews M, Davis ID, Gedye C, Parente P, Svobodova S, Chia J, Browne K, Campbell IG, Phillips WA, Voskoboinik I, Cebon JS (2013) Human perforin mutations and susceptibility to multiple primary cancers. Oncoimmunology 2:e24185
95. Trentham DE (1977) Autoimmunity to type II collagen an experimental model of arthritis. J Exp Med 146:857–868
96. Van den Broek ME, Kägi D, Ossendorp F, Toes R, Vamvakas S, Lutz WK, Melief CJ, Zinkernagel RM, Hengartner H (1996) Decreased tumor surveillance in perforin-deficient mice. J Exp Med 184:1781–1790
97. Vastert SJ, van Wijk R, D'Urbano LE, de Vooght KMK, de Jager W, Ravelli A, Magni-Manzoni S, Insalaco A, Cortis E, van Solinge WW, Prakken BJ, Wulffraat NM, de Benedetti F, Kuis W (2010) Mutations in the perforin gene can be linked to macrophage activation syndrome in patients with systemic onset juvenile idiopathic arthritis. Rheumatology 49:441–449

98. Voskoboinik I, Dunstone MA, Baran K, Whisstock JC, Trapani JA (2010) Perforin: structure, function, and role in human immunopathology. Immunol Rev 235:35–54
99. Voskoboinik I, Sutton VR, Ciccone A, House CM, Chia J, Darcy PK, Yagita H, Trapani JA (2007) Perforin activity and immune homeostasis: the common A91V polymorphism in perforin results in both presynaptic and postsynaptic defects in function. Blood 110:1184–1190
100. Voskoboinik I, Thia M-C, Fletcher J, Ciccone A, Browne K, Smyth MJ, Trapani JA (2005) Calcium-dependent plasma membrane binding and cell lysis by perforin are mediated through its C2 domain: a critical role for aspartate residues 429, 435, 483, and 485 but not 491. J Biol Chem 280:8426–8434
101. Wakeland EK, Liu K, Graham RR, Behrens TW (2001) Delineating the genetic basis of systemic lupus erythematosus. Immunity 15:397–408
102. Walsh CM, Matloubian M, Liu CC, Ueda R, Kurahara CG, Christensen JL, Huang MT, Young JD, Ahmed R, Clark WR (1994) Immune function in mice lacking the perforin gene. Proc Natl Acad Sci USA 91:10854–10858
103. Yañez DM, Manning DD, Cooley AJ, Weidanz WP, van der Heyde HC (1996) Participation of lymphocyte subpopulations in the pathogenesis of experimental murine cerebral malaria. J Immunol 157:1620–1624
104. Yawalkar N, Schmid S, Braathen LR, Pichler WJ (2001) Perforin and granzyme B may contribute to skin inflammation in atopic dermatitis and psoriasis. Br J Dermatol 144:1133–1139
105. Yoon J-W, Jun H-S (2005) Autoimmune destruction of pancreatic beta cells. Am J Ther 12:580–591
106. Young JD, Damiano A, DiNome MA, Leong LG, Cohn ZA (1987) Dissociation of membrane binding and lytic activities of the lymphocyte pore-forming protein (perforin). J Exp Med 165:1371–1382
107. Young LH, Joag SV, Zheng LM, Lee CP, Lee YS, Young JD (1990) Perforin-mediated myocardial damage in acute myocarditis. Lancet 336:1019–1021
108. Youssef A-R, Otley C, Mathieson PW, Smith RM (2004) Role of CD4+ and CD8+ T cells in murine skin and heart allograft rejection across different antigenic desparities. Transpl Immunol 13:297–304
109. Zhang Y (2008) I-TASSER server for protein 3D structure prediction. BMC Bioinformatics 9:40
110. Zur Stadt U, Beutel K, Weber B, Kabisch H, Schneppenheim R, Janka G (2004) A91V is a polymorphism in the perforin gene not causative of an FHLH phenotype. Blood 104:1909 (author reply 1910)
111. Zur Stadt U, Rohr J, Seifert W, Koch F, Grieve S, Pagel J, Strauss J, Kasper B, Nürnberg G, Becker C, Maul-Pavicic A, Beutel K, Janka G, Griffiths G, Ehl S, Hennies HC (2009) Familial hemophagocytic lymphohistiocytosis type 5 (FHL-5) is caused by mutations in Munc18-2 and impaired binding to syntaxin 11. Am J Hum Genet 85:482–492
112. Zur Stadt U, Schmidt S, Kasper B, Beutel K, Diler AS, Henter J-I, Kabisch H, Schneppenheim R, Nürnberg P, Janka G, Hennies HC (2005) Linkage of familial hemophagocytic lymphohistiocytosis (FHL) type-4 to chromosome 6q24 and identification of mutations in syntaxin 11. Hum Mol Genet 14:827–834

Chapter 12
The Role of MACPF Proteins in the Biology of Malaria and Other Apicomplexan Parasites

Joana Tavares, Rogerio Amino and Robert Ménard

Abstract Apicomplexans are eukaryotic parasites of major medical and veterinary importance. They have complex life cycles through frequently more than one host, interact with many cell types in their hosts, and can breach host cell membranes during parasite traversal of, or egress from, host cells. Some of these parasites make a strikingly heavy use of the pore-forming MACPF domain, and encode up to 10 different MACPF domain-containing proteins. In this chapter, we focus on the two most studied and medically important apicomplexans, *Plasmodium* and *Toxoplasma*, and describe the known functions of their MACPF polypeptide arsenal. Apicomplexan MACPF proteins appear to be involved in a variety of membrane-damaging events, making them an attractive model to dissect the structure-function relationships of the MACPF domain.

Keywords APC-β · Apicomplexa · *Plasmodium* · *Plasmodium* perforin-like proteins · SPECT2 · TgPLP1 · *Toxoplasma*

Abbreviations

APC-β	Apicomplexan perforin-like protein C-terminal β-sheet
CAD1	Constitutively activated cell death 1
CCP	Complement control protein
CDCs	Cholesterol-dependent cytolysins
EPM	Erythrocyte plasma membrane
FIMAC	Factor I membrane attack complex
HPM	Hepatocyte plasma membrane
LDL-A	Low density lipoprotein receptor domain A
MACPF	Membrane-attack complex/perforin
MACPF^{api}	MACPF domains of apicomplexans

J. Tavares · R. Amino · R. Ménard (✉)
Institut Pasteur, Unité de Biologie et Génétique du Paludisme, Département de Parasitologie et Mycologie, 28 rue du Docteur Roux 75015 Paris, France
e-mail: rmenard@pasteur.fr

G. Anderluh and R. Gilbert (eds.), *MACPF/CDC Proteins - Agents of Defence, Attack and Invasion*, Subcellular Biochemistry 80, DOI: 10.1007/978-94-017-8881-6_12,

MAOP	Membrane-attack ookinete protein
PLP	Perforin-like protein
PPLP	*Plasmodium* perforin-like protein
PV	Parasitophorous vacuole
PVM	Parasitophorous vacuole membrane
RBC	Red blood cell
SCR	Short consensus repeats
SPECT	Sporozoite microneme protein essential for cell traversal
TSR	Thrombospondin type I repat
WT	Wild type

Introduction

The membrane attack complex/perforin (MACPF) domain is conserved in many kingdoms of life. In vertebrates, it is present in the five components of the complement membrane attack complex (C6, C7, C8α, C8β, and C9) and perforin. The MACPF domain is found in proteins that confer plant immunity, such as the constitutively activated cell death 1 (CAD1) protein of *Arabidopsis thaliana*, and in cytolytic toxins from venomous sea anemones [25, 33, 42]. Bacteria also express MACPF domain-containing proteins, including non-pathogenic species such as most *Bacteroides* spp. of the microbiota of the human intestinal tract [43], and pathogenic species such as *Chlamydia* [29] or the insect pathogen *P. luminescens* [32]. Protozoa also produce MACPF proteins, including free-living predatory organisms, such as *Paramecium* and *Tetrahymena*, and parasites of the *Apicomplexa* phylum [20].

Apicomplexan parasites are an interesting case with respect to the MACPF domain. These organisms, some of which are of major medical and veterinary importance, convert between multiple forms during their life cycle, interact with many cell types in their hosts, and can breach host cell membranes from both sides. MACPF proteins are widespread in apicomplexan parasites, with many genera expressing multiple (up to seven) different versions [20]. Apicomplexan MACPF proteins thus constitute a valuable system to investigate the function of the MACPF domain. In addition, since all MACPF proteins studied so far are critical to parasite infectivity, they are attractive targets for anti-parasite control strategies.

Here, we overview the conservation of the MACPF domain in apicomplexans and the structural singularities of MACPF proteins in these parasites, and focus on the known contributions of the MACPF proteins in the life cycles of *Plasmodium* and *Toxoplasma*.

MACPF Proteins in *Apicomplexa* Parasites

Apicomplexans

The *Apicomplexa* phylum of protozoa contains almost exclusively parasitic organisms, which frequently exhibit complex life cycles involving more than one host. Some infect blood cells and rely on blood feeding invertebrates for transmission, such as *Plasmodium,* the agent of malaria, or *Theileria* and *Babesia*, which cause severe diseases in cattle [6]. Others invade the intestinal tract of animals and are transmitted via the fecal-oral route, such as *Toxoplasma* and *Cryptosporidium*, which cause severe pathologies and death in immuno-compromised patients [15, 39].

Most apicomplexan parasites are obligatorily intracellular, i.e., multiply only inside host cells, and continuously switch between an intracellular and an extracellular position in their hosts. Their extracellular stages (zoites) are typically polarized and motile cells. They use a sub-membrane actin-myosin motor to power a unique type of substrate-dependent motility, termed gliding motility, which propels them at high speeds (up to 5 μm/sec) without overt change in parasite shape [23]. In most cases apicomplexan zoites invade host cells inside a parasitophorous vacuole (PV), formed upon invagination of the host cell plasma membrane around the entering zoite [4, 8]. Inside the PV, the parasite multiplies into daughter zoites, which then egress the host cell and go on to invade new host cells [5, 12]. Some apicomplexan stages, including in *Plasmodium*, can traverse host cells, i.e., breach their plasma membrane and glide through their cytosol before egressing, without PV formation. Motility, cell traversal and cell invasion require the discharge of apical secretory organelles called micronemes, which are a defining feature of the phylum [37].

Apicomplexan MACPF Proteins

All genomes of apicomplexan parasites that have been sequenced so far encode MACPF domains, except that of *Cryptosporidium,* an early-branching clade in the phylum. Remarkably, among studied apicomplexans those transmitted by insects, e.g. *Plasmodium*, *Theileria* and *Babesia*, appear to produce more MACPF proteins (Fig. 12.1). *Plasmodium*, transmitted by *Anopheles* mosquitoes, expresses five MACPF proteins, while *Theileria* and *Babesia*, which are transmitted by ticks, produce at least six and seven MACPF proteins, respectively. Unlike *Plasmodium* (Haemosporididae), *Theileria* and *Babesia* (Piroplasmidae) escape from the enclosing PV shortly after invasion and replicate free within the host cell cytoplasm.

Structural homology modeling of the MACPF domain in PLP1 of *T. gondii* [19] confirmed that the characteristic structural features of MACPF domains, which

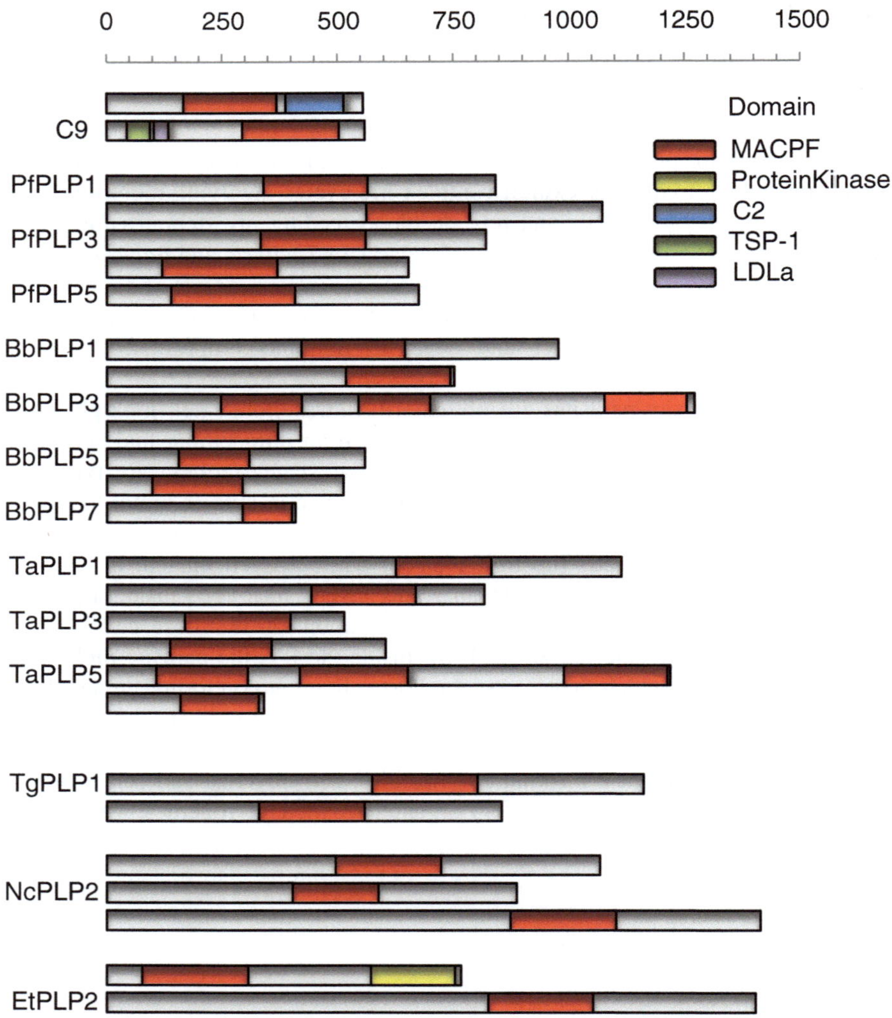

Fig. 12.1 Apicomplexan PLPs. Schematic representation of the genes encoding MACPF domains in *Plasmodium falciparum* (Pf), *Babesia bovis* (Bb), *Theileria annulata* (Ta), *Toxoplasma gondii* (Tg), *Neospora caninum* (Nc) and *Eimeria tennela (*Et). The MACPF domain is represented in *red*. Other classical domains such as C2 (*blue*), Thrombospondin type I repeat (TSP-1, *green*) and Low Density Lipoprotein receptor domain A (LDL-A, *purple*) are found, respectively, in perforin and complement factor 9. In apicomplexan PLPs only a kinase domain (*yellow*) is found in *Eimeria* PLP1

comprise a twisted, four-stranded β-sheet placed on top of two helical clusters [13, 32], are conserved in the MACPF domains of apicomplexans (MACPF^{api}). The helical clusters of MACPF^{api} domains exhibit the pattern of alternating hydrophobic and hydrophilic amino acids found in other MACPF domains [19], which convert into amphipathic β-hairpins and insert into the target membrane during pore-formation. The MACPF^{api} domains also contain a conserved insertion of a pair of anti-parallel α-helices, which is absent from canonical MACPF domains. This insertion is modeled to reside on the external face of the MACPF^{api} domain in a region that contains important residues for oligomerization of membrane-bound perforin monomers [19, 20]. MACPF domains exhibit limited sequence similarity but contain a signature [YW]-G-[TS]-H-[FY]-x(6)-G-G motif [32, 36], while MACPF^{api} domains display a signature pattern of W-x(2)-[FL]-[FI]-x(2)-[FY]-G-T-H-x(7)-G-G.

The MACPF domain is commonly found associated with N- and C-terminal domains, such as TSR, LDLRA, Sushi/CCP/SCR, FIMAC or C2, which are thought to control and/or target MACPF function. The MACPF^{api} domains characterized to date all share similar domain architectures with an N-terminal region, followed by a centrally located MACPF^{api} domain, and a C-terminal β-pleated sheet-rich domain unique to the apicomplexan phylum. However, the N-terminal region of MACPF^{api} proteins is not conserved across the phylum, varying considerably in length and sequence, suggesting that they might have unique functions in each MACPF^{api} protein. The C-terminal region of *Toxoplasma* PLP1, which consists of a β-sheet-rich domain, is reminiscent of C-terminal domains in other MACPF proteins, which play critical roles in binding to lipids or protein receptors in target membranes [35]. These C-terminal domains typically consist of β-pleated sheets, e.g., an EGF-like domain followed by a TSR in C8α or the β-sandwich immunoglobulin domain in cholesterol-dependent cytolysins (CDCs) [28]. The C-terminal domains of MACPF^{api} proteins contain multiple copies of a domain predicted to possess five β-strands [20]. This domain, named Apicomplexan Perforin-like protein C-terminal β-sheet (APC-β) domain, is $\sim$55-residue long, is generally present in three tandem copies, is found associated with MACPF^{api} domains across the apicomplexan phylum, and appears to be unique to that phylum [20].

MACPF Proteins in *Plasmodium*

Plasmodium *Life Cycle: In and Out and Through Host Cells*

Malaria, caused by *Plasmodium*, remains the most deadly parasitic disease in humans (WHO 2011 report, [24]), causing up to one million deaths per year. The *Plasmodium* parasite is transmitted during the bite of an *Anopheles* mosquito. During its life, the parasite adopts distinct forms adapted to the various tissues encountered and parasitized in both hosts, as schematized in Fig. 12.2. Central to the parasite life and causing the symptoms in the mammalian host are the cycles of parasite multiplication inside RBC. The parasite invades a RBC inside a PV and either produces 10–30 new parasites during asexual multiplication or differentiates into sexual forms. In either case, the parasite(s) ultimately leave the RBC. Sexual

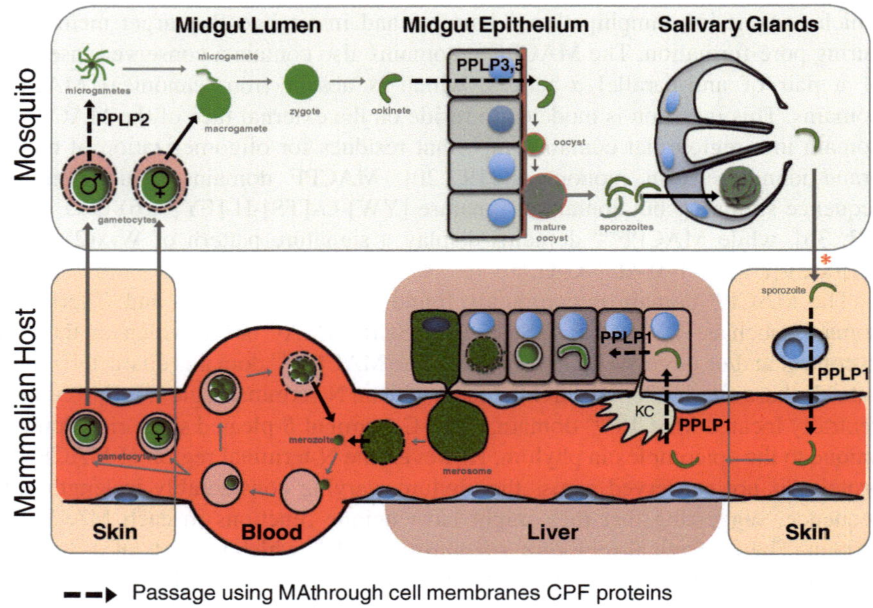

Fig. 12.2 *Plasmodium* life cycle through the mosquito vector and the mammalian host, and parasite-induced disruption of host cell membranes. During a blood meal, an infected anopheline mosquito injects sporozoites into the skin (*red asterisk*). The motile sporozoites traverse cells in the skin using PPLP1/SPECT2 to reach blood vessels, finding their way to the liver. In the liver sinusoids, sporozoites use PPLP1 to resist clearance by Kupffer cells and cross the sinusoids, getting access to hepatocytes. After traversing several hepatocytes, a sporozoite invades a final hepatocyte inside a PV, where it transforms into tens of thousands of merozoites. Merozoites egress hepatocytes inside merosomes, which are vesicles bound by the hepatocyte membrane shed into the sinusoid circulation that protects them from local macrophages (Kupffer cells, KC) in the liver sinusoids. Merozoites invade RBC inside a PV and asexually multiply, generating 10–30 new merozoites at each cycle and causing the symptoms of the disease. Intra-RBC parasites can also differentiate into gametocytes and, when ingested by the mosquito, into gametes. One female macrogamete and eight male microgametes egress from RBC; egress of the male microgametes from the RBC membrane requires PPLP2. Fertilization in the mosquito gut lumen/blood meal generates a zygote, which transforms into a motile ookinete. The ookinete traverses epithelial cells, using at least PPLP3 and PPLP5, and lodges underneath the basal lamina that separates the gut epithelium from the mosquito body cavity/circulation. Inside an oocyst, thousands of sporozoites are formed, which are released in the cavity, attach to the salivary glands, traverse cells in a MACPF-independent way and finally reach the salivary cavities and ducts, where they await transmission to a new host

reproduction takes the parasite inside a mosquito, successively in the midgut lumen, across the midgut epithelium, inside the mosquito circulation and through the salivary gland system. Once injected in the skin of the host during the mosquito bite, the parasite rapidly invades capillaries in the dermis and egresses sinusoids in the liver and eventually invades hepatocytes inside a PV, where it re-transforms into the RBC-infecting form.

MACPF Proteins in Plasmodium

Five genes encoding a MACPF domain are present in all *Plasmodium* genomes sequenced so far: *P. falciparum*, *P. vivax* and *P. knowlesi*, which cause malaria in humans, and *P. yoelli, P. berghei* and *P. chabaudi*, which infect rodents. Their products were initially called *Plasmodium* perforin-like proteins (PPLPs) [21]. The role that these proteins play in the parasite life cycle was addressed in *P. berghei*, a species that is particularly amenable to genetic manipulation [26]. Parasites in which the PPLP1-, PPLP2-, PPLP3- and PPLP5-encoding genes were inactivated were selected and cloned (only erythrocytic stages of *Plasmodium* can be genetically manipulated). This indicates that these genes are not important for parasite multiplication in, and egress from, erythrocytes, in agreement with gene expression analysis in *P. yoelii* and *P. falciparum* showing that the family is not expressed in blood-stage parasites [21]. Nonetheless, inactivation of the PPLP4-encoding gene, which is mostly expressed in the sexual and/or ookinete stages [14, 30] generated mutants that could not be cloned, raising the possibility that PPLP4 might act during parasite multiplication in the blood [11]. All characterized MACPF proteins locate to parasite micronemes [10, 17, 18, 21].

PPLP2: Male Gamete Egress from Erythrocytes

To egress erythrocytes, gametocytes must rupture both the parasitophorous vacuole membrane (PVM) and the erythrocyte plasma membrane (EPM) to fertilize as gametes in the mosquito midgut lumen (Fig. 12.2). In *P. berghei*, male, but not female gametocytes produce PPLP2. *P. berghei* PPLP2(-) parasites generate normally fertile female gametes, but infertile male gametes [9]. In the wild-type (WT), eight male flagellated gametes (one axoneme and one nucleus surrounded by a flagellar membrane) normally break out from the cell to rapidly attach to erythrocytes, forming exflagellation centers. In contrast, PPLP2(-) male gametocytes, while rupturing the PVM, fail to rupture the EPM, forming flagella that remain bundled and trapped in the intact EPM. However, the fertilization block is not complete. Ookinetes are occasionally formed, which in turn do not display any detectable defect in crossing the mosquito midgut epithelium, showing that the contribution of PPLP2 is restricted to egress of the male gametocyte.

PPLP3 and PPLP5: Ookinete Traversal of Insect Cells

Fertilization in the mosquito midgut generates a motile zygote, called ookinete, which needs to cross the midgut epithelium to reach the basal lamina, where it transforms into an oocyst. The ookinete wounds and traverses midgut epithelial cells [34], usually killing the traversed cell [45]. *P. berghei* parasites deficient in PPLP3/MAOP (membrane-attack ookinete protein) [18] or in PPLP5 [10] display the same phenotype: ookinetes are normally formed but fail to cross midgut epithelial cells and do not generate oocysts. In both cases, mutant ookinetes remain attached to the microvilli of the apical side of the midgut epithelium, sometimes causing the epithelial cell membrane to invaginate, without breaking [18]. Interestingly, injection of PPLP5-deficient ookinetes in the mosquito hemocoel (body cavity), i.e., on the other side of the midgut epithelium, restores parasite infectivity in subsequent stages, showing that PPLP5 activity is restricted to ookinete passage through the midgut epithelium barrier.

PPLP1: Sporozoite Traversal of Mammalian Cells

The oocyst, which derives from the ookinete, generates thousands of sporozoites, the only parasite stage that infects cells in both the mosquito and mammalian hosts. The oocyst releases sporozoites in the mosquito hemocoel, which invade the salivary gland cells to reach the salivary cavities and ducts. Sporozoites then reside extracellularly in the mosquito salivary system, where they await transmission during a mosquito bite [27].

The motile sporozoite can breach the plasma membrane of, and traverse host cells, including macrophages [41], hepatocytes and epithelial cell types [22]. PPLP1/SPECT2 (sporozoite protein essential for cell traversal 2) was the first MACPF-containing product characterized in *Plasmodium* [17, 21]. SPECT2 is expressed only in sporozoites and specifically in those present in the mammalian host, and is the sole PPLP produced by sporozoites. *P. berghei* SPECT2(-) parasites are normally infective until sporozoites have invaded the mosquito salivary glands [17]. They fail to wound hepatocytes and other host cell types in vitro and are poorly infective after injection into mice [17]. Intravital imaging showed that SPECT2 was important to (i) traverse host cells and freely locomote in the skin [1], (ii) traverse and resist killing by Kupffer cells, the resident macrophages in the liver sinusoids [40], and (iii) cross the sinusoidal barrier by breaching through endothelial cells and/or Kupffer cells [40]; Fig. 12.3). The SPECT2-mediated host cell traversal activity thus serves multiple purposes during the sporozoite journey, including progression through cellular barriers in the skin and across endothelia and preventing phagocytosis by host immune cells.

TgPLP1 of *Toxoplasma gondii*: Tachyzoite Egress from Host Cells

Toxoplasma causes congenital birth defects and severe encephalitis in immunocompromised individuals. Its life cycle consists of a sexual phase in cats and an asexual phase that can occur in virtually any warm-blooded animal. In the latter, *Toxoplasma* sporozoites invade intestinal epithelial cells, where they transform

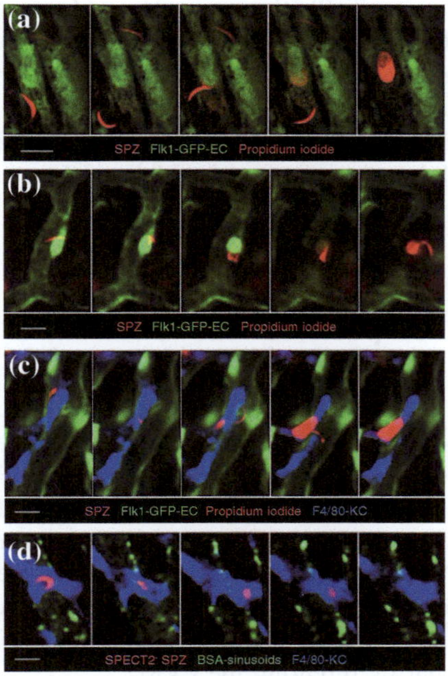

Fig. 12.3 Host cell traversal (CT) by *Plasmodium* sporozoites in the liver. Sporozoites are capable of cell traversal, i.e., breaching host cell plasma membranes, gliding through the cytosol and egressing. Sporozoite CT requires PPLP1/SPECT2 and can be detected by the incorporation of membrane-impermeant dye on the extracellular medium or alternatively by the loss of intracellular contents. **a** Time-lapse microscopy of a sporozoite (SPZ-*red*) traversing a primary liver sinusoidal GFP$^+$ endothelial cell (EC-*green*) isolated from a *flk1-gfp* mouse in the presence of propidium iodide (PI) and triggering the specific decrease of the GFP fluorescence intensity of EC and the simultaneous incorporation of PI in the EC (*red*). **b** Intravital imaging in the liver of mice showing the fading of the GFP and the incorporation of PI by the EC when a sporozoite (SPZ-*red*) crosses the sinusoid (Flk1-GFP-EC, *green*) and enters the *black* parenchyma. **c** During their journey in the sinusoids, motile sporozoites (SPZ-*red*) frequently interact and wound Kupffer cells (KC, resident macrophages) as detected by the incoporation of PI (*red)* in F4/80$^+$ cells (*blue*). **d** In contrast, PPLP1/SPECT2-deficient sporozoites (SPECT2$^-$ SPZ) associate with and are cleared by F4/80$^+$-KC (*blue*)

into tachyzoites, a rapidly multiplying parasite stage that lacks cell traversal ability. Tachyzoites spread via the blood stream to all organs and tissues of the body, where they invade host cells and convert into bradyzoites, the dormant stage that forms tissue cysts and can reactivate infection upon host immune depression.

The *Toxoplasma gondii* genome encodes two MACPF proteins, TgPLP1 and TgPLP2. Only TgPLP1 was functionally characterized. TgPLP1(-) tachyzoites are impaired in egressing from host cells and rupturing the PVM [19]. Interestingly, in cells containing both TgPLP1(-) and WT tachyzoites in distinct PVs, TgPLP1(-) tachyzoites readily egress their PV via the activity of TgPLP1 produced by WT tachyzoites, which shows that TgPLP1 can function from either side of the PVM.

Like all MACPF proteins, TgPLP1 contains N- and C-terminal domains flanking the MACPF domain. Biochemical analysis of TgPLP1 showed that it forms large, multimeric membrane-embedded complexes, and that recombinant TgPLP1 was sufficient for membrane disruption [31]. Structure-function analysis also showed that each of the central MACPF domain and the ß-sheet-rich C-terminal domain, but not the N-terminal domain, were required for activity, and that each of the N- and C-terminal domains were sufficient for membrane association.

Conclusions and Perspectives

It appears from the initial characterization of the roles of MACPF^{api} proteins that they mediate many, but not all host membrane-disruption events imposed by the life cycles of apicomplexan parasites. In *Plasmodium*, MACPF proteins are mainly used for cell traversal by the ookinete and the sporozoite, although the sporozoite appears to traverse salivary gland cells in a MACPF-independent way. A MACPF protein used by the *P. berghei* male, but not female gamete permits egress from erythrocytes, by rupturing the EPM. Merozoites, however, do not use a MACPF protein for egressing host cells. In contrast, *T. gondii* tachyzoites use a MACPF protein for egressing, but by rupturing the PVM.

Since the merozoite appears to be the only *Plasmodium* stage that does not use a MACPF to egress the host cell, it is interesting to note that merozoites egress hepatocytes by a complex mechanism that separates the rupturing of the PVM and of the host cell/hepatocyte membrane (HPM) [38]. After the rupture of the PVM, and the release of thousands of merozoites in the hepatocyte cytosol, the HPM initially remains intact. Merozoite-filled vesicles bound by the HPM (merosomes) bud off from infected hepatocytes into the sinusoid circulation. The merosome/hepatocyte membrane shields merozoites from local macrophages patrolling in the liver sinusoids, and only ruptures when merosomes have reached the lungs, where released merozoites safely invade erythrocytes [3].

Why does the ookinete use two, and possibly three, MACPF proteins to cross a single epithelial cell in the insect when the sporozoite use only one to traverse multiple cell types in different tissues in the mammal? Do the two or three MACPF proteins secreted by the ookinete act as a complex, analogous to the interaction of C6-C9 in the membrane attack complex? If so, what specific function is conferred by multiple MACPF domains in the pore-forming complex of the ookinete? The PPLP1/SPECT2 of the sporozoite might also act as a complex with non-MACPF proteins, including SPECT that has a similar contribution to SPECT2 in sporozoite infectivity [1, 16, 17]. Reconstitution of the entire pore-forming machineries at each stage will be essential for understanding their mode of action.

How do apicomplexan MACPF proteins act? Formation of large pore might be sufficient for a motile parasite (egressing or traversing) to squeeze in and move through. Pore formation may also permit the translocation of other effectors, e.g., membrane-damaging proteases or lipases (in the case of egress) or molecules facilitating the degradation of the cortical actin network (in the case of traversal). The pore may also cause ion fluxes that influence parasite egress. For example, ATP or K^+ efflux or Ca^{2+} influx might induce host cell contraction, initiate parasite motility, or activate parasite (SUB1, SERAs; [2, 44]) or host (calpains; [7]) proteases.

The receptor(s) recognized by apicomplexan MACPF proteins on the various membranes parasites encounter in insect and mammalian hosts are unknown. *Toxoplasma* PLP1 acts on either side of the PVM, and *Plasmodium* sporozoites need to rupture the membrane of the traversed cell from the outside and the inside. Both the N-terminal and C-terminal domains of TgPLP1 bind to membranes, although only the latter is required for MACPF activity while the former may regulate MACPF activity via a processing event [31]. Understanding how MACPF activity is regulated might help explain how *Plasmodium* sporozoites, which invade host cells by secreting micronemes and thus MACPF proteins, prevent the rupture of the nascent PVM.

Further studies in apicomplexans promise to yield interesting insights into the MACPF domain. The genetic tractability of these parasites, along with the diversity of membrane disruption events they cause, should help dissect the structure-function relationships and adaptive potential of the fascinating MACPF fold.

References

1. Amino R, Giovannini D, Thiberge S, Gueirard P, Boisson B, Dubremetz JF, Prévost MC, Ishino T, Yuda M, Ménard R (2008) Host cell traversal is important for progression of the malaria parasite through the dermis to the liver. Cell Host Microbe 3:88–96
2. Arastu-Kapur S, Ponder EL, Fonović UP, Yeoh S, Yuan F, Fonović M, Grainger M, Phillips CI, Powers JC, Bogyo M (2008) Identification of proteases that regulate erythrocyte rupture by the malaria parasite *Plasmodium falciparum*. Nat Chem Biol 4:203–213
3. Baer K, Klotz C, Kappe SH, Schnieder T, Frevert U (2007) Release of hepatic *Plasmodium yoelii* merozoites into the pulmonary microvasculature. PLoS Pathog 3:e171
4. Bargieri D, Lagal V, Tardieux I, Ménard R (2012) Host cell invasion by apicomplexans: what do we know? Trends Parasitol 28:131–135
5. Blackman MJ (2008) Malarial proteases and host cell egress: an 'emerging' cascade. Cell Microbiol 10:1925–1934
6. Chalmers RM, Giles M (2010) Zoonotic cryptosporidiosis in the UK—challenges for control. J Appl Microbiol 109:1487–1497
7. Chandramohanadas R, Davis PH, Beiting DP, Harbut MB, Darling C, Velmourougane G, Lee MY, Greer PA, Roos DS, Greenbaum DC (2009) Apicomplexan parasites co-opt host calpains to facilitate their escape from infected cells. Science 324:794–797
8. Cowman AF, Berry D, Baum J (2012) The cellular and molecular basis for malaria parasite invasion of the human red blood cell. J Cell Biol 198:961–971
9. Deligianni E, Morgan RN, Bertuccini L, Wirth CC, Silmon de Monerri NC, Spanos L, Blackman MJ, Louis C, Pradel G, Siden-Kiamos I (2013) A perforin-like protein mediates disruption of the erythrocyte membrane during egress of *Plasmodium berghei* male gametocytes. Cell Microbiol 15:1438–1455
10. Ecker A, Pinto SB, Baker KW, Kafatos FC, Sinden RE (2007) *Plasmodium berghei*: *Plasmodium* perforin-like protein 5 is required for mosquito midgut invasion in *Anopheles stephensi*. Exp Parasitol 116:504–508

11. Ecker A, Bushell ES, Tewari R, Sinden RE (2008) Reverse genetics screen identifies six proteins important for malaria development in the mosquito. Mol Microbiol 70:209–220
12. Graewe S, Rankin KE, Lehmann C, Deschermeier C, Hecht L, Froehlke U, Stanway RR, Heussler V (2011) Hostile takeover by *Plasmodium*: reorganization of parasite and host cell membranes during liver stage egress. PLoS Pathog 7:e1002224
13. Hadders MA, Beringer DX, Gros P (2007) Structure of C8alpha-MACPF reveals mechanism of membrane attack in complement immune defense. Science 317:1552–1554
14. Hall N, Karras M, Raine JD, Carlton JM, Kooij TW, Berriman M, Florens L, Janssen CS, Pain A, Christophides GK, James K, Rutherford K, Harris B, Harris D, Churcher C, Quail MA, Ormond D, Doggett J, Trueman HE, Mendoza J, Bidwell SL, Rajandream MA, Carucci DJ, Yates JR 3rd, Kafatos FC, Janse CJ, Barrell B, Turner CM, Waters AP, Sinden RE (2005) A comprehensive survey of the Plasmodium life cycle by genomic, transcriptomic, and proteomic analyses. Science 307:82–86
15. Innes EA (2010) A brief history and overview of *Toxoplasma gondii*. Zoonoses Public Health 57:1–7
16. Ishino T, Yano K, Chinzei Y, Yuda M (2004) Cell-passage activity is required for the malarial parasite to cross the liver sinusoidal cell layer. PLoS Biol 2:E4
17. Ishino T, Chinzei Y, Yuda M (2005) A *Plasmodium sporozoite* protein with a membrane attack complex domain is required for breaching the liver sinusoidal cell layer prior to hepatocyte infection. Cell Microbiol 7:199–208
18. Kadota K, Ishino T, Matsuyama T, Chinzei Y, Yuda M (2004) Essential role of membrane-attack protein in malarial transmission to mosquito host. Proc Natl Acad Sci USA 101:16310–16315
19. Kafsack BF, Pena JD, Coppens I, Ravindran S, Boothroyd JC, Carruthers VB (2009) Rapid membrane disruption by a perforin-like protein facilitates parasite exit from host cells. Science 323:530–533
20. Kafsack BF, Carruthers VB (2010) Apicomplexan perforin-like proteins. Commun Integr Biol 3:18–23
21. Kaiser K, Camargo N, Coppens I, Morrisey JM, Vaidya AB, Kappe SH (2004) A member of a conserved Plasmodium protein family with membrane-attack complex/perforin (MACPF)-like domains localizes to the micronemes of sporozoites. Mol Biochem Parasitol 133:15–26
22. Mota MM, Pradel G, Vanderberg JP, Hafalla JC, Frevert U, Nussenzweig RS, Nussenzweig V, Rodríguez A (2001) Migration of *Plasmodium sporozoites* through cells before infection. Science 291:141–144
23. Münter S, Sabass B, Selhuber-Unkel C, Kudryashev M, Hegge S, Engel U, Spatz JP, Matuschewski K, Schwarz US, Frischknecht F (2009) *Plasmodium sporozoite* motility is modulated by the turnover of discrete adhesion sites. Cell Host Microbe 6:551–562
24. Murray CJ, Rosenfeld LC, Lim SS, Andrews KG, Foreman KJ, Haring D, Fullman N, Naghavi M, Lozano R, Lopez AD (2012) Global malaria mortality between 1980 and 2010: a systematic analysis. Lancet 379:413–431
25. Oshiro N, Kobayashi C, Iwanaga S, Nozaki M, Namikoshi M, Spring J, Nagai H (2004) A new membrane-attack complex/perforin (MACPF) domain lethal toxin from the nematocyst venom of the Okinawan sea anemone Actineria villosa. Toxicon 43:225–228
26. Philip N, Orr R, Waters AP (2013) Transfection of rodent malaria parasites. Methods Mol Biol 923:99–125
27. Pimenta PF, Touray M, Miller L (1994) The journey of malaria sporozoites in the mosquito salivary gland. J Eukaryot Microbiol 41:608–624
28. Polekhina G, Giddings KS, Tweten RK, Parker MW (2005) Insights into the action of the superfamily of cholesterol-dependent cytolysins from studies of intermedilysin. Proc Natl Acad Sci USA 102:600–605
29. Ponting CP (1999) Chlamydial homologues of the MACPF (MAC/perforin) domain. Curr Biol 9:R911–913

30. Raibaud A, Brahimi K, Roth CW, Brey PT, Faust DM (2006) Differential gene expression in the ookinete stage of the malaria parasite Plasmodium berghei. Mol Biochem Parasitol 150:107–113
31. Roiko MS, Carruthers VB (2013) Functional dissection of toxoplasma gondii perforin-like protein 1 reveals a dual domain mode of membrane binding for cytolysis and parasite egress. J Biol Chem 288:8712–8725
32. Rosado CJ, Buckle AM, Law RH, Butcher RE, Kan WT, Bird CH, Ung K, Browne KA, Baran K, Bashtannyk-Puhalovich TA, Faux NG, Wong W, Porter CJ, Pike RN, Ellisdon AM, Pearce MC, Bottomley SP, Emsley J, Smith AI, Rossjohn J, Hartland EL, Voskoboinik I, Trapani JA, Bird PI, Dunstone MA, Whisstock JC (2007) A common fold mediates vertebrate defense and bacterial attack. Science 317:1548–1551
33. Rosado CJ, Kondos S, Bull TE, Kuiper MJ, Law RH, Buckle AM, Voskoboinik I, Bird PI, Trapani JA, Whisstock JC, Dunstone MA (2008) The MACPF/CDC family of pore-forming toxins. Cell Microbiol 10:1765–1774
34. Shahabuddin M, Pimenta PF (1998) *Plasmodium gallinaceum* preferentially invades vesicular ATPase-expressing cells in *Aedes aegypti* midgut. Proc Natl Acad Sci USA 95:3385–3389
35. Shatursky O, Bayles R, Rogers M, Jost BH, Songer JG, Tweten RK (2000) *Clostridium perfringens* beta-toxin forms potential-dependent, cation-selective channels in lipid bilayers. Infect Immun 68:5546–5551
36. Slade DJ, Lovelace LL, Chruszcz M, Minor W, Lebioda L, Sodetz JM (2008) Crystal structure of the MACPF domain of human complement protein C8 alpha in complex with the C8 gamma subunit. J Mol Biol 379:331–342
37. Singh S, Chitnis CE (2012) Signalling mechanisms involved in apical organelle discharge during host cell invasion by apicomplexan parasites. Microbes Infect 14:820–824
38. Sturm A, Amino R, van de Sand C, Regen T, Retzlaff S, Rennenberg A, Krueger A, Pollok JM, Menard R, Heussler VT (2006) Manipulation of host hepatocytes by the malaria parasite for delivery into liver sinusoids. Science 313:1287–1290
39. Suarez CE, Noh S (2011) Emerging perspectives in the research of bovine babesiosis and anaplasmosis. Vet Parasitol 180:109–125
40. Tavares J, Formaglio P, Thiberge S, Mordelet E, Van Rooijen N, Medvinsky A, Ménard R, Amino R (2013) Role of host cell traversal by the malaria sporozoite during liver infection. J Exp Med 210:905–915
41. Vanderberg JP, Chew S, Stewart MJ (1990) *Plasmodium sporozoite* interactions with macrophages in vitro: a videomicroscopic analysis. J Protozool 37:528–536
42. Voskoboinik I, Smyth MJ, Trapani JA (2006) Perforin-mediated target-cell death and immune homeostasis. Nat Rev Immunol 6:940–952
43. Xu Q, Abdubek P, Astakhova T, Axelrod HL, Bakolitsa C, Cai X, Carlton D, Chen C, Chiu HJ, Clayton T, Das D, Deller MC, Duan L, Ellrott K, Farr CL, Feuerhelm J, Grant JC, Grzechnik A, Han GW, Jaroszewski L, Jin KK, Klock HE, Knuth MW, Kozbial P, Krishna SS, Kumar A, Lam WW, Marciano D, Miller MD, Morse AT, Nigoghossian E, Nopakun A, Okach L, Puckett C, Reyes R, Tien HJ, Trame CB, van den Bedem H, Weekes D, Wooten T, Yeh A, Zhou J, Hodgson KO, Wooley J, Elsliger MA, Deacon AM, Godzik A, Lesley SA, Wilson IA (2010) Structure of a membrane-attack complex/perforin (MACPF) family protein from the human gut symbiont *Bacteroides thetaiotaomicron*. Acta Crystallogr, Sect F: Struct Biol Cryst Commun 66:1297–1305
44. Yeoh S, O'Donnell RA, Koussis K, Dluzewski AR, Ansell KH, Osborne SA, Hackett F, Withers-Martinez C, Mitchell GH, Bannister LH, Bryans JS, Kettleborough CA, Blackman MJ (2007) Subcellular discharge of a serine protease mediates release of invasive malaria parasites from host erythrocytes. Cell 131:1072–1083
45. Zieler H, Dvorak JA (2000) Invasion in vitro of mosquito midgut cells by the malaria parasite proceeds by a conserved mechanism and results in death of the invaded midgut cells. Proc Natl Acad Sci USA 97:11516–11521

Chapter 13
Chlamydial MACPF Protein CT153

Lacey D. Taylor and David E. Nelson

Abstract Chlamydiae are obligate intracellular bacterial parasites that infect a wide range of metazoan hosts. Some *Chlamydia* species are important causes of chronic inflammatory diseases of the ocular, genital and respiratory tracts in humans. Genes located in a variable region of chlamydial genomes termed the plasticity zone are known to be key determinants of pathogenic diversity. The plasticity zone protein CT153, present only in select species, contains a membrane attack complex/perforin (MACPF) domain, which may mediate chlamydial interactions with the host cell. CT153 is present throughout the *C. trachomatis* developmental cycle and is processed into polypeptides that interact with membranes differently than does the parent protein. Chlamydiae interact extensively with membranes from the time of invasion until they eventually exit host cells, so numerous roles for a MACPF protein in pathogenesis of these pathogens are conceivable. Here, we present an overview of what is known about CT153 and highlight potential roles of a MACPF family protein in a group of pathogens whose intracellular development is marked by a series of interactions with host cell membranes and organelles. Finally, we identify new strategies for identifying CT153 functions made feasible by the recent development of a basic toolset for genetic manipulation of chlamydiae.

Keywords CT153 · *Chlamydia trachomatis* · Plasticity zone · MACPF domain · MIR domain

---

L. D. Taylor (✉)
Laboratory of Intracellular Parasites, Rocky Mountain Laboratories, National Institute of Allergy and Infectious Disease, National Institutes of Health, 903 S. 4th Street, Hamilton MT 59840, USA
e-mail: ltaylor113@hotmail.com

D. E. Nelson
Department of Microbiology and Immunology, Indiana University School of Medicine MS411B, 635 North Barnhill Drive, Indianapolis IN 46202, USA

G. Anderluh and R. Gilbert (eds.), *MACPF/CDC Proteins - Agents of Defence, Attack and Invasion*, Subcellular Biochemistry 80, DOI: 10.1007/978-94-017-8881-6_13,

Abbreviations

CDC	Cholesterol dependent cytolysins
EB	Elementary body
LGV	Lymphogranuloma venereum
MACPF	Membrane attack complex/perforin
MIR	Mannosyltransferase, inositol 1,4,5-triphosphate receptor, ryanodine receptor
PLD	Phospholipase D
RB	Reticulate body
TgPLP1	*Toxoplasma gondii* PLP1

Chlamydial Species and Disease

Organisms in the Chlamydiaceae are all Gram-negative obligate intracellular parasites that must infect metazoan host organisms to survive and replicate. The genus *Chlamydia* contains nine cultivated species at present: *Chlamydia trachomatis*, *C. pneumoniae*, *C. muridarum*, *C. caviae*, *C. pecorum*, *C. felis*, *C. suis*, *C. abortus*, and *C. psittaci* although recent cultivation-independent studies suggest more species exist [16]. *C. trachomatis* and *C. pneumoniae* are human pathogens responsible for chronic inflammatory diseases of mucosal surfaces. *C. pneumoniae* causes atypical pneumonia and may promote atherosclerosis in humans. The genomes of all *C. pneumoniae* isolates are almost identical, and with exception of duplication of the tyrP gene, few genetic differences have been identified that contribute to pathogenic diversity of the existing isolates of this species [21]. The genomes of *C. trachomatis* strains are also highly similar, but strain subgroups differ in their normal tissue tropisms and cause distinct diseases. *C. trachomatis* serovars A–C infect the conjunctiva and elicit trachoma, which can cause blindness. Serovar D–K strains infect columnar epithelial cells of the urogenital tract and trigger acute and chronic inflammatory responses, which can manifest as urethritis in men and pelvic inflammatory disease and infertility in women. *C. trachomatis* serovars L1–L3 cause lymphogranuloma venereum (LGV). Unlike the also sexually transmitted serovar D–K strains, LGV strains disseminate to lymph nodes where they induce massive inflammatory responses and tissue destruction [63–65].

Chlamydial Developmental Cycle

Despite major differences in host range, tissue tropism, and disease pathology, all characterized *Chlamydia* spp. share a similar bi-phasic developmental cycle that includes transition between two easily distinguishable morphological forms

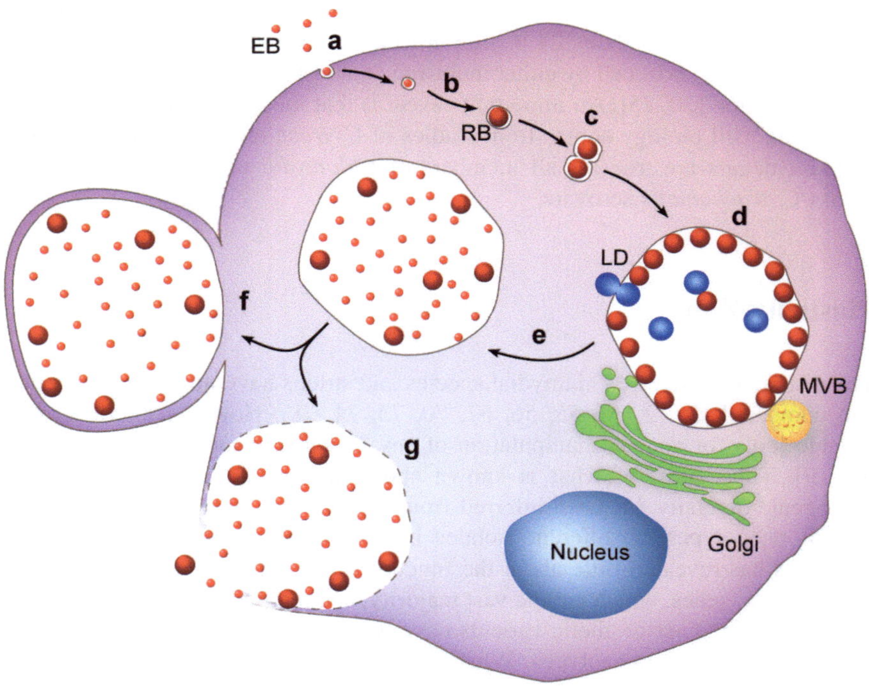

Fig. 13.1 Overview of the chlamydial developmental cycle. *a* EBs attach to cells, induce uptake at the plasma membrane and remain confined to membrane bound vacuole (inclusion) for the entire developmental cycle. Early after infection, *b* EBs differentiate to RBs, and *c* multiple nascent inclusions fuse to form a single vacuole. *d* RBs closely associate with the inclusion membrane to secrete effector proteins into the host cell cytosol and to parasitize host-derived nutrients from intracellular vacuoles and organelles. By mid-cycle, RBs continue to replicate by binary fission and *e* asynchronously differentiate to EBs. Late in development, chlamydiae orchestrate the release of organisms from the cell by *f* extrusion of the plasma membrane-encased inclusion or by *g* lytic rupture of the inclusion and cellular membranes. *EB* elementary bodies. *RB* reticulate bodies. *MVB* multivesicular bodies. *LD* lipid droplets

(Fig. 13.1). The infectious metabolically dormant elementary body (EB) invades host cells and remains inside a membrane vesicle called an inclusion. EBs differentiate into metabolically active non-infectious reticulate bodies (RBs) that grow and replicate by binary fission inside the inclusion [44]. Following infection, the inclusion rapidly dissociates from the endocytic pathway, circumvents fusion with lysosomes, and migrates along microtubules to the microtubule organizing center to reside in the peri-Golgi region of the cell [23, 28, 66]. The inclusion selectively interacts with host trafficking pathways and organelles to acquire host-derived lipids and possibly other nutrients [62]. In the presence of the cytokine interferon-gamma, penicillin or during tryptophan starvation chlamydiae can enter a persistent growth state in which RBs transform into enlarged non-replicating forms [5]. Chlamydiae can rapidly resume the normal developmental cycle upon removal of these stresses. *C. trachomatis* RBs begin to asynchronously differentiate to EBs starting about 24 h after infection, and these EBs are eventually released from the host cell by either inclusion and host cell lysis or extrusion of the intact inclusion [30, 75]. An important caveat is that most of what is known about chlamydial cell biology comes from studies of *C. trachomatis* serovar L2, not all of these details are true for all members of *Chlamydia* nor are they conserved across *C. trachomatis* serovars.

Plasticity Zone

The genomes of diverse chlamydial species and strains have now been sequenced and compared [2, 12, 36, 55, 56, 67, 70, 73, 74, 78]. However, there were no reliable tools for genetic manipulation of any *Chlamydia* spp. until recently [19, 37, 79]. Thus, most of what is known about the genetic basis of chlamydial pathogenic diversity has been inferred from comparisons of genomes in the context of phenotypes of different isolates in culture and in vivo. Unfortunately, genomics has revealed little about the functions of highly conserved Chlamydiaceae-specific genes, which are the vast majority, and there have been few instances where specific genes or allelic variants have been unambiguously linked to strain- and species-specific phenotypes. What we can be sure of is that chlamydial genomes have a high degree of similarity in terms of size and synteny, and most genes are represented in all chlamydial species. This indicates chlamydiae contain a highly conserved core genome that mediates common functions. Genetic differences obviously must underlie differences in behavior of chlamydiae. The difficulty has been identifying the genes most relevant to chlamydial pathogenic diversity.

The most variable region of chlamydial genomes is limited to ~50 kbp near the replication terminus and was termed the plasticity zone by Read and colleagues [55]. Characterized genes in the plasticity zone encode strain- and species-variable alleles that have been associated with in vivo tissue tropism and immune avoidance [9, 42, 47]. For example, presence of tryptophan biosynthesis and putative cytotoxin genes in the plasticity zone has been correlated with in vivo infection tropism and evasion of host immune effectors [6, 9, 11, 42, 46, 47]. Chlamydial phospholipase D (PLD) superfamily genes (*ct154–ct158*) located in the plasticity zones of some Chlamydiaceae are believed to be involved with chlamydial acquisition and processing of host-derived lipids [52]. The plasticity zone PLDs may be important for survival late in the developmental cycle and during recovery from interferon-mediated persistent infections [45]. At least one of these proteins, CT156/Lda-1, localizes to host cell lipid droplets [38]. Precisely how these proteins function is unclear as their structures and enzymology have not been thoroughly characterized.

CT153

The *C. trachomatis* plasticity zone gene *ct153* encodes a protein with substantial homology to membrane attack complex/perforin (MACPF) domain-containing proteins. Full-length *ct153* orthologs are found in all *C. trachomatis*, *C. muridarum*, *C. felis* and *C. psittaci* strains sequenced to date [2, 12, 55, 70, 73, 78]. Orthologs are also present in the plasticity zones of *C. pneumoniae* and *C. caviae*, but these have accumulated multiple frameshift mutations, whereas no genes with significant homology to *ct153* are present in *C. abortus* [36, 56, 67, 74]. In *C. trachomatis*, *ct153* orthologs are highly conserved and are located immediately upstream from the plasticity zone PLD genes suggesting the proteins share concomitant functions [12, 50, 55, 70, 73]. There is no evidence of recent horizontal gene transfer of *ct153* between *C. trachomatis* or the closely related species, *C. muridarum* and *C. felis*, and the phylogenetically distant *C. psittaci*, indicating that CT153 may have been essential in the last common ancestor of these strains. Why CT153 is dispensable in some species that diverged from this ancestor is unknown but suggests *ct153* has a niche-specific function. No obvious phenotype or characteristic neatly sorts *Chlamydia* species by presence of this gene so these comparisons have not suggested a function for CT153. Chlamydiae also interact with multiple host and host-derived phospholipid membranes (plasma membrane, early endosome membrane, inclusion membrane, exocytic vesicles, lipid droplets) that could conceivably be targeted by MACPF proteins, which further complicates assigning potential roles to CT153.

Members of the MACPF domain family of proteins share a canonical domain architecture of an N-terminal region of variable length followed by a membrane-spanning domain and C-terminal motifs that typically mediate binding of target membranes [22]. CT153 exhibits a similar domain arrangement. The CT153 N-terminal sequence is unique to *Chlamydia* and contains no conserved domains. The C-terminal amino acid residues 427–621 of CT153 share homology with the MACPF domain [50]. In lytic components of the human immune system, perforin and complement 9, this domain maps to the membrane-spanning regions of the proteins [49]. The crystal structures of prokaryotic and eukaryotic MACPF domain-containing proteins show the MACPF domain is structurally homologous to pore-forming cholesterol dependent cytolysins (CDC) of Gram-positive bacteria [25, 60, 68]. The domain signature is a conserved core fold that controls conformational rearrangements of alpha helices into the membrane-spanning beta-hairpins causing membrane insertion of the protein. CT153 also contains a C-terminal mannosyltransferase, inositol 1,4,5-triphosphate receptor, ryanodine receptor (MIR) domain (amino acids 759–786) that may confer a ligand transferase function or an ability to recognize membranes [51].

Proteins containing the MACPF domain are broadly represented in metazoans and prokaryotes, and their diverse ranges of functions are connected by the ability of the domain to interact with cellular membranes. MACPF/CDC domain superfamily proteins are secreted as monomers that oligomerize to form lytic pores and function in non-lytic membrane interactions [61, 76]. Some of these proteins can transiently perturb membrane morphology without destroying membrane integrity [22]. For example, MACPF domain-containing proteins from *Toxoplasma* and *Plasmodium* facilitate parasite egress from cells and allow parasites to transverse host cells, respectively [32, 33, 35, 59]. *Listeria monocytogenes* listeriolysin O, a secreted cytolysin, and listerial phospholipases mediate pathogen escape from the nascent membrane bound vacuole and disrupt host membranes during intercellular dissemination [17, 69]. Presence of a MACPF domain in CT153 and proximity of this gene to PLDs in the plasticity zone suggests CT153 interacts with host cell membranes and is involved with membrane or lipid modification [50].

CT153 in Chlamydial Pathogenesis

Chlamydiae have evolved sophisticated ways of manipulating host cells to establish their intracellular niche, access nutrients and escape at the end of the developmental cycle. Species variable alleles in the plasticity zone play important roles in chlamydial niche specificity and disease diversity. These aspects of chlamydial biology imply roles for CT153 in chlamydial pathogenesis. A recent study showed CT153 is expressed by all 15 human *C. trachomatis* serovars and provided the first characterization of a chlamydial MACPF domain-containing protein [72]. CT153 orthologs from other chlamydial species have not been studied extensively; therefore this review will focus on the potential functions of *C. trachomatis* CT153.

Early Host Cell Interactions

MACPF proteins interact with membranes, and one of the first barriers chlamydial EBs face is crossing the host cell plasma membrane during pathogen-mediated endocytosis. Proteins that mediate chlamydial attachment and entry are often enriched in EBs and Taylor et al. showed the full-length form of CT153 (p91) predominates in *C. trachomatis* EBs, which was confirmed by shotgun proteomic analysis of EB proteins in another study [39]. Most MACPF domain-containing proteins are secreted as soluble polypeptides that insert into target membranes [1], which suggests if CT153 functions in chlamydial entry it would need to be exposed on the EB surface or secreted into host cells. Anti-CT153 polyclonal sera did not react with or neutralize unfixed EBs indicating CT153 is not displayed on the EB surface. However, other effectors that mediate chlamydial attachment and entry are also not displayed on free EBs and are only secreted after EBs attach to host cells. The actin recruiting effector protein, TARP, is secreted from a pre-formed type III secretion apparatus on EBs following a second irreversible step in pathogen-mediated endocytosis [14]. Chlamydial polymorphic outer membrane proteins, a large family of autotransporter proteins, may also only translocate their effector domains at specific times in the developmental cycle [39, 48]. CT153 does not appear to have canonical type II or type III secretion signals or characteristics of other chlamydial autotransporter proteins, thus if it is secreted from EBs this occurs via an unknown pathway. We are unaware of attempts to directly address this question using immunoelectron microscopy or surrogate secretion systems. Human perforin molecules induce vesiculation on the plasma membrane of eukaryotic cells triggering an endocytosis-like event [53]. If CT153 is secreted or exposed during entry it might similarly induce membrane invaginations that facilitate EB uptake. Mechanisms of chlamydial entry vary even among *C. trachomatis* isolates, and not all chlamydiae encode CT153, so this protein would be more likely to play an accessory, as opposed to a primary, role in entry.

CT153 rapidly undergoes host-mediated or autoproteolytic cleavage during EB invasion of host cells [72]. Infection-dependent proteolysis of CT153 separates the N-terminal p41 from the C-terminal p57 fragment containing the MACPF and MIR domains, and this is independent of chlamydial transcription and translation. In contrast to p41, the p91 and p57 are stable up to 6 h post-infection indicating distinct functions for the intact and processed polypeptides in early host cell interactions. Some MACPF proteins require cleavage for activation [77], but it is unknown if p91 processing is necessary for activation of CT153.

Inclusion Fusion

When cells are infected with multiple *C. trachomatis* EBs, the individual inclusions eventually fuse to form a single vacuole [8]. In contrast, *C. caviae* inclusions do not fuse and display a multilobed inclusion structure [58]. IncA, which forms SNARE-like complexes on the cytosolic face of inclusion membranes for both species, has been implicated in homotypic inclusion membrane fusion for *C. trachomatis* [18, 71]. Differences in fusogenicity between the two species indicate additional *C. trachomatis*-specific factors mediate inclusion fusion. The presence of CT153 in *C. trachomatis* and absence in *C. caviae* suggest CT153 participates in fusion of *C. trachomatis* inclusions. In this scenario, IncA could tether adjacent inclusions in close proximity to one another while insertion of CT153 into inclusion membranes could destabilize the bilayers to promote fusion.

Mid-Developmental Cycle and Nutrient Acquisition

Transcriptome analysis of the *C. trachomatis* developmental cycle showed *ct153* transcription initiates early in infection, a time in which RBs are actively replicating, and peaks when RBs begin to differentiate to EBs [7]. During these time points, cleavage of CT153 results in a p41 N-terminal fragment and a p57 polypeptide containing the C-terminal MACPF and MIR domains similar to those seen following early infection [72]. The full-length and processed forms of CT153 exhibit differences in their ability to interact with membranes. The p41 and p57 are associated with the insoluble fractions of *C. trachomatis*-infected cells, and p91 is found in both insoluble and soluble membrane fractions [72]. These differences point to specialized functions for each form of the protein. The pore-forming activity of *Toxoplasma gondii* PLP1 (TgPLP1) is controlled by proteolytic cleavage of the N-terminal portion of the protein during trafficking [59]. Thus, proteolysis of CT153 most likely occurs after the protein is secreted or once it is associated with its target membrane and may regulate the function of CT153. Soluble MACPF domain protein monomers assemble as oligomers to form pores in membranes. In SDS-PAGE gels of *C. trachomatis*-infected cells, CT153 migrated primarily as p91, p57 and p41 polypeptides and as high molecular weight complexes that were also observed in gels of purified recombinant protein (Taylor, unpublished data). The presence of highly stable complexes resistant to SDS and boiling suggests higher ordered CT153 oligomeric structures are formed during infection, but this needs to be confirmed by further studies.

Confocal and transmission electron microscopy studies showed CT153 is expressed by RBs that are closely positioned adjacent to the inclusion. These RBs had increased expression of Hsp60, a chlamydial chaperonin that is upregulated by stress [4, 54]. This suggests RBs that express CT153 may have specialized metabolic capabilities. CT155, a plasticity zone PLD, localizes to similar RBs [45], consistent with the hypothesis that CT153 and the plasticity zone PLDs are functionally related. In addition to roles in parasite and protein translocation across membranes [32, 40] members of the MACPF family participate in cytolytic functions and have the ability to disrupt membrane stability [20, 22, 26]. RBs in close proximity to the inclusion membrane respond to cytosolic signals, acquire host nutrients, and secrete proteins into the host cell cytosol. A chlamydial membrane-spanning pore protein could mediate these activities. Thus, CT153, along with the plasticity zone PLD proteins, might participate in nutrient acquisition or membrane modification unique to *C. trachomatis*.

Chlamydiae have undergone reductive genomic evolution and depend upon their host for many metabolites [41]. *C. trachomatis* encodes several nutrient uptake systems and acquires nutrients by redirecting transport vesicles and intracellular organelles [62, 70]. Host-derived lipids are captured via multiple routes [27, 82]. *Chlamydia* induces Golgi fragmentation and intercepts Golgi-derived exocytic vesicles from which cholesterol, sphingomyelin and possibly other nutrients are acquired [10, 24, 29]. Inclusions sequester lipid droplets and fuse with multivesicular bodies, which both contain sphingomyelin and lysobiphosphatidic acid [3, 15, 57]. Presence of CT153 and the plasticity zone PLDs is positively correlated with the ability to recruit lipid droplets [56, 70, 72]. The inclusion membrane protein, IncA, Lda proteins 1–3, CT153 and the PLD paralogs have all been implicated in lipid droplet acquisition [15, 38, 45, 72], but how *C. trachomatis* targets and assimilates the organelles is unknown. Lipid droplets accumulate at the periphery of the inclusion. Entire organelles are transported across the inclusion at sites enriched with IncA and Lda proteins and are consumed in the inclusion lumen by RBs [15]. *Listeria* and *Rickettsia* employ MACPF domain-containing proteins and phospholipases to modify host membranes [17, 69, 81], and as in the delivery of granzymes by perforin, MACPF proteins can translocate other effector proteins across membranes [40, 43]. RBs expressing CT153 are positioned at sites of lipid droplet translocation and are closely associated with intra-inclusion lipid droplets [72]. Therefore CT153 may play a role in the capture and degradation of these organelles by remodeling membranes at the site of lipid droplet translocation or by serving as a conduit to transport lipid-modifying enzymes.

Exit from the Host Cell

Chlamydiae exit the host cell either by a process termed extrusion, an exocytosis-like event, or by cellular lysis [30, 75]. The *chlamydia*-containing inclusion pinches off and exits the cell enveloped in a layer of the eukaryotic plasma membrane leaving the host cell membrane intact during extrusion. Extrusion requires formation of actin coats on the inclusion membrane and is dependent upon chlamydial protein synthesis suggesting this is a pathogen-directed process mediated through host cell signaling pathways [13, 30]. The second exit pathway, lysis, consists of sequential rupture of the inclusion, cellular and plasma membranes that kills the host cell. Inclusion lysis can be blocked by cysteine protease inhibitors, and rupture of the plasma membrane is dependent on extracellular calcium influx. It has yet to be determined if lysis or extrusion is more relevant to EB escape from epithelial cells in vivo. It is also unclear how broadly the extrusion mechanism is conserved or if it is correlated with presence of a CT153 ortholog.

The mode of chlamydial exit from the host cell is most likely *Chlamydia*-directed since the process for either pathway begins from within the inclusion [30]. The mechanisms the organism uses to decide which pathway to follow likely involve intricate interactions between the inclusion and cell that are established late in the developmental cycle. Levels of CT153 in the inclusion lumen may be a contributing factor to the decision to exit by extrusion or lysis. The amount of CT153 in *C. trachomatis*-infected cells steadily increases during intracellular development [72]. Levels are low early in infection and the inclusion membrane is stable at that time. As the developmental cycle progresses, infectious EBs are formed and remaining RBs continue to express CT153. Accumulation of secreted CT153 might cue lytic exit. If this hypothesis is true we would expect strains lacking CT153 to primarily exit via extrusion.

Numerous intracellular parasites utilize pore-forming proteins and phospholipases to escape from vacuoles, and there are multiple examples where MACPF-domain containing proteins of other pathogens mediate exit from host cells [31, 34]. *T. gondii* egress from the cell is achieved by sequential rupture of the parasite vacuole followed by the eukaryotic plasma membrane [35]. Permeabilization of the parasitophorous vacuolar membrane is mediated by pores of the MACPF domain-containing protein TgPLP1 [35, 59]. Analogous to the action of TgPLP1 in escape of *T. gondii*, CT153 could form pores in the inclusion membrane. These pores could directly disrupt the membrane or provide a passage for other effector proteins involved in host cell lysis.

Escape by cellular lysis requires the pathogen to breach the inclusion and plasma membranes. It is unknown if the same or different mechanisms are used to rupture these two membranes. Rupture of the inclusion is dependent on host or chlamydial cysteine proteases [30]. Protease activity could be a necessary prerequisite for insertion or activation of CT153 in membranes leading to pore-formation. These pores could function to destabilize the membrane or serve to transport other effectors, such as phospholipases or proteases, to disrupt the inclusion membrane or additional cellular targets. The nuclear compartment ruptures prior to the plasma membrane after inclusion lysis suggesting soluble factors released from the inclusion disrupt additional membranes [30]. Soluble CT153 released from the ruptured inclusion could act on cellular membranes and cause weakening of the plasma membrane leading to lysis.

Future Directions

A major limitation in determining the function of CT153 is lack of an assay to measure if and when this protein is functional. To date our efforts to demonstrate pore-forming ability of CT153 in sheep red blood cells lysis assays using protocols published for other MACPF family proteins have been unsuccessful. Reasons these assays may have failed include that critical chlamydial cofactors required for activity of CT153 were absent or that unprocessed CT153 is inactive. These assays might be more successful with mixtures of recombinant processed proteins if only processed CT153 forms pores. It needs to be clarified if full-length and processed forms of CT153 oligomerize in vitro and in vivo.

Other limitations in ascertaining CT153 function are that little is known about the localization of this protein at any point in the chlamydial developmental cycle and if full-length and processed forms are similarly localized. For example, if full-length CT153 was exclusively found in or on RB while processed fragments accumulated in the inclusion membrane this might indicate the processed form plays a role in fusion of the inclusion with incoming vesicles or lysis of the inclusion. One way to assess this could be to label infected cells simultaneously with antibodies targeting an epitope spanning the processing junction (thus specific for the full-length protein) and a second antibody targeting the N- or C-terminal portion of CT153. Alternately, the recent demonstration that green fluorescent protein and other recombinant transgenes can be expressed from the chlamydial plasmid [80] suggests expression of fluorescently tagged full-length and truncated versions could be used to localize full-length and processed forms of CT153 and to identify key residues that mediate localization.

The developmental cycle of the Chlamydiaceae involves sequential interactions with multiple host cell membranes, which has made it difficult to differentiate which of these interactions could be mediated by CT153. The recent development of rudimentary tools for forward and reverse genetic analysis of *Chlamydia* spp. gives hope that a definitive function of CT153 can be identified by these approaches. Although we have as yet been unable to isolate a CT153 nonsense mutant using the reverse genetic TILLING approach developed by Kari et al., these efforts have yet to reach saturation, and we now have a large collection of viable CT153 missense mutant strains confirming CT153 is mutable. Other genetic methods for interrogating CT153 function might be to use mutant screens to identify isolates with phenotypes consistent with what would be predicted for loss of CT153. For example, if CT153 plays a role in inclusion lysis, isolates that have enlarged inclusions and/or only exit via extrusion might have mutations in CT153 and/or accessory proteins. Although a negative result is difficult to interpret in a surrogate system, an especially promising approach might be to test if CT153 can complement the exit defect of a *T. gondii* MACPF mutant.

Work on the chlamydial MACPF domain protein, CT153, is beginning to reveal the importance of this protein in chlamydial biology and fundamental questions regarding the role of CT153 in chlamydial pathogenesis remain to be answered. Demonstration of pore-formation by CT153 and defining required cognate receptors, cofactors and target membranes will be central to elucidating the structure and function of the protein.

References

1. Anderluh G, Lakey JH (2008) Disparate proteins use similar architectures to damage membranes. Trends Biochem Sci 33:482–490
2. Azuma Y, Hirakawa H, Yamashita A, Cai Y, Rahman MA, Suzuki H, Mitaku S, Toh H, Goto S, Murakami T, Sugi K, Hayashi H, Fukushi H, Hattori M, Kuhara S, Shirai M (2006) Genome sequence of the cat pathogen, *Chlamydophila felis*. DNA Res 13:15–23
3. Beatty WL (2006) Trafficking from CD63-positive late endocytic multivesicular bodies is essential for intracellular development of *Chlamydia trachomatis*. J Cell Sci 119:350–359
4. Beatty WL, Morrison RP, Byrne GI (1994) Immunoelecton microscopic quantitation of differential levels of chlamydial proteins in a cell culture model of persistent *Chlamydia trachomatis* infection. Infect Immun 62:4059–4062
5. Beatty WL, Morrison RP, Byrne GI (1995) Reactivation of persistent *Chlamydia trachomatis* infection in cell culture. Infect Immun 63:199–205
6. Belland RJ, Scidmore MA, Crane DD, Hogan DM, Whitmire W, McClarty G, Caldwell HD (2001) *Chlamydia trachomatis* cytotoxicity associated with complete and partial cytotoxin genes. Proc Natl Acad Sci USA 98:13984–13989
7. Belland RJ, Zhong G, Crane DD, Hogan D, Sturdevant D, Sharma J, Beatty WL, Caldwell HD (2003) Genomic transcriptional profiling of the developmental cycle of *Chlamydia trachomatis*. Proc Natl Acad Sci USA 100:8478–8483
8. Blyth WA, Taverne J (1972) Some consequences of the multiple infection of cell cultures by TRIC organisms. J Hyg 70:33–37

9. Caldwell HD, Wood H, Crane D, Bailey R, Jones RB, Mabey D, Maclean I, Mohammed Z, Peeling R, Roshick C, Schachter J, Solomon AW, Stamm WE, Suchland RJ, Taylor L, West SK, Quinn TC, Belland RJ, McClarty G (2003) Polymorphisms in *Chlamydia trachomatis* tryptophan synthase genes differentiate between genital and ocular isolates. J Clin Invest 111:1757–1769
10. Carabeo RA, Mead DJ, Hackstadt T (2003) Golgi-dependent transport of cholesterol to the *Chlamydia trachomatis* inclusion. Proc Natl Acad Sci USA 100:6771–6776
11. Carlson JH, Hughes S, Hogan D, Cieplak G, Sturdevant DE, McClarty G, Caldwell HD, Belland RJ (2004) Polymorphisms in the *Chlamydia trachomatis* cytotoxin locus associated with ocular and genital isolates. Infect Immun 72:7063–7072
12. Carlson JH, Porcella SF, McClarty G, Caldwell HD (2005) Comparative genomic analysis of *Chlamydia trachomatis* oculotropic and genitotropic strains. Infect Immun 73:6407–6418
13. Chin E, Kirker K, Zuck M, James G, Hybiske K (2012) Actin recruitment to the *Chlamydia* inclusion is spatiotemporally regulated by a mechanism that requires host and bacterial factors. PLoS One 7:e46949
14. Clifton DR, Fields KA, Grieshaber NA, Dooley CA, Fischer ER, Mead DJ, Carabeo RA, Hackstadt T (2004) A chlamydial type III translocated protein is tyrosine-phosphorylated at the site of entry and associated with recruitment of actin. Proc Natl Acad Sci USA 101:10166–10171
15. Cocchiaro JL, Kumar Y, Fischer ER, Hackstadt T, Valdivia RH (2008) Cytoplasmic lipid droplets are translocated into the lumen of the *Chlamydia trachomatis* parasitophorous vacuole. Proc Natl Acad Sci USA 105:9379–9384
16. Corsaro D, Greub G (2006) Pathogenic potential of novel Chlamydiae and diagnostic approaches to infections due to these obligate intracellular bacteria. Clin Microbiol Rev 19:283–297
17. Cossart P, Vicente MF, Mengaud J, Baquero F, Perez-Diaz JC, Berche P (1989) Listeriolysin O is essential for virulence of *Listeria monocytogenes*: direct evidence obtained by gene complementation. Infect Immun 57:3629–3636
18. Delevove C, Nilges M, Dautry-Varsat A, Subtil A (2004) Conservation of the biochemical properties of IncA from *Chlamydia trachomatis* and *Chlamydia caviae*: oligomerization of IncA mediates interaction between facing membranes. J Biol Chem 279:46896–46906
19. DeMars R, Weinfurter J (2008) Interstrain gene transfer in *Chlamydia trachomatis* in vitro: mechanism and significance. J Bacteriol 190:1605–1614
20. Geoffroy C, Gaillard JL, Alouf JE, Berche P (1987) Purification, characterization, and toxicity of the sulfhydryl-activated hemolysin listeriolysin O from *Listeria monocytogenes*. Infect Immun 55:1641–1646
21. Geiffers J, Durling L, Ouellette SP, Rupp J. Maass M, Byrne GI, Caldwell HD, Belland RJ (2003) Genotypic differences in the *Chlamydia pneumonia* tyrP locus related to vascular tropism and pathogenicity. J Infect Dis 188:1085–1093
22. Gilbert RJ, Mikelj M, Dalla Serra M, Froelich CJ, Anderluh G (2013) Effects of MACPF/CDC proteins on lipid membranes. Cell Mol Life Sci 70:2083–2098
23. Grieshaber SS, Grieshaber NA, Hackstadt T (2003) *Chlamydia trachomatis* uses host cell dynein to traffic to the microtubule-organizing center in a p50 dynamitin-independent process. J Cell Sci 116:3793–3802
24. Hackstadt T, Rockey DD, Heinzen RA, Scidmore MA (1996) *Chlamydia trachomatis* interrupts an exocytic pathway to acquire endogenously synthesized sphingomyelin in transit from the Golgi apparatus to the plasma membrane. EMBO J 15:964–977
25. Hadders MA, Beringer DX, Gros P (2007) Structure of C8 alpha-MACPF reveals mechanism of membrane attack in complement immune defense. Science 317:1552–1554
26. Hadding U, Muller-Eberhard HJ (1969) The ninth component of human complement: isolation, description and mode of action. Immunology 16:719–735
27. Hatch GM, McClarty G (1998) Phospholipid composition of purified *Chlamydia trachomatis* mimics that of the eukaryotic host cell. Infect Immun 66:3727–3735

28. Heinzen RA, Scidmore MA, Rockey DD, Hackstadt T (1996) Differential interaction with endocytic and exocytic pathways distinguish parasitophorous vacuoles for *Coxiella burnetti* and *Chlamydia trachomatis*. Infect Immun 64:796–809
29. Heuer D, Lipinski AR, Machuy N, Karlas A, Wehrens A, Siedler F, Brinkmann V, Meyer TF (2009) *Chlamydia* causes fragmentation of the Golgi compartment to ensure reproduction. Nature 457:731–735
30. Hybiske K, Stephens RS (2007) Mechanisms of host cell exit by the intracellular bacterium *Chlamydia*. Proc Natl Acad Sci USA 104:11430–11435
31. Hybiske K, Stephens RS (2008) Exit strategies of intracellular pathogens. Nat Rev Microbiol 6:99–110
32. Ishino T, Chinzei Y, Yuda M (2005) A *Plasmodium sporozoite* protein with a membrane attack complex domain is required for breaching the liver sinusoidal cell layer prior to hepatocyte infection. Cell Microbiol 7:199–208
33. Kadota K, Ishino T, Matsuyama T, Chinzei Y, Yuda M (2004) Essential role of membrane-attack protein in malarial transmission to mosquito host. Proc Natl Acad Sci USA 101:16310–16315
34. Kafsack BF, Carruthers VB (2010) Apicomplexan perforin-like proteins. Commun Integr Biol 3:18–23
35. Kafsack BF, Pena JD, Coppens I, Ravindran S, Boothroyd JC, Carruthers VB (2009) Rapid membrane disruption by a perforin-like protein facilitates parasite exit from host cells. Science 323:530–533
36. Kalman S, Mitchell W, Marathe R, Lammel C, Fan J, Hyman RW, Olinger L, Grimwood J, Davis RW, Stephens RS (1999) Comparative genomes of *Chlamydia pneumoniae* and *C. trachomatis*. Nat Genet 21:385–389
37. Kari L, Goheen MM, Randall LB, Taylor LD, Carlson JH, Whitmire WM, Virok D, Rajaram K, Endresz V, McClarty G, Nelson DE, Caldwell HD (2011) Generation of targeted *Chlamydia trachomatis* null mutants. Proc Natl Acad Sci USA 108:7189–7193
38. Kumar Y, Cocchiaro J, Valdivia RH (2006) The obligate intracellular pathogen *Chlamydia trachomatis* targets host lipid droplets. Curr Biol 16:1646–1651
39. Liu X, Afrane M, Clemmer DE, Zhong G, Nelson DE (2010) Identification of *Chlamydia trachomatis* outer membrane complex proteins by differential proteomics. J Bacteriol 192:2852–2860
40. Madden JC, Ruiz N, Caparon M (2001) Cytolysin-mediated translocation (CMT): a functional equivalent of type III secretion in Gram-positive bacteria. Cell 104:143–152
41. McClarty G (1999) Chlamydial metabolism as inferred from the complete genome sequence. In: Stephens RS (ed) *Chlamydia*: intracellular biology, pathogenesis, and immunity. ASM Press, Washington
42. McClarty G, Caldwell HD, Nelson DE (2007) Chlamydial interferon gamma immune evasion influences infection tropism. Curr Opin Microbiol 10:47–51
43. Metkar SS, Wang B, Aguilar-Santelises M, Raja SM, Uhlin-Hansen L, Podack E, Trapani JA, Froelich CJ (2002) Cytotoxic cell granule-mediated apoptosis: perforin delivers granzyme B-seglycin complexes into target cells without plasma membrane pore formation. Immunity 16:417–428
44. Moulder JW (1991) Interaction of chlamydiae and host cells in vitro. Microbiol Rev 55:143–190
45. Nelson DE, Crane DD, Taylor LD, Dorward DW, Goheen MM, Caldwell HD (2006) Inhibition of chlamydiae by primary alcohols correlates with the strain-specific complement of plasticity zone phospholipase D genes. Infect Immun 74:73–80
46. Nelson DE, Taylor LD, Shannon JG, Whitmire WM, Crane DD, McClarty G, Su H, Kari L, Caldwell HD (2007) Phenotypic rescue of *Chlamydia trachomatis* growth in IFN-gamma treated mouse cells by irradiated *Chlamydia muridarum*. Cell Microbiol 9:2289–2298

47. Nelson DE, Virok DP, Wood H, Roshick C, Johnson RM, Whitmire WM, Crane DD, Steele-Mortimer O, Kari L, McClarty G, Caldwell HD (2005) Chlamydial IFN-gamma immune evasion is linked to host infection tropism. Proc Natl Acad Sci USA 102:10658–10663
48. Nelson DE (2012) The Chlamydial cell envelope. In: Tan M, Bavoil P (eds) Intracellular pathogens I: chlamydiales. ASM Press, Washington
49. Peitsch MC, Amiguet P, Guy R, Brunner J, Maizel JV Jr, Tschopp J (1990) Localization and molecular modeling of the membrane-inserted domain of the ninth component of human complement and perforin. Mol Immunol 27:589–602
50. Ponting CP (1999) Chlamydial homologues of the MACPF (MAC/perforin) domain. Curr Biol 9:R911–R913
51. Ponting CP (2000) Novel repeats in ryanodine and IP3 receptors and protein O-mannosyltransferases. Trends Biochem Sci 25:48–50
52. Ponting CP, Kerr ID (1996) A novel family of phospholipase D homologues that includes phospholipid synthases and putative endonucleases: identification of duplicated repeats and potential active site residues. Protein Sci 5:914–922
53. Praper T, Sonnen AF, Kladnik A, Andrighetti AO, Viero G, Morris KJ, Volpi E, Lunelli L, Dalla Serra M, Froelich CJ, Gilbert RJ, Anderluh G (2011) Perforin activity at membranes leads to invaginations and vesicle formation. Proc Natl Acad Sci USA 108:21016–21021
54. Raulston JE (1997) Response of *Chlamydia trachomatis* serovar E to iron restriction in vitro and evidence for iron-regulated chlamydial proteins. Infect Immun 65:4539–45447
55. Read TD, Brunham RC, Shen C, Gill SR, Heidelberg JF, White O, Hickey EK, Peterson J, Utterback T, Berry K, Bass S, Linher K, Weidman J, Khouri H, Craven B, Bowman C, Dodson R, Gwinn M, Nelson W, DeBoy R, Kolonay J, McClarty G, Salzberg SL, Eisen J, Fraser CM (2000) Genome sequences of *Chlamydia trachomatis* MoPn and *Chlamydia pneumoniae* AR39. Nucleic Acids Res 28:1397–1406
56. Read TD, Myers GS, Brunham RC, Nelson WC, Paulsen IT, Heidelberg J, Holtzapple E, Khouri H, Federova NB, Carty HA, Umayam LA, Haft DH, Peterson J, Beanan MJ, White O, Salzberg SL, Hsia RC, McClarty G, Rank RG, Bavoil PM, Fraser CM (2003) Genome sequence of *Chlamydophila caviae* (*Chlamydia psittaci* GPIC): examining the role of niche-specific genes in the evolution of the Chlamydiaceae. Nucleic Acids Res 31:2134–2147
57. Robertson DK, Gu L, Rowe RK, Beatty WL (2009) Inclusion biogenesis and reactivation of persistent *Chlamydia trachomatis* requires host cell sphingolipid biosynthesis. PLoS Pathog 5:e1000664
58. Rockey DD, Fischer ER, Hackstadt T (1996) Temporal analysis of the developing *Chlamydia psittaci* inclusion by use of fluorescence and electron microscopy. Infect Immun 64:4269–4278
59. Roiko MS, Carruthers VB (2013) Functional dissection of *Toxoplasma gondii* perforin-like protein 1 reveals a dual domain mode of membrane binding for cytolysis and parasite egress. J Biol Chem 288:8712–8725
60. Rosado CJ, Buckle AM, Law RH, Butcher RE, Kan WT, Bird CH, Ung K, Browne KA, Baran K, Bashtannyk-Puhalovich TA, Faux NG, Wong W, Porter CJ, Pike RN, Ellisdon AM, Pearce MC, Bottomley SP, Emsley J, Smith AI, Rossjohn J, Hartland EL, Voskoboinik I, Trapani JA, Bird PI, Dunstone MA, Whisstock JC (2007) A common fold mediates vertebrate defense and bacterial attack. Science 317:1548–1551
61. Rosado CJ, Kondos S, Bull TE, Kuiper MJ, Law RH, Buckle AM, Voskoboinik I, Bird PI, Trapani JA, Whisstock JC, Dunstone MA (2008) The MACPF/CDC family of pore-forming toxins. Cell Microbiol 10:1765–1774
62. Saka HA, Valdivia RH (2010) Acquisition of nutrients by Chlamydiae: unique challenges of living in an intracellular compartment. Curr Opin Microbiol 13:4–10
63. Schachter J (1978) Chlamydial infections (first of three parts). N Engl J Med 298:428–434
64. Schachter J (1978) Chlamydial infections (second of three parts). N Engl J Med 298:490–495
65. Schachter J (1978) Chlamydial infections (third part of three parts). N Engl J Med 298:540–549

66. Scidmore MA, Fischer ER, Hackstadt T (2003) Restricted fusion of *Chlamydia trachomatis* vesicles with endocytic compartments during the initial stages of infection. Infect Immun 71:973–984
67. Shirai M, Hirakawa H, Kimoto M, Tabuchi M, Kishi F, Ouchi K, Shiba T, Ishii K, Hattori M, Kuhara S, Nakazawa T (2000) Comparison of whole genome sequences of *Chlamydia pneumoniae* J138 from Japan and CWL029 from USA. Nucleic Acids Res 28:2311–2314
68. Slade DJ, Lovelace LL, Chruzcz M, Minor W, Lebioda L, Sodetz JM (2008) Crystal structure of the MACPF domain of human complement protein C8 alpha in complex with the C8 gamma subunit. J Mol Biol 379:331–342
69. Smith GA, Marquis H, Jones S, Johnston NC, Portnoy DA, Goldfine H (1995) The two distinct phospholipases C of *Listeria monocytogenes* have overlapping roles in escape from a vacuole and cell-to-cell spread. Infect Immun 63:4231–4237
70. Stephens RS, Kalman S, Lammel C, Fan J, Marathe R, Aravind L, Mitchell W, Olinger L, Tatusov RL, Zhao Q, Koonin EV, Davis RW (1998) Genome sequence of an obligate intracellular pathogen of humans: *Chlamydia trachomatis*. Science 282:754–759
71. Suchland RJ, Rockey DD, Bannantine JP, Stamm WE (2000) Isolates of *Chlamydia trachomatis* that occupy nonfusogenic inclusions lack IncA, a protein localized to the inclusion membrane. Infect Immun 68:360–367
72. Taylor LD, Nelson DE, Dorward DW, Whitmire WM, Caldwell HD (2010) Biological characterization *Chlamydia trachomatis* plasticity zone MACPF domain family protein CT153. Infect Immun 78:2691–2699
73. Thomson NR, Holden MT, Carder C, Lennard N, Lockey SJ, Marsh P, Skipp P, O'Connor CD, Goodhead I, Norbertzcak H, Harris B, Ormond D, Rance R, Quail MA, Parkhill J, Stephens RS, Clarke IN (2008) *Chlamydia trachomatis*: genome sequence analysis of lymphogranuloma venereum isolates. Genome Res 18:161–171
74. Thomson NR, Yeats C, Bell K, Holden MT, Bentley SD, Livingstone M, Cerdeno-Tarraga AM, Harris B, Doggett J, Ormond D, Mungall K, Clarke K, Feltwell T, Hance Z, Sanders M, Quail MA, Price C, Barrell BG, Parkhill J, Longbottom D (2005) The *Chlamydophila abortus* genome sequence reveals an array of variable proteins that contribute to interspecies variation. Genome Res 15:629–640
75. Todd WJ, Caldwell HD (1985) *Chlamydia trachomatis* host cell interactions: Ultrastructural studies on the mechanism of release of a biovar II strain from HeLa 229 cells. J Infect Dis 151:1037–1044
76. Tweten RK (2005) Cholesterol-dependent cytolysins, a family of versatile pore-forming toxins. Infect Immun 73:6199–6209
77. Uellner R, Zvelebil MJ, Hopkins J, Jones J, MacDougall LK, Morgan BP, Padock E, Waterfield MD, Griffiths GM (1997) Perforin is activated by a proteolytic cleavage during biosynthesis which reveals a phospholipid-binding C2 domain. EMBO J 16:7287–7296
78. Voigt A, Schofl G, Saluz HP (2012) The *Chlamydia psittaci* genome: a comparative analysis of intracellular pathogens. PLoS One 7:e35097
79. Wang Y, Kahane S, Cutcliffe LT, Skilton RJ, Lambden PR, Clarke IN (2011) Development of a transformation system for *Chlamydia trachomatis*: restoration of glycogen biosynthesis by acquisition of a plasmid shuttle vector. PLoS Pathog 7:e1002258
80. Wang Y, Cutcliffe LT, Skilton RJ, Persson K, Bjartling C, Clarke IN (2013) Transformation of a plasmid-free, genital tract isolate of *Chlamydia trachomatis* with a plasmid vector carrying a deletion in CDS6 revealed that this gene regulated inclusion phenotype. Pathog Dis 67:100–103
81. Whitworth T, Popov VL, Yu XJ, Walker DH, Bouyer DH (2005) Expression of the *Rickettsia prowazekii* pld or tlyC gene in *Salmonella enterica* serovar Typhimurium mediates phagosomal escape. Infect Immun 73:6668–6673
82. Wylie JL, Hatch GM, McClarty G (1997) Host cell phospholipids are trafficked to and then modified by *Chlamydia trachomatis*. J Bacteriol 179:7233–7242

Chapter 14
Fungal MACPF-Like Proteins and Aegerolysins: Bi-component Pore-Forming Proteins?

Katja Ota, Matej Butala, Gabriella Viero, Mauro Dalla Serra, Kristina Sepčić and Peter Maček

Abstract Proteins with membrane-attack complex/perforin (MACPF) domains are found in almost all kingdoms of life, and they have a variety of biological roles, including defence and attack, organism development, and cell adhesion and signalling. The distribution of these proteins in fungi appears to be restricted to some Pezizomycotina and Basidiomycota species only, in correlation with another group of proteins with unknown biological function, known as aegerolysins. These two protein groups coincide in only a few species, and they might operate in concert as cytolytic bi-component pore-forming agents. Representative proteins here include pleurotolysin B, which has a MACPF domain, and the aegerolysin-like protein pleurotolysin A, and the very similar ostreolysin A, which have been purified from oyster mushroom (*Pleurotus ostreatus*). These have been shown to act in concert to perforate natural and artificial lipid membranes with high cholesterol and sphingomyelin content. The aegerolysin-like proteins provide the membrane cholesterol/sphingomyelin selectivity and recruit oligomerised pleurotolysin B molecules, to create a membrane-inserted pore complex. The resulting protein structure has been imaged with electron microscopy, and it has a 13-meric rosette-like structure, with a central lumen that is ~4–5 nm in diameter. The opened transmembrane pore is non-selectively permeable for ions and smaller neutral solutes, and is a cause of cytolysis of a colloid-osmotic type. The biological significance of these proteins for the fungal life-style is discussed.

Keywords Aegerolysins · Fungi · Membrane-attack complex/perforin · Ostreolysin A · *Pleurotus ostreatus* · Pleurotolysin A · Pleurotolysin B

K. Ota · M. Butala · K. Sepčić · P. Maček (✉)
Department of Biology Biotechnical Faculty, University of Ljubljana,
Večna pot 111, 1000 Ljubljana, Slovenia
e-mail: peter.macek@bf.uni-lj.si

G. Viero · M. Dalla Serra
Istituto di Biofisica, Consiglio Nazionale delle Ricerche and Fondazione
Bruno Kessler, via alla Cascata 56/C, 38123 Trento, Italy

G. Anderluh and R. Gilbert (eds.), *MACPF/CDC Proteins - Agents of Defence, Attack and Invasion*, Subcellular Biochemistry 80, DOI: 10.1007/978-94-017-8881-6_14,

Abbreviations

3D	Three-dimensional
diΦPC	diphytanoylphosphatidylcholine
EGFP	Enhanced green fluorescent protein
Hi	*H. irregulare*
LUV	Large unilamellar vesicles
MACPF	Membrane-attack complex/perforin
Mp	*M. perniciosa*
OlyA	Ostreolysin A
PLM	Planar lipid membranes
PlyA	Pleurotolysin A
PlyA2	Pleurotolysin A2
PlyB	Pleurotolysin B
Ss	*S. stellatus*
Tv	*T. versicolor*

Introduction

Fungi are currently assigned to 8–10 phyla (traditionally Chytridiomycota, Zygomycota, Ascomycota and Basidiomycota only), and they are an extremely variable group of eukaryotic organisms that inhabit all niches of our planet. The number of fungal species, which are commonly known as yeasts, moulds and mushrooms, has been estimated to be at least in the seven-digit range (for review, see [3]) (Fig. 14.1a). Beyond their ecological impact in nature as a major group of decomposers of dead organic matter, fungi have also evolved hemibiotrophic and biotrophic life-styles, and they are thus also notorious plant, animal and human pathogens (for review, see [23]). Fungi have an enormous molecular inventory potential, including antibiotics, cellulases and lignolytic enzymes for example. They have had very significant economic impact on human society, as they have been traditionally exploited in human and animal nutrition, and in medicine. They have also attracted a lot of attention in traditional and modern biotechnological applications [12, 44] and as emerging threats to animal, plant and ecosystem health [18]. As one consequence, several fungal genome projects have been started, and annotations of several of these have contributed markedly to information on the biology and evolution of medically, industrially and environmentally important fungi [19, 21].

Here, we focus on a small group of fungal proteins that have a predicted membrane-attack complex/perforin (MACPF) domain (Prosite PS51412, Pfam PF01823). We define these in conjunction with the implication of their damage to cellular membranes and model lipid membranes. This fungal protein group is represented by pleurotolysin B (PlyB), from the edible oyster mushroom *Pleurotus*

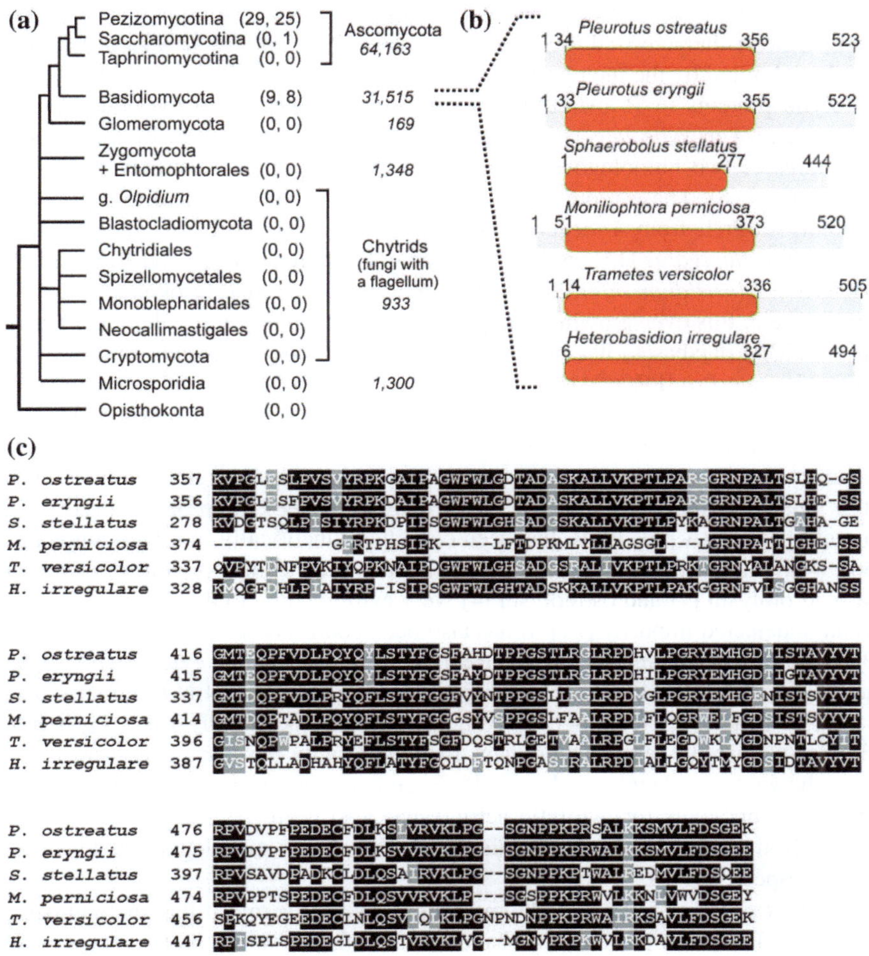

Fig. 14.1 Distribution of aegerolysins and proteins with the MACPF domain in fungi, and structure of pleurotolysin B (PlyB) and PlyB-like proteins. **a** Fungal taxonomic classification and data on species number are from Bass and Richards [3]. In *brackets* number of species with one or more predicted or confirmed aegerolysin-like protein, and number of proteins with a predicted MACPF domain, respectively. The total number of well-described fungal species is in italics. Data were collected from NCBI, UniProt, and JGI nucleotide and protein databases. **b** Sequence location of the MACPF domain in PlyB (*P. ostreatus*, NCBI: BAD66667) and PlyB-like proteins (*P. eryngii*, NCBI: BAI45248; *S. stellatus*, JGI: Protein ID 808130; *M. perniciosa*, NCBI: XP_002389006; *T. versicolor*, NCBI: EIW53798; *H. irregulare*, JGI: Protein ID 426372), as indicated. **c** Clustal W [30] amino-acid alignment of the C-terminal region, following the MACPF domain as shown in (**b**). The incomplete amino-acid sequence of an *A. cylindracea* PlyB-like protein that was deduced from expressed sequence tags (NCBI EST database) and from proteome and transcriptome data [68] is not shown. *Black shading*, identical amino acids; *grey shading*, similar amino acids

ostreatus. Historically, pleurotolysin was first described by Bernheimer and Avigad [7], who characterised this mushroom haemolysin as a ~12-kDa protein that can be inhibited by the membrane lipid sphingomyelin. In 2002, reinvestigation of cytolytic proteins in *P. ostreatus*, as compared to those from the edible poplar mushroom, *Agrocybe aegerita*, resulted in partial determination of the primary structures of two homologous, cytolytic, ~15-kDa proteins: ostreolysin and aegerolysin [4, 6]. These proteins were shown to be highly identical to a predicted protein, Aa-Pri1, from *A. aegerita* [17], and also similar to Asp-haemolysin from *Aspergillus fumigatus* [15], and the non-haemolytic proteins P16 and P18 from the bacterium *Clostridium bifermentans* [2]. With reference to aegerolysin, as the Aa-Pri1 protein, this ever-growing group of proteins (currently more than 350 entries in the NCBI databases) has been named as the aegerolysin protein family (Pfam PF06355, InterPro IPR009413). Their structure, putative functional characteristics, and occurrence across the tree of life have been reviewed elsewhere [5, 39]. However, it was shown for the first time by Tomita et al. [65] that an ostreolysin isoform from *P. ostreatus*, known as pleurotolysin A (PlyA), is not cytolytic unless it is supplemented with another non-cytolytic, ~59-kDa protein, PlyB. This protein pair, PlyA plus PlyB, has been classified functionally as a two-component pore-forming protein. Indeed, it is clear now that both of the previous preparations of pleurotolysin [7] and ostreolysin [4] were cytolytic due to the presence of PlyB as an undetected impurity [41]. It has also been confirmed recently that two native PlyA isoforms and a recombinant ostreolysin A (OlyA) are not cytolytic per se, but are essential for the cytolytic activity of native PlyB and its N-terminally truncated, i.e., matured, form [41], as originally suggested for PlyB and PlyA [65]. Herein, we use the term ‘ostreolysin’ for the partially purified ostreolysin A that was shown to be cytolytic due to PlyB contamination. The necessity for two protein components for cytolytic activity has also been reported for erylysins A and B from *Pleurotus eryngii*, which are very similar orthologues of PlyA and PlyB, respectively [62].

Current searches of nucleotide and protein databases reveal that the number of ~59-kDa PlyB-like proteins that have been confirmed at the protein level or are predicted is low (6–8 proteins, to date), and that they are distributed exclusively in some species of the fungal class Agaricomycetes. In contrast, aegerolysin-like proteins similar to PlyA and OlyA are much more widely distributed in bacteria, fungi and plants, and one representative each has also been reported in a lepidopteran (*Chrysodeixis includens*) [69] and in a virus species (*Trichoplusia ni ascovirus 2c*). In view of the saprophytic way of life of the *Pleurotus* species, the biological roles of PlyB-like proteins and aegerolysin proteins remain enigmatic. The in vitro cytolytic activity of these combined A- and B-type proteins is also intriguing, because these mushroom proteins are not known to serve as defensive or attack agents against other co-habiting organisms.

Here, we summarise recent data on the distribution of PlyB-like proteins and aegerolysins, including their structural features and functions, with a focus on their permeabilising of natural cellular membranes and model lipid membranes. We also discuss some previously controversial data on the putative functions of aegerolysins, their lipid-binding characteristics, and their pore formation, and we propose mechanisms of pore formation by PlyB-like proteins in combination with their species-specific aegerolysin counterparts. We also provide data on the electrophysiological characteristics of the transmembrane channels that are induced by purified native PlyB and OlyA in planar lipid membranes (PLMs). Furthermore, we suggest the molecular mechanism of formation of the transmembrane pore complex formed by PlyB and OlyA (or PlyA).

Occurrence in Fungi of Aegerolysins and Proteins with an MACPF Domain, Including PlyB-Like Proteins

Fungal genome sequencing projects have provided a steadily increasing number of novel putative proteins that have been assigned to the aegerolysin protein family and to the proteins with a predicted MACPF domain. Current data from a bioinformatics survey (March, 2013) on the distribution of these proteins across the fungal taxa are summarised in Fig. 14.1a. Although the great majority of the fungal species classified belong to the Ascomycota and Basidiomycota phyla, it is interesting that some of the other fungal groups completely lack both aegerolysins and MACPF-domain-bearing proteins. The estimate of the number of fungal species that have proteins with the MACPF domain (33 species) shown in Fig. 14.1a matches the total number of these proteins (including the PlyB-like proteins shown in Fig. 14.1b) that have recently been predicted in fungi (31 proteins) [20]. Although these estimates may not be accurate, it appears that each of these species can code for a low number (1–2) of different proteins with the MACPF domain. Proteins with the MACPF domain are relatively widely distributed in all of the kingdoms across living organisms [20], including the fungal taxa, as summarised in Fig. 14.1a. However, in contrast to some vertebrate proteins, the functions and biological roles of these fungal MACPF-domain-containing proteins are completely unknown in fungi. So far, only PlyB-like proteins from the genus *Pleurotus* have been confirmed at the protein level, and their cytolytic activity has been documented. In addition, based on functional studies of *A. aegerita* proteins [4] and on *Agrocybe cylindracea* transcriptome data [63], aegerolysins (Aa-Pri1, and similar proteins) and orthologues of PlyB these species can be included in the list. BLAST searches in nucleotide and protein databases using PlyB as the query show that the amino-acid sequence of the C-terminal region of PlyB and of PlyB-like proteins (Fig. 14.1c) is unique, and this makes these proteins distinctive from all other known pore-forming proteins, including those with the MACPF domain and the cholesterol-dependent cytolysins (CDCs) [9, 13, 20, 22, 25, 26, 52]. At present, very few fungal species are known to have a matched aegerolysin and PlyB-like protein combination (see Fig. 14.1b). These species are restricted to the families of Pleurotaceae (*P. ostreatus*, *P. eryngii*), Geastraceae (*Sphaerobolus stellatus*), Marasmiaceae (*Moniliophtora perniciosa*),

Polyporaceae (*Trametes versicolor*), Bondarzewiaceae (*Heterobasidion irregulare*), and Strophariaceae (*A. aegerita* and *A. cylindracea*). This list should probably also include *Pleurotus nebrodensis* [34].

Structure of Fungal Cytolytic PlyB-Like Proteins and Aegerolysins

The structures, putative functions and biological roles of members of the aegerolysin family have been extensively reviewed elsewhere [5, 39]. Figure 14.2 illustrates the alignment of the amino-acid sequences for selected aegerolysins only, as: (i) those already shown to participate with PlyB-like proteins in cytolysis (OlyA, PlyA, EryA); (ii) one that is very likely to be implicated in cytolysis (Aa-Pri1 from *A. aegerita*; Aa); and (iii) those that have not yet been proven at the protein level, but could be predicted to participate with their predicted PlyB-like counterparts in cytolysis, based on structural homology: the *S. stellatus* (Ss), *M. perniciosa* (Mp), *T. versicolor* (Tv) and *H. irregulare* (Hi) aegerolysins. The primary structures of these selected aegerolysins are very similar; e.g., PlyA and OlyA differ in only two amino-acid substitutions. However, one visible exception is the predicted *T. versicolor* aegerolysin, with a propeptide that precedes the core sequence. Similarly, there are also propeptide sequences in some Ascomycota aegerolysins. It is not known whether the propeptide is post-translationally cleaved during the process of protein maturation. Although the number of sequences collected in Fig. 14.2 is relatively few, and thus may not be conclusive, at least some of the characteristic amino-acid motifs can be deduced that might make this group distinctive from other aegerolysins: (i) the N-terminal sequence AYAQWVIIIIIHNVG, with a few conservative substitutions (the predicted *T. versicolor* protein is again an exception); and (ii) two amino-acid motifs with a conserved cysteine residue in each (CGR, CPWG). In comparison, in the N-terminal part of the other aegerolysins, there are varied polar and charged amino-acid substitutions, which makes this part of them less hydrophobic. Also, the CGR and/or CPWG motifs are not conserved in the other aegerolysins, and in some aegerolysin sequences, additional cysteines are inserted elsewhere. In this respect, it is of note that thiol-reactive reagents can totally impair the cytolytic activity of ostreolysin [4], and of OlyA in combination with PlyB [40]. The primary structures of PlyB and its orthologues are also very similar to each other, with regard to the position of the MACPF domain and the C-terminal region, as discussed above. At present, the nucleotide/protein sequencing data are not sufficient to make any reliable analysis of the primary structures of this protein group. Of note, in PlyB-like proteins, the number of cysteine residues that are regularly located in the C-terminal region is lower (1–2) than in the other fungal proteins with MACPF domains (5–8; our unpublished data).

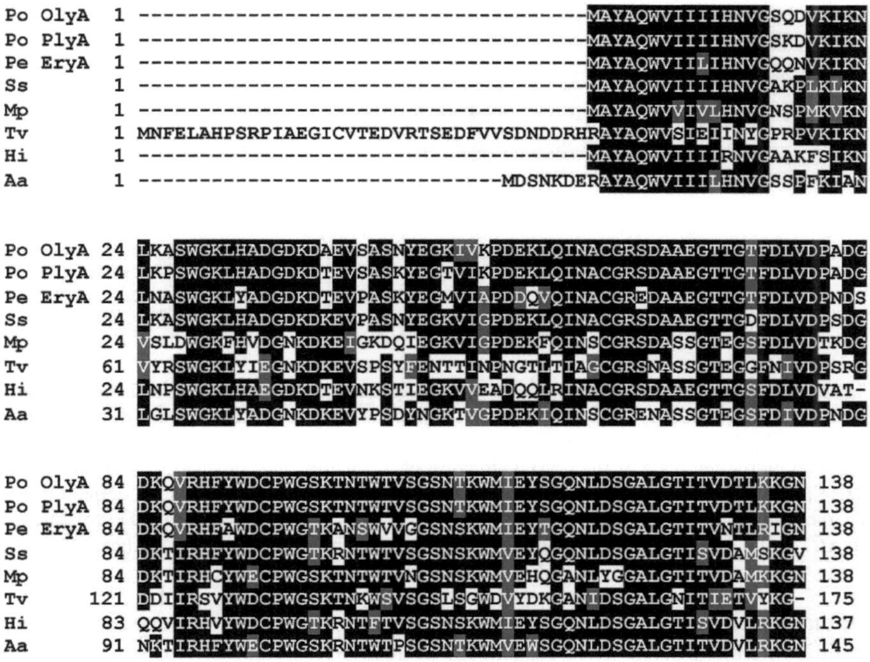

Fig. 14.2 Amino-acid alignment of aegerolysins from fungal species with PlyB or PlyB-like proteins. See Fig. 14.1b, including aegerolysin, Aa-Pri1, from *A. aegerita*. Po OlyA, *P. ostreatus* OlyA6 (NCBI: AGH25589); Po PlyA, *P. ostreatus* pleurotolysin A (NCBI: BAD66668); Pe EryA, *P. eryngeii* erylysin A (NCBI: BAI45247); Ss, *S. stellatus* (JGI:ID 267302); Mp, *M. perniciosa* aegerolysin (NCBI:XP_002389486); Tv, *T. versicolor* aegerolysin (JGI:ID 134289); Hi, *H. irregulare* aegerolysin (JGI:ID 148469); Aa, *A. aegerita* Aa-Pri1 (NCBI:AAC02265). Sequences were aligned with ClustalW. *Black shading*, identical amino acids; *grey shading*, similar amino acids

To date, the three-dimensional (3D) structures of PlyB and PlyB-like proteins have not been solved, and there is only one known crystal structure of an aegerolysin family member [58]: the Cry34A protein from *Bacillus thuringiensis* [32, 59]. As 3D structural data are lacking, protein 3D structural homology modelling might be helpful to obtain some insights into the protein structure and function. Figure 14.3 shows the 3D structural models of the full-length precursor of PlyB and an OlyA isoform, OlyA6, which have been used in functional studies [40, 41]. Searches for proteins with similar 3D structures have shown that this modelled PlyB is most similar to mouse perforin (PDB: 3NSJ) with regard to the MACPF domain structure. The homology modelling predicts a β-sandwich fold for OlyA6, a structure that is most similar to that of a sea anemone actinoporin, fragaceatoxin C (PDB: 3LIM). Thermal and pH-induced denaturation of ostreolysin have suggested that the tertiary structure of this β-structured protein is relatively stable [6]. In contrast, PlyB is highly unstable in solution and is prone to cleavage of an

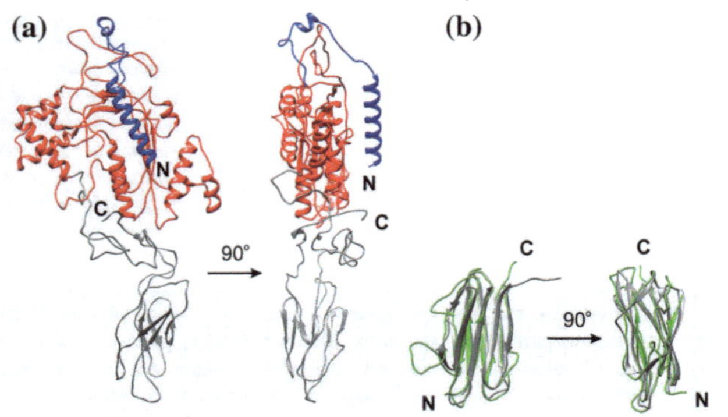

Fig. 14.3 Three-dimensional structure models of PlyB and OlyA6. The models of the full-length PlyB (NCBI:BAD66667) and OlyA6 were obtained from the iterative threading assembly refinement by I-TASSER [54] and rendered by CHIMERA [42]. The OlyA6 3D structure was modelled on the template of the crystal structure of a *Bacillus thuringiensis* aegerolysin, 149B1 Cry34 [58]. **a** The PlyB model, with the N-terminal α-helix in *blue*, the MACPF domain in *red*, and the C-terminal region in *grey* (see also Fig. 14.1b, c). **b** Superimposed 3D structures of OlyA6 (*green*) and 149B1 Cry34 (*gray*). The N-termini and C-termini are as indicated

internal peptide bond [41, 65]. Interestingly, this cleavage does not significantly diminish its haemolytic activity, and it has also been noted for erylysin B from *P. eryngii* [62].

The predicted 3D structural similarity between PlyB and perforin and other MACPF/CDC proteins suggests that the PlyB MACPF domain might be a prerequisite for the formation of pores across cellular membranes and model lipid membranes. The actual 3D structure of the PlyB C-terminal region is, however, uncertain, and it is clearly different from that of perforin and the other MACPF/CDC proteins discussed here. The C-terminally located C2 domain in perforin [52] and C-terminal domain 4 in CDCs [25] are the membrane-binding parts of these proteins. At present, it is not known whether the C-terminal region of PlyB is involved in the attaching of the molecule to susceptible membranes.

Lipid-Binding and Pore-Forming Characteristics of PlyB, OlyA and the PlyA Aegerolysins

Contrary to numerous database entries on predicted and proven aegerolysins, and despite their apparently wide distribution in various organisms, only a few of them have been functionally characterised. Asp-haemolysin from the mould *A. fumigatus* (syn. *Neosartorya fumigata*) was the first aegerolysin-like protein to be described [15, 55]. Functionally, it was isolated and characterised as a haemolytic protein [14] that binds to oxidised-LDL and lysophospholipids [29]. In a recent study, a similar protein from *Aspergillus terreus*, recombinant terrelysin, was suggested to be haemolytic [38], while an aegerolysin, known as RahU (the PA0122 protein), from the bacterium *Pseudomonas aeruginosa* is similar to Asp-haemolysin, and it can interact with oxidised-LDL and lysophospholipids, although it is not haemolytic [47]. The *C. bifermentans* P16 and P18 aegerolysin-like proteins have also been shown not to be haemolytic [2].

The apparent functional diversity of this group of proteins is illustrated further by the aegerolysin-like proteins known as the insecticidal 14 kDa crystal (Cry34) toxins, from *Bacillus thuringiensis* [16, 24]. These proteins have been shown to permeabilise lipid vesicles made of phosphatidylserine/phosphatidylcholine and to open ion channels, with electric conductances ranging from ~10 to 300 pS in PLMs [36]. These Cry34 proteins were classified as binary crystal toxins, as their membrane-damaging activities are maximal when combined with the *B. thuringiensis* 44 kDa Cry35 proteins [36]. The insecticidal activity of these Cry34Ab1/Cry35Ab1 proteins that are used for in-plant control of corn rootworm [58] was suggested to be different from other *B. thuringiensis* crystal toxins [32]. In contrast, the basidiomycetous aegerolysins PlyA, OlyA and EryA have been specifically shown not to be cytolytic per se; moreover, for OlyA, it has been confirmed that it does not bind lysophospholipids [41]. It was previously reported that the cytolytic activity induced by ostreolysin (as mixed with the PlyB impurity) might be inhibited by lysophospholipids and free fatty acids [60]. However, it was shown later that the inhibition was caused indirectly by partitioning of the lysophospholipids into the lipid membranes, which consequently resulted in diminished binding of ostreolysin to the membranes [10].

Studies of the lipid selectivity of the native ostreolysin plus PlyB mixture using cellular membranes or model lipid membranes have shown that these proteins bind exclusively to membranes rich in cholesterol (or other sterols) only if the sterols are combined with sphingomyelin [10, 49, 50, 61]. The binding of these proteins and their lipid-vesicle permeabilisation were significantly reduced when sphingomyelin was substituted by the fully saturated phosphatidylcholine in the binary lipid mixture, and it was abolished by the addition of unsaturated glycerophosphatides to cholesterol/sphingomyelin membranes [61]. In contrast, PlyA and PlyB have been classified as sphingomyelin-specific two-component cytolysins [65]. However, in this study of Tomita et al. [65], there were no membranes with a control lipid composition without cholesterol, as all of the tested lipid mixtures contained cholesterol plus a variable lipid. Recent re-examination of the lipid selectivity of recombinant OlyA and PlyB, individually and combined, have confirmed the exclusive specificity of OlyA for cholesterol/sphingomyelin-rich membranes, which are typical of lipid rafts in cellular membranes. Also, when OlyA6 was assayed on a glycan array of >300 different human oligosaccharides, it did not interact with any of them [41], which thus confirms the lipid binding preference of OlyA6. It has also been shown that in contrast to OlyA, PlyB on its own can bind to lipid vesicles of variable lipid compositions, although with relatively low affinities compared to OlyA6 [41]. In this particular study, it was also noted that OlyA6 and PlyB do not interact with each other in solution. For both the

PlyA/PlyB and OlyA/PlyB pairs, it was shown that they can form large sodium dodecyl sulphate (SDS)-resistant oligomeric structures in susceptible membranes [41, 65]. Moreover, based on the kinetics of haemolysis provoked by varied proportions of PlyB and PlyA, a PlyB/PlyA stoichiometry of 1/3 was suggested as optimal for pore formation [65]. This specific study also suggested that PlyA is firmly bound in the complex and cannot be dissociated from the pore complex with SDS. In another study, however, is could not be shown that there was OlyA in the SDS-resistant protein-lipid complexes [41]. This discrepancy might be accounted for by differences in the experimental details, such as the membrane composition, the protocols for solubilisation of the complex with SDS, and the antibodies used in the Western blotting of these proteins. However, labelling of the on-membrane complex of PlyB and OlyA, N-terminally tagged with hexahistidine, and electron microscopy confirmed that OlyA is a constituent of the rosette-like protein-lipid complex [40] (see below). It is not known, however, what the component stoichiometry in the pore complex is.

The PlyA/PlyB pore complex on erythrocyte membranes has been imaged as a ring-like structure with electron microscopy [65]. More details on the oligomeric architecture of these complexes have been provided by recent electron microscopy imaging of recombinant OlyA6 and Δ48PlyB (N-terminally truncated PlyB, corresponding to mature PlyB), both solely or as combined proteins, on large unilamellar vesicles (LUVs) made from a total erythrocyte lipid extract [41], as shown in Fig. 14.4. OlyA alone induced reshaping of the LUVs, which were seen as membrane invaginations and evaginations (Fig. 14.4a), and the evaginations could even pinch-off from the LUV membrane. Of note, Δ48PlyB alone can sporadically form ring-like, and even rosette-like, structures in the presence of the LUVs (Fig. 14.4b), while treatment of the LUVs with both OlyA6 and Δ48PlyB resulted in abundant rosette-like oligomers on the membranes (Fig. 14.4c). These structures have a radial symmetry, and are 19.7 nm in diameter and ~9 nm in height (seen in Fig. 14.4d), and they are mostly 13-mers and similar to the rosette-like oligomers that are produced by Δ48PlyB only. Moreover, these rosette-like oligomers on the top of two opposing LUVs were seen to dimerise, to form associated rosette-like structures (Fig. 14.4d, right), which might be the reason for the significant aggregation of the LUVs that was observed with the OlyA/PlyB combination. Of note here, the size of the single rosette-like structures under electron microscopy is in reasonable agreement with that estimated for the SDS-resistant oligomers of OlyA/PlyB [41] and PlyA/PlyB [65]. It was also noted that the dimensions of the monomers that formed rosette-like oligomers resemble those of the bacterial CDC pores that are formed by perfringolysin O and pneumolysin [41].

Further evidence for the membrane-pore-forming activity of PlyB and OlyA extends to experiments on PLMs, as shown in Fig. 14.5. The mixture of the native PlyB/OlyA (molar ratio, ~1:10) induced a step-wise increase in the transmembrane current across PLMs prepared from erythrocyte lipids and PLMs made from combined cholesterol and sphingomyelin, all of which were supplemented with diphytanoylphosphatidylcholine (diΦPC), to stabilise the PLMs. Pure diΦPC PLMs were not affected by PlyB/OlyA, but the mixed proteins produced pore openings in

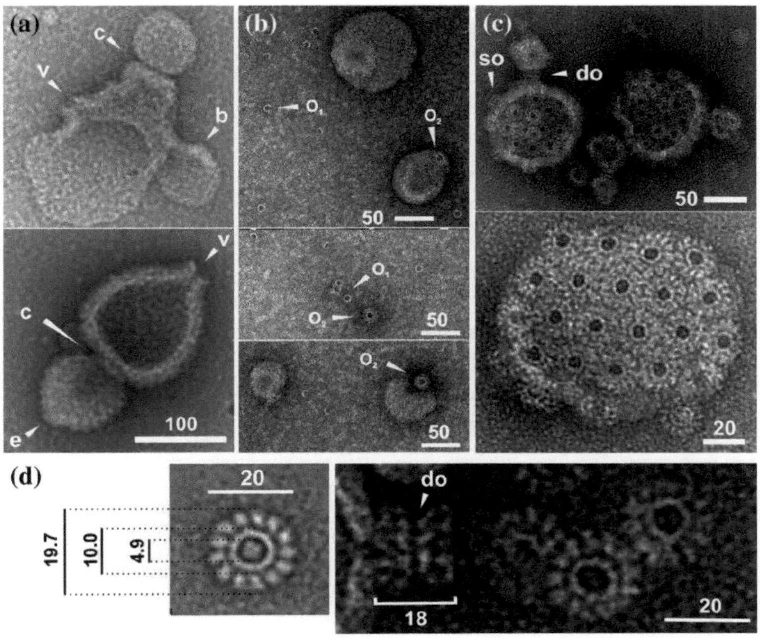

Fig. 14.4 Electron microscopy of Δ48PlyB and OlyA6 on lipid-vesicle membranes. Representative electron microscopy images of erythrocyte-lipid large unilamellar vesicles (LUVs) treated with OlyA6 and Δ48PlyB. OlyA6 and Δ48PlyB, individually or combined, were incubated with LUVs in 140 mM NaCl, 20 mM Tris-HCl, 1 mM EDTA, pH 8.0, and transferred to carbon electron microscopy grids, with 2 % uranyl acetate for contrast. Scale bars are given in nm. **a** LUVs treated with 4 μM OlyA6, showing vesicle budding (b), vesicle constriction (c), evagination (e), and 'volcano-like' structure (v). **b** LUVs treated with 4 μM Δ48PlyB, showing *ring*-like structures (o_1) and rosette-like structures (o_2). **c** *Top* LUVs treated with the combination of 533 nM OlyA6 and 266 nM Δ48PlyB, showing dimeric rosette-like oligomer (do) and single rosette-like oligomer (so). *Bottom* Single LUV treated with the combination of 8 μM OlyA6 and 4 μM Δ48PlyB, showing the hexagonal array of rosette-like oligomers obtained. **d** *Left* Averaged image (n = 9) of the 13-mer rosette-like oligomer formed on LUVs treated with OlyA6 and Δ48PlyB. *Right* detail of a dimeric rosette-like oligomers (do), as ~18 nm in height, that connects two opposing LUV membranes (see also **c**). *Bars* and *numbers* indicate dimensions in nm. Reproduced, with permission, from Ref. [41] © 2013 Elsevier Masson SAS

PLMs composed of diΦPC and at least 26 mol% cholesterol (data not shown). This suggests again a ubiquitous role for cholesterol in the OlyA/PlyB pore-forming activity. Furthermore, this also supports the view that sphingomyelin can be substituted by another membrane lipid, such as either a fully saturated glycerophosphatide or diΦPC. As the cholesterol/diΦPC mixture, which has a preference to form a liquid-disordered lipid phase [35, 66], is as operational as that of cholesterol/sphingomyelin, this also raises the question whether the formation of the liquid-ordered lipid phase is a prerequisite for the pore-forming activity of these proteins, as has been suggested previously [49, 61]. Statistical analyses showed that the

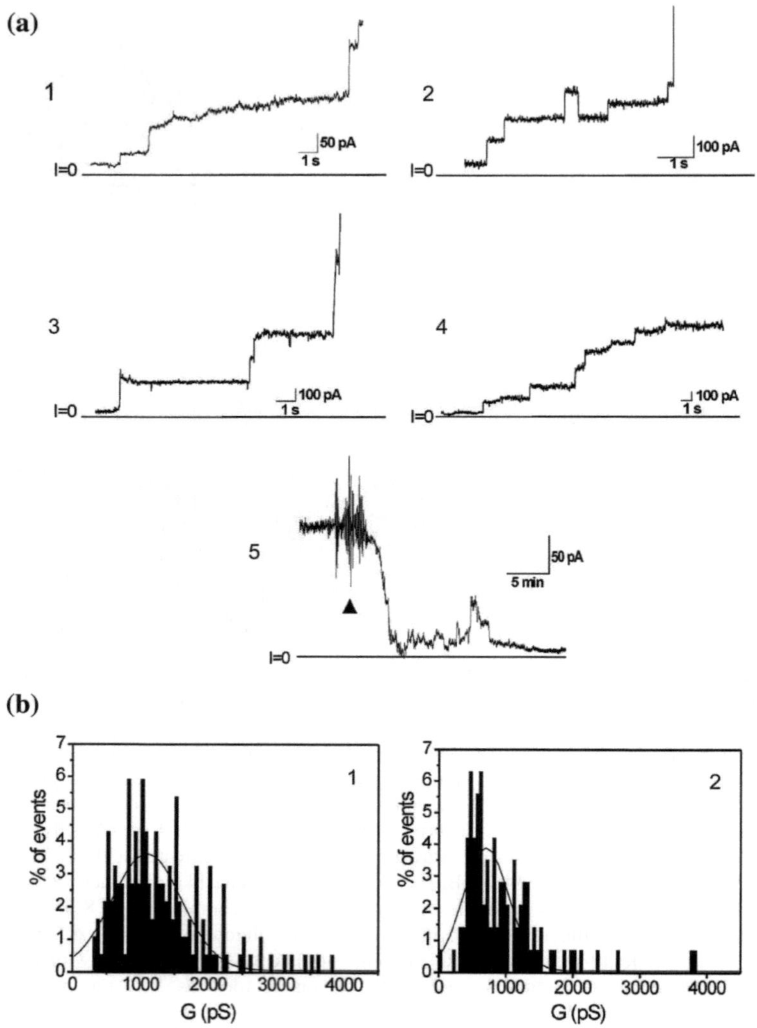
(a)
1
50 pA
1 s
I=0
2
100 pA
1 s
I=0
3
100 pA
1 s
I=0
4
100 pA
1 s
I=0
5
50 pA
5 min
I=0
(b)
% of events
G (pS)
1
2
0
1000
2000
3000
4000

◀ **Fig. 14.5** Electrophysiological characteristics of the transmembrane pores. Pores were formed by mixed native pleurotolysin B plus ostreolysin A in planar lipid membranes using techniques described elsewhere [11]. **a** Representative transmembrane current increases induced by 0.08 μM PlyB plus 0.82 μM OlyA (*a1*) or 0.17 μM PlyB plus 1.67 μM OlyA (*a2–a4*) added to the *cis*-side of a stable preformed lipid bilayer composed of 20 % (w/w) total sheep erythrocyte lipids and 80 % (w/w) di-phytanoyl-phosphatidylcholine (diΦPC) (*a1*) or cholesterol/sphingomyelin/diΦPC in a 1/1/1 molar ratio (*a2–a4*). The bathing solutions were symmetrical and composed of 100 mM (*a1–a3*) or 20 mM (*a4*) KCl, 10 mM Tris-HCl, 0.1 mM EDTA (pH 8.0). The holding potential was either +40 mV (*a1*, *a2* and *a4*) or +100 mV (*a3*). The blocking effect of La^{3+} (1 mM, indicated by *arrow*) added on the *cis*-side (*a5*). The holding potential, membrane lipid composition, PlyB and OlyA concentrations and buffer were as in *a4*. **b** Conductance distributions and Gaussian fits of the pores formed by 0.17 μM PlyB plus 1.67 μM OlyA in planar lipid bilayers composed of cholesterol/sphingomyelin/diΦPC in a 1/1/1 molar ratio. Conductances ($n = 143-186$) were randomly collected from traces obtained in 9–12 independent experiments, and the number of single events of pore opening were plotted *versus* conductance, using a bin-size of 50 pS. Bathing solutions were symmetrical and composed of 100 mM (*b1*) or 20 mM (*b2*) KCl, 10 mM Tris-HCl, 0.1 mM EDTA (pH 8.0). Mean conductances were 1,100 pS (*b1*) and 700 pS (*b2*); the holding potential was +100 mV

distribution of the PlyB/OlyA pore conductance is relatively homogeneous, with a mean of 1,100 ± 600 pS and 700 ± 300 pS using bathing solutions containing 100 mM (Fig. 14.5b1) and 20 mM KCl (Fig. 14.5b2), respectively. In comparison, measurements of single-channel conductance and transmembrane macroscopic current revealed that PlyB/OlyA form ion-conducting pores with broader and skewed conductance distributions in N18 neuroblastoma (~600 pS) and CHO-K1 (~485 pS) cell membranes [57]. Despite these differences in the pores conductance the current-voltage characteristics of the pores were linear and symmetrical, which also suggests that the pores, either created in the planar lipid membranes or natural membranes, are not voltage-gated and are relatively large; these pores are not ion-selective. Of note, these characteristics are in accordance with the dimensions of the central-most inner ring of about 4.9 nm that was seen in the electron microscopy images of the rosette-like structures in Fig. 14.4d, and is in accordance with a previous estimation of the hydrodynamic radius of the pores produced by ostreolysin (containing PlyB as an impurity) or purified PlyB/OlyA [57]. These characteristics of the pores are also in agreement with the suggested colloid-osmotic type of cytolysis that is produced by these proteins [60].

Is the Pore-Forming Activity of PlyB and OlyA (or PlyA) Biologically Relevant?

The pore-forming activities of *P. ostreatus* PlyB and OlyA or PlyA, and also the haemolytic activity of the *A. aegerita* proteins [4], are not understood in terms of the biology of these mushrooms. Both of these species are widely cultivated as popular edible mushrooms, and they are known not to be poisonous when thermally pre-treated. However, raw *P. ostreatus* basidiocarps have been reported to cause sporadic intoxication in humans and animals upon ingestion of large quantities of these fresh mushrooms [1]. This toxicity was associated with the mushroom proteins, which provoked contraction of non-vascular tracheal smooth muscle in vitro [56]. Intravenous application of partially purified ostreolysin containing PlyB was reported to be lethal in mice, due to hyperkalaemia (50 % lethal dose, ~1.2 mg/kg), which could result from intravasal cytolysis caused by these proteins [70]. These proteins also induced sustained contraction of porcine coronary arteries and endothelial dysfunction in middle-sized and large vessels, which is also in agreement with their pore-forming activity [28]. Recent toxicity assays have shown that the PlyB/OlyA combination is not harmful to the development of *Arabidopsis thaliana* seedlings, to the eggs and larvae of the nematode *Caenorhabditis elegans* [41], or to *Drosophila* eggs (our unpublished data). These results were interpreted in terms of the absence of either sterol or sphingomyelin in the cellular membranes of these model organisms [41]. These observations, and the intracellular location of PlyB and OlyA (or PlyA), imply that these proteins have roles in fungal development rather than in their defence, as was originally proposed for Aa-Pri1 [17] and other aegerolysins [5, 39].

It has been shown using Western blotting that the biosynthesis of *P. ostreatus* ostreolysin and *A. agrocybe* Aa-Pri1 starts early in hyphal development, and peaks during the formation of fruiting bodies [4, 67]. Such an expression pattern was also supported by an analysis of *P. ostreatus* transcripts during the growth of liquid-cultured mycelia and fruiting stages [31]. Another extensive analysis of expressed sequence tags (ESTs) of *P. ostreatus* revealed high redundancy of aegerolysin transcripts of PriA, a PlyA/OlyA isoform, in primordial and young fruiting bodies, with lower redundancy in the aged mushrooms [27]. In this study by Joh et al. [27], however, no PlyB transcripts were detected. In fruiting bodies, ostreolysin (mainly as OlyA) was immuno-localised predominantly in the basidiomata growing zones, and the gills and spores [67]. A large-scale study of gene-expression transcripts of PlyB/PlyA/OlyA orthologues showed that an aegerolysin (Aa-Pria1) and a PlyB-like protein are up-regulated during the formation of fruiting bodies in *A. cylindracea* [63]. This is consistent with the appearance of haemolytic activity during the mushroom development, as also noted for both *P. ostreatus* and *A. aegerita* [4]. Detailed analysis of the differential expression of developmentally regulated genes in *M. perniciosa*, which is a causal agent of the Witches' broom disease of *Theobroma cacao* [37], demonstrated that the relatively low basal transcription of a PlyB-like protein in young 'white' mycelia significantly increased at the 'redish-pink' developmental stage of the mycelia (15.2-fold), and slightly decreased towards the development of the primordia, with a return to almost that of the 'white' mycelia in basidiomata (1.6-fold higher). Meanwhile, the transcription levels of the aegerolysin orthologues were low in the early developmental stages, but suddenly increased to ~90-fold in basidiomata [43]. Hence, in *P. perniciosa* at least, the expression levels of PlyB and aegerolysins do not appear to be co-regulated; however, the translated proteins probably coincide in the primordia, and particularly in the basidiomata.

Accumulated data on the regulation of the expression and turn-over of PlyB-like proteins and aegerolysins in fungi are still fragmentary. Elucidation of their roles in fungal physiology will be of interest, as all of the species with PlyB-like proteins plus aegerolysins are very important economically, as lignin and cellulose decomposers, and at least one is a well-known plant pathogen. Specifically, genera *Pleurotus*, *Agrocybe*, *Trametes* (syn. *Phomes*), and *Sphaerobolus* are white rotting fungi that promote the decay of wood lignin, which can thus cause marked economic losses. *Heterobasidion* spp. can cause huge damage to conifer woods through what is known as 'annosum root rot' disease. Meanwhile, *P. perniciosa* is a well-known hemibiotrophic pathogen of the cacao tree (*T. cacao*), which can reduce cacao crops by up to 90 % [45].

Concluding Remarks

The MACPF domain appears to be widely spread across all kingdoms of life, including fungi, and it can be combined with other structural and functional protein domains [20]; (see Section "Occurrence in Fungi of Aegerolysins and Proteins with an MACPF Domain, Including PlyB-Like Proteins"). However, combination of this domain with the specific C-terminal region in PlyB-like proteins, and in addition to this, the co-occurrence of PlyB-like proteins and aegerolysins (as shown in Fig. 14.1) is unique, and has been found for only a few Basidiomycota species to date. Such a low distribution of these proteins among the fungal taxa cannot be ascribed simply to a lack of the genomic data in the other fungal groups. Further in-depth comparative analyses of the genomic data, including intronic, intergenic, and regulatory sequences, should provide novel data on the phylogeny of PlyB-like proteins, particularly in fungi, and their relationships with the other eukaryotic proteins with MACPF domains. In comparison, the distribution of aegerolysin-like proteins appears to be much wider among evolutionarily distant organisms. However, their roles in these organisms appear even more enigmatic. Aegerolysin-like proteins might even have evolved for different purposes. Their roles might be pleiotropic in the organism, as suggested for the *P. aeruginosa* RahU aegerolysin [46, 48], and also varied in evolutionarily distant organisms. Their conserved β-sandwich core and repertoire of variable substitutions on loops might provide variable binding sites that support their variable functions, which include binding to lipids and cooperation with the PlyB-like proteins.

The accumulated data on the functional characteristics of OlyA/PlyA and PlyB include their selectivity for lipids and binding to lipid membranes, their modelled 3D structures, and their assumed structural and functional analogies with the well-studied MACPF/CDC pore-forming proteins [20, 52, 53]. Also considering in particular the role of their MACPF domains, an overall hypothetical model of PlyB and OlyA (or PlyA) pore-formation in the lipid bilayer is presented schematically in Fig. 14.6. We suggest that both the monomeric proteins OlyA and PlyA, with

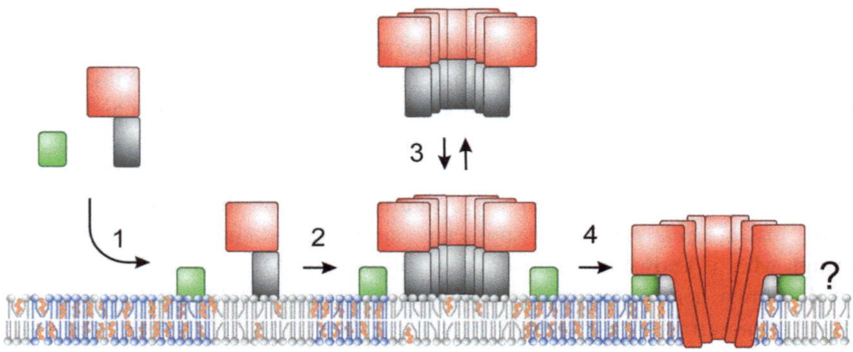

Fig. 14.6 Hypothetical model of the mechanism of pore formation by PlyB and OlyA (or PlyA) on lipid membranes. In the lipid bilayer, the lipids sphingomyelin (*blue*) and cholesterol (*orange*) are shown clustered, and surrounded by glycerophosphatides (*grey*). *Step 1* Monomeric OlyA (*green*) binds preferentially to cholesterol/sphingomyelin-rich membrane regions; monomeric PlyB, depicted with the MACPF domain (*red*) and C-terminal region (*grey*), binds with low affinity to the bilayer surface and oligomerises, to form the presumed pre-pore complex (*step 2*). *Step 3* The oligomerised PlyB complex can dissociate from the surface of the bilayer or associate with OlyA in *step 4*, to form the transmembrane pore-complex. The location and number of OlyA monomers in the final complex is not certain (i.e., question mark)

their high selectivity and affinity for cholesterol/sphingomyelin-enriched membrane nanodomains, and PlyB, with its lower lipid-selectivity and lower affinity for the bilayer lipids, can bind superficially to exposed membrane surfaces. The attachment of the PlyB molecules, which presumably occurs through the C-terminal domain, then orients the protein molecules and enables their oligomerisation to form the pre-pore complex (detected as PlyB SDS-resistant oligomers; [41]). This complex thus appears likely to represent the pre-pore complex that can also dissociate from the bilayer surface, as seen in electron microscopy images (see Fig. 14.4). In the next step, several surface-attached OlyA monomers associate successively with the PlyB complex, which induces cooperative conformational changes in the PlyB molecules, and promotes the extension of certain MACPF domain structural elements for their protrusion through the membrane bilayer, thus forming the transmembrane pores. Location of the OlyA molecules and their numbers in the pore complexes remain uncertain.

An interesting extension of the observed selectivity of ostreolysin and OlyA for membranes enriched in cholesterol/sphingomyelin (>30 mol% cholesterol; [61]) might allow the use of this protein for selective labelling of membrane nanodomains and microdomains that are enriched in cholesterol and sphingomyelin (e.g., as for those classified as lipid rafts, caveolae, detergent-resistant membranes). Previous studies have shown that labelling of these membrane structures with ostreolysin might be specific and different from the labelling with other lipid-selective proteins [10, 33, 51], such as with the cholesterol-specific domain 4 of perfringolysin, the lipid-raft-specific cholera toxin subunit B, and the sphingomyelin-specific pore-forming toxins such as lysenin and equinatoxin (for review, see [64]). Moreover, an ortholog of *P.ostreatus* pleurotolysin A and ostreolysin A from *P.eryngii*, named pleurotolysin A2 (PlyA2), when fused with enhanced green fluorescent protein (EGFP) bound to cholesterol/sphingomyelin but not to cholesterol/phosphatidylcholine lipid vesicles. The fusion protein, PlyA2-EGFP, was used to label cell surface. The surface binding of the fluorescent probe was abolished after methyl-β-cyclodextrin as well as sphingomyelinase treatment of the cells, removing cholesterol and sphingomyelin, respectively. It was suggested that PlyA2-EGFP specifically binds cell surface cholesterol/sphingomyelin rafts [8].

In conclusion, fungal PlyB-like proteins with the MACPF domain appear non-lytic on their own but need the assistance of specific aegerolysins to be cytolytic, pore-forming proteins. Compared to other MACPF/CDC proteins, their mechanisms of membrane permeabilisation appear to be specific, through the combination with aegerolysin molecules for the stabilisation of a pre-pore PlyB complex on the membranes, and in the further processes that lead to the formation of the final transmembrane pore complex, which is permeable for solutes below $\sim$4–5 nm in diameter. The solving of their crystal structure and further structure-functional studies are needed in order to be able to understand their mechanisms of membrane permeabilisation and the nature of these protein complexes in the membranes. The biological relevance of the pore-forming activity of these proteins in fungi also remains to be understood fully. Further studies on these proteins should result in the definition of their unprecedented molecular function(s) and their biological significance.

Acknowledgments GV and MDS thank the Laboratory of Biomolecular Sequence and Structure Analysis for Health (LaBSSAH) for technical support. The Slovenian authors were supported by the Slovenian Research Agency (contracts P1-0207 and J1-4305). The authors thank Christopher Berrie for editorial revision of the manuscript.

References

1. Al-Deen IH, Twaij HA, Al-Badr AA, Istarabadi TA (1987) Toxicologic and histopathologic studies of *Pleurotus ostreatus* mushroom in mice. J Ethnopharmacol 21:297–305
2. Barloy F, Lecadet MM, Delécluse A (1998) Cloning and sequencing of three new putative genes from *Clostridium bifermentans* CH18. Gene 211:293–299
3. Bass D, Richards TA (2011) Three reasons to re-evaluate fungal diversity 'on earth and in the ocean'. Fungal Biol Rev 25:159–164
4. Berne S, Križaj I, Pohleven F, Turk T, Maček P, Sepčić K (2002) *Pleurotus* and *Agrocybe* hemolysins, new proteins hypothetically involved in fungal fruiting. Biochim Biophys Acta 1570:153–159
5. Berne S, Lah L, Sepčić K (2009) Aegerolysins: structure, function, and putative biological role. Protein Sci 18:694–706
6. Berne S, Sepčić K, Anderluh G, Turk T, Maček P, Poklar Ulrih N (2005) Effect of pH on the pore forming activity and conformational stability of ostreolysin, a lipid raft-binding protein from the edible mushroom *Pleurotus ostreatus*. Biochemistry 44:11137–11147

7. Bernheimer AW, Avigad LS (1979) A cytolytic protein from the edible mushroom, *Pleurotus ostreatus*. Biochim Biophys Acta 585:451–461
8. Bhat HB, Kishimoto T, Abe M, Makino A, Inaba T, Murate M, Dohmae N, Kurahashi A, Nishibori K, Fujimori F, Greimel P, Ishitsuka R, Kobayashi T (2013) Binding of a pleurotolysin ortholog from *Pleurotus eryngii* to sphingomyelin and cholesterol-rich membrane domains. J Lipid Res 54:2933–2943
9. Bubeck D, Roversi P, Donev R, Morgan BP, Llorca O, Lea SM (2011) Structure of human complement C8, a precursor to membrane attack. J Mol Biol 405:325–330
10. Chowdhury HH, Rebolj K, Kreft M, Zorec R, Maček P, Sepčić K (2008) Lysophospholipids prevent binding of a cytolytic protein ostreolysin to cholesterol-enriched membrane domains. Toxicon 51:1345–1356
11. Dalla Serra M, Menestrina G (2000) Characterisation of molecular properties of pore-forming toxins with planar lipid bilayers. In: Holst O (ed) Bacterial toxins: methods and protocols. Humana Press, Totowa
12. Dobson GD, van West P, Gadd GM (2007) Exploitation of fungi. Cambridge University Press, Cambridge
13. Dunstone MA, Tweten RK (2012) Packing a punch: the mechanism of pore formation by cholesterol dependent cytolysins and membrane attack complex/perforin-like proteins. Curr Opin Struct Biol 22:342–349
14. Ebina K, Ichinowatari S, Yokota K (1985) Studies on toxin of *Aspergillus fumigatus*. XXII. Fashion of binding of Asp-hemolysin to human erythrocytes and Asp-hemolysin-binding proteins of erythrocyte membranes. Microbiol Immunol 29:91–101
15. Ebina K, Sakagami H, Yokota K, Kondo H (1994) Cloning and nucleotide sequence of cDNA encoding Asp-hemolysin from *Aspergillus fumigatus*. Biochim Biophys Acta 1219:148–150
16. Ellis RT, Stockhoff BA, Stamp L, Schnepf HE, Schwab GE, Knuth M et al (2002) Novel *Bacillus thuringiensis* binary insecticidal crystal proteins active on Western corn rootworm, *Diabrotica virgifera virgifera* LeConte. Appl Environ Microbiol 68:1137–1145
17. Fernandez Espinar MT, Labarère J (1997) Cloning and sequencing of the *Aa-Pri1* gene specifically expressed during fruiting initiation in the edible mushroom *Agrocybe aegerita*, an analysis of the predicted amino-acid sequence. Curr Genet 32:420–424
18. Fisher MC, Henk D, Briggs CJ, Brownstein JS, Madoff LC, McCraw SL et al (2012) Emerging fungal threats to animal, plant and ecosystem health. Nature 484:186–194
19. Galagan JE, Henn MR, Ma L-J, Cuomo CA, Birren B (2005) Genomics of the fungal kingdom: insights into eukaryotic biology. Genome Res 15:1620–1631
20. Gilbert RJ, Mikelj M, Dalla Serra M, Froelich CJ, Anderluh G (2013) Effects of MACPF/CDC proteins on lipid membranes. Cell Mol Life Sci 70:2083–2098
21. Grigoriev IV, Nordberg H, Shabalov I, Aerts A, Cantor M, Goodstein D et al (2012) The genome portal of the Department of Energy Joint Genome Institute. Nucleic Acids Res 40(Database issue):D26–D32
22. Hadders MA, Beringer DX, Gros P (2007) Structure of C8α-MACPF reveals mechanism of membrane attack in complement immune defense. Science 317:1552–1554
23. Heitman J (2011) Microbial pathogens in the fungal kingdom. Fungal Biol Rev 25:48–60
24. Herman RA, Scherer PN, Young DL, Mihaliak CA, Meade T, Woodsworth AT et al (2002) Binary insecticidal crystal protein from *Bacillus thuringiensis*, Strain PS149B1: effects of individual protein components and mixtures in laboratory bioassays. J Econ Entomol 95:635–639
25. Heuck AP, Moe PC, Johnson BB (2010) The cholesterol-dependent cytolysin family of Gram-positive bacteria. In: Harris JR (ed) Cholesterol binding and cholesterol transport proteins. Subcellular biochemistry, Springer, Netherlands
26. Hotze EM, Tweten RK (2012) Membrane assembly of the cholesterol-dependent cytolysin pore complex. Biochim Biophys Acta 1818:1028–1038
27. Joh J-H, Lee S-H, Lee J-S, Kim K-H, Jeong S-J, Youn W-H et al (2007) Isolation of genes expressed during the developmental stages of the oyster mushroom, *Pleurotus ostreatus*, using expressed sequence tags. FEMS Microbiol Lett 276:19–25

28. Juntes P, Rebolj K, Sepčić K, Maček P, Žužek MC, Cestnik V et al (2009) Ostreolysin induces sustained contraction of porcine coronary arteries and endothelial dysfunction in middle- and large-sized vessels. Toxicon 54:784–792
29. Kudo Y, Fukuchi Y, Kumagai T, Ebina K, Yokota K (2001) Oxidized low-density lipoprotein-binding specificity of Asp-hemolysin from *Aspergillus fumigatus*. Biochim Biophys Acta 1568:183–188
30. Larkin MA, Blackshields G, Brown NP, Chenna R, McGettigan PA, McWilliam H et al (2007) Clustal W and Clustal X version 2.0. Bioinformatics 23:2947–2948
31. Lee SH, Kim BG, Kim KJ, Lee JS, Yun DW, Hahn JH et al (2002) Comparative analysis of sequences expressed during the liquid-cultured mycelia and fruit body stages of *Pleurotus ostreatus*. Fungal Gen Biol 35:115–134
32. Li H, Olson M, Lin G, Hey T, Tan SY, Narva KE (2013) *Bacillus thuringiensis* Cry34Ab1/Cry35Ab1 interactions with Western corn rootworm midgut membrane binding sites. PLoS ONE 8(1):e53079
33. Lokar M, Kabaso D, Resnik N, Sepčić K, Kralj-Iglič V, Veranič P et al (2012) The role of cholesterol-sphingomyelin membrane nanodomains in the stability of intercellular membrane nanotubes. Int J Nanomedicine 7:1891–1902
34. Lv H, Kong Y, Yao Q, Zhang B, Leng FW, Bian HJ et al (2009) Nebrodeolysin, a novel hemolytic protein from mushroom *Pleurotus nebrodensis* with apoptosis-inducing and anti-HIV-1 effects. Phytomedicine 16:198–205
35. Marsh D (2010) Liquid-ordered phases induced by cholesterol: A compendium of binary phase diagrams. Biochim Biophys Acta Biomembranes 1798:688–699
36. Masson L, Schwab G, Mazza A, Brousseau R, Potvin L, Schwartz JL (2004) A novel *Bacillus thuringiensis* (PS149B1) containing a Cry34Ab1/Cry35Ab1 binary toxin specific for the Western corn rootworm *Diabrotica virgifera virgifera* LeConte forms ion channels in lipid membranes. Biochemistry 43:12349–12357
37. Mondego JMC, Carazzolle MF, Costa GGL, Formighieri EF, Parizzi LP, Rincones J et al (2008) A genome survey of *Moniliophthora perniciosa* gives new insights into Witches' broom disease of cacao. BMC Genom 9:548
38. Nayak AP, Blachere F, Hettick J, Lukomski S, Schmechel D, Beezhold DH (2011) Characterization of recombinant terrelysin, a hemolysin of *Aspergillus terreus*. Mycopathologia 171:23–34
39. Nayak AP, Green BJ, Beezhold DH (2013) Fungal hemolysins. Med Mycol 51:1–16
40. Ota K (2013) Functional characterization of fungal two-component pore-forming proteins. Dissertation, University of Ljubljana
41. Ota K, Leonardi A, Mikelj M, Skočaj M, Wohlschlager T, Künzler M et al (2013) Membrane cholesterol and sphingomyelin, and ostreolysin A are obligatory for pore-formation by a MACPF/CDC-like pore-forming protein, pleurotolysin B. Biochimie 80:1855–1864
42. Pettersen EF, Goddard TD, Huang CC, Couch GS, Greenblatt DM, Meng EC et al (2004) UCSF Chimera-a visualization system for exploratory research and analysis. J Comput Chem 25:1605–1612
43. Pires ABL, Gramacho KP, Silva DC, Goes-Neto A, Silva MM, Muniz-Sobrinho JS et al (2009) Early development of *Moniliophthora perniciosa* basidiomata and developmentally regulated genes. BMC Microbiol 9:158
44. Pointing S, Hyde K (2001) Bio-exploitation of filamentous Fungi. In: Fungal diversity research series, University of Hong Kong, Hong Kong
45. Purdy L, Schmidt R (1996) Status of cacao Witches' broom: biology, epidemiology, and management. Ann Rev Phytopathol 34:573–594
46. Rao J, DiGiandomenico A, Artamonov M, Leitinger N, Amin AR, Goldberg JB (2011) Host derived inflammatory phospholipids regulate rahU (PA0122) gene, protein, and biofilm formation in *Pseudomonas aeruginosa*. Cell Immunol 270:95–102
47. Rao J, DiGiandomenico A, Unger J, Bao Y, Polanowska-Grabowska RK, Goldberg JB (2008) A novel oxidized low-density lipoprotein-binding protein from *Pseudomonas aeruginosa*. Microbiol 154:654–665

48. Rao J, Elliott MR, Leitinger N, Jensen RV, Goldberg JB, Amin AR (2011) RahU: An inducible and functionally pleiotropic protein in *Pseudomonas aeruginosa* modulates innate immunity and inflammation in host cells. Cell Immunol 270:103–113
49. Rebolj K, Bakrač B, Garvas M, Ota K, Šentjurc M, Potrich C et al (2010) EPR and FTIR studies reveal the importance of highly ordered sterol-enriched membrane domains for ostreolysin activity. Biochim Biophys Acta 1798:891–902
50. Rebolj K, Ulrih-Poklar N, Maček P, Sepčić K (2006) Steroid structural requirements for interaction of ostreolysin, a lipid-raft binding cytolysin, with lipid monolayers and bilayers. Biochim Biophys Acta 1758:1662–1670
51. Resnik N, Sepčić K, Plemenitaš A, Windoffer R, Leube R, Veranič P (2011) Desmosome assembly and cell-cell adhesion are membrane raft-dependent processes. J Biol Chem 286:1499–1507
52. Rosado CJ, Buckle AM, Law RH, Butcher RE, Kan WT, Bird CH et al (2007) A common fold mediates vertebrate defense and bacterial attack. Science 317:1548–1551
53. Rosado CJ, Kondos S, Bull TE, Kuiper MJ, Law RH, Buckle AM et al (2008) The MACPF/CDC family of pore-forming toxins. Cell Microbiol 10:1765–1774
54. Roy A, Kucukural A, Zhang Y (2010) I-TASSER: a unified platform for automated protein structure and function prediction. Nat Protoc 5:725–738
55. Sakaguchi O, Shimada H, Yokota K (1975) Proceedings: purification and characteristics of hemolytic toxin from *Aspergillus fumigatus*. Jap J Med Sci & Biol 28:328–331
56. Schachter EN, Zuskin E, Goswami S, Castranova V, Arumugam U, Whitmer M et al (2005) Pharmacological study of oyster mushroom (*Pleurotus ostreatus*) extract on isolated guinea pig trachea smooth muscle. Lung 183:63–71
57. Schlumberger S, Kristan-Črnigoj K, Ota K, Frangež R, Molgó J, Sepčić K, Benoit E, Maček P (2013) Permeability characteristics of cell-membrane pores induced by ostreolysin A/pleurotolysin B, binary pore-forming proteins from the oyster mushroom. FEBS Lett 588:35–40
58. Schnepf HE (2013) Modified *Cry34* proteins. Patent, U.S.A 7939651
59. Schnepf HE, Lee S, Dojillo J, Burmeister P, Fencil K, Morera L et al (2005) Characterization of Cry34/Cry35 binary insecticidal proteins from diverse *Bacillus thuringiensis* strain collections. Appl Environ Microbiol 71:1765–1774
60. Sepčić K, Berne S, Potrich C, Turk T, Maček P, Menestrina G (2003) Interaction of ostreolysin, a cytolytic protein from the edible mushroom *Pleurotus ostreatus*, with lipid membranes and modulation by lysophospholipids. Eur J Biochem 270:1199–1210
61. Sepčić K, Berne S, Rebolj K, Batista U, Plemenitaš A, Šentjurc M et al (2004) Ostreolysin, a pore-forming protein from the oyster mushroom, interacts specifically with membrane cholesterol-rich lipid domains. FEBS Lett 575:81–85
62. Shibata T, Kudou M, Hoshi Y, Kudo A, Nanashima N, Miyairi K (2010) Isolation and characterization of a novel two-component hemolysin, erylysin A and B, from an edible mushroom, *Pleurotus eryngii*. Toxicon 56:1436–1442
63. Shim SM, Kim SB, Kim HY, Rho H-S, Lee HS, Lee MW et al (2006) Genes expressed during fruiting body formation of *Agrocybe cylindracea*. Mycobiol 34:209–213
64. Skočaj M, Bakrač B, Križaj I, Maček P, Anderluh G, Sepčić K (2013) The sensing of membrane microdomains based on pore-forming toxins. Curr Med Chem 20:491–501
65. Tomita T, Noguchi K, Mimuro H, Ukaji F, Ito K, Sugawara-Tomita N et al (2004) Pleurotolysin, a novel sphingomyelin-specific two-component cytolysin from the edible mushroom *Pleurotus ostreatus*, assembles into a transmembrane pore complex. J Biol Chem 279:26975–26982
66. Veatch SL, Gawrisch K, Keller SL (2006) Closed-loop miscibility gap and quantitative tie-lines in ternary membranes containing diphytanoyl PC. Biophys J 90:4428–4436
67. Vidic I, Berne S, Drobne D, Maček P, Frangež R, Turk T et al (2005) Temporal and spatial expression of ostreolysin during development of the oyster mushroom (*Pleurotus ostreatus*). Mycol Res 109:377–382

68. Wang M, Gu B, Huang J, Jiang S, Chen Y, Yin Y et al (2013) Transcriptome and proteome exploration to provide a resource for the study of *Agrocybe aegerita*. PLoS ONE 8(2):e56686
69. Zhang S, Clark KD, Strand MR (2011) The protein P23 identifies capsule-forming plasmatocytes in the moth *Pseudoplusia includens*. Develop Comp Immunol 35:501–510
70. Žužek MC, Maček P, Sepčić K, Cestnik V, Frangež R (2006) Toxic and lethal effects of ostreolysin, a cytolytic protein from edible oyster mushroom (*Pleurotus ostreatus*), in rodents. Toxicon 48:264–271

Chapter 15
Fluorescence Imaging of MACPF/CDC Proteins: New Techniques and Their Application

Michael J. Senior and Mark I. Wallace

Abstract Structural and biochemical investigations have helped illuminate many of the important details of MACPF/CDC pore formation. However, conventional techniques are limited in their ability to tackle many of the remaining key questions, and new biophysical techniques might provide the means to improve our understanding. Here we attempt to identify the properties of MACPF/CDC proteins that warrant further study, and explore how new developments in fluorescence imaging are able to probe these properties.

Keywords Artificial lipid bilayers · Fluorescence microscopy · Pore-forming toxins · Single-molecule fluorescence · Super-resolution imaging

Abbreviations

Badan	6-bromoacetyl-2-dimethylaminonaphthalene
CDC	Cholesterol-dependent cytolysin
ER	Endoplasmic reticulum
FIONA	Fluorescence imaging with one-nanometre accuracy
FPs	Fluorescent proteins
FRET	Förster resonance energy transfer
FWHM	Full width at half maximum
ILY	Intermedilysin
LLO	Listeriolysin O
MACPF	Membrane attack complex/perforin
NBD	7-nitro-2-1,3-benzoxadiazol-4-yl
PALM	Photoactivated localization microscopy
PDMS	Poly(dimethyl) siloxane
PFO	Perfringolysin O
PLY	Pneumolysin

M. J. Senior · M. I. Wallace (✉)
Department of Chemistry, Oxford University, 12 Mansfield Rd, Oxford, OX1 3TA, UK
e-mail: mark.wallace@chem.ox.ac.uk

G. Anderluh and R. Gilbert (eds.), *MACPF/CDC Proteins - Agents of Defence, Attack and Invasion*, Subcellular Biochemistry 80, DOI: 10.1007/978-94-017-8881-6_15,

PSF	Point spread function
SHRImP	Single-molecule high-resolution imaging with photobleaching
SIM	Structured illumination microscopy
smFRET	Single-molecule FRET
STED	Stimulated emission depletion
STORM	Stochastic optical reconstruction microscopy
TIRF	Total internal reflection fluorescence
μsSMT	Microsecond single-molecule tracking

Introduction

Our current understanding of MACPF/CDC proteins can be attributed to the effective and complementary application of a range of structural and functional techniques (e.g. [7, 16, 18, 24, 37, 38, 40, 63, 65, 68, 70, 76, 77, 85, 91, 93]). Although previous work has revealed much about the behavior of these proteins [15, 26–28, 39, 69], there remain many essential features that we know surprisingly little about. As a reader of this chapter, it is likely that you need little convincing of the significance of these proteins for both our immune system and bacterial pathogenesis [17, 48, 95]. We hope you would therefore agree that an incomplete description of how these proteins work is unacceptable. So how might we rectify the gaps in our knowledge? This chapter aims to introduce several recent advances in fluorescence imaging that, in our opinion, have the potential to overcome many of the obstacles that prevent a fuller understanding of the MACPF/CDC proteins. We focus on fluorescence in accordance with our expertise. Fluorescence imaging affords high resolution, excellent molecular specificity and the ability to study both in vitro and in vivo systems. Our approach is to pose a set of questions that outline the gaps in our understanding, before exploring a number of new techniques and addressing how they might be used to answer these questions.

Our task of identifying the key biological questions is greatly aided by a number of recent MACPF/CDC reviews [15, 26–28, 39, 69]. These questions can be broadly separated into two categories (Fig. 15.1). Firstly, there are those questions that relate to the molecular mechanism by which the proteins achieve their function. Secondly, there are questions that relate to the cellular function of the proteins; for example, the effects that the protein has on its target cell or its relationship with the cell that produced it. Following this distinction, the techniques reviewed in this chapter are introduced according to the type of question that they can help to answer. A natural consequence is that, when investigating the cellular function of the MACPF/CDC proteins, experiments are ideally performed in vivo. In comparison, investigations of molecular mechanism benefit from a reduction in system complexity in vitro to the point where biological function is retained, but the effects of each component are easily studied in isolation.

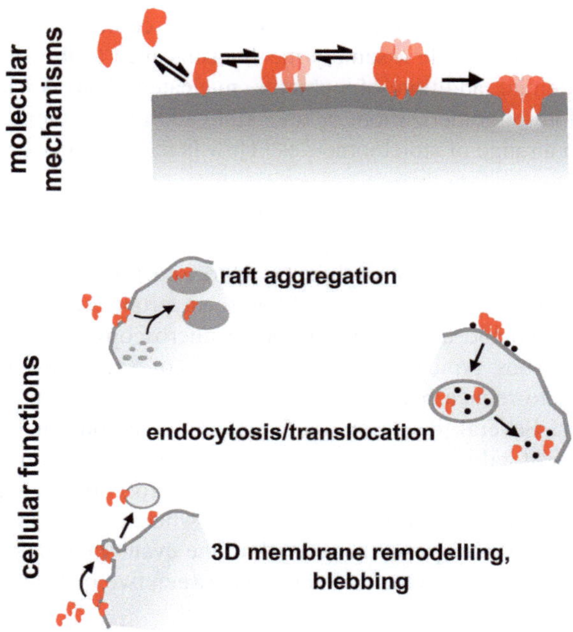

Fig. 15.1 Molecular mechanism and cellular function. Molecular mechanism: the assembly of individual complexes on a target membrane from monomers in solution through the prepore step to the pore [26, 39]. Cellular function: the effects on the target cell arising from the action of the MACPF/CDC protein, including but not limited to lipid raft aggregation [22], translocation of substrates via endocytosis [52, 91] and 3D membrane effects such as blebbing and invaginations [25, 47, 62, 64, 88]

Molecular Mechanism

What are the Dynamic Steps During Pore Formation?

So far, our understanding of the mechanism of MACPF/CDC pore formation is based largely on knowing the discrete stages of assembly: unbound monomer, membrane-bound monomer, prepore, and membrane-inserted pore [26, 39, 41, 78, 93]. Structural studies [26, 68, 70, 93] and bulk fluorescence measurements [14, 37, 38, 40, 65, 76, 77] have revealed a great deal about this mechanism, but we currently lack a dynamic picture of the molecular steps of pore assembly. Arguably, without such observations MACPF/CDC pore formation cannot be comprehensively described. For example, recent work on the CDC perfringolysin O (PFO) by Tweten and colleagues has supported a mechanism involving initial cholesterol recognition by residues in loop L1 [18], with subsequent structural changes in domain 3 coupled to membrane binding by the highly conserved undecapeptide region [14], allowing interaction between monomers and oligomerization. By directly visualising the pore assembly process, we could robustly discriminate between a variety of possible assembly mechanisms such as this one. Is assembly driven by sequential addition of subunits nucleated from a single activated monomer, random oligomerization, or reversible addition of monomers? Why do pores form with a range of stoichiometries? How is the process of oligomerization regulated?

Our recent single-molecule fluorescence measurements of pore formation in the *Staphylococcus aureus* β-barrel pore-forming toxin α-hemolysin [92] suggest one method by which such questions might be addressed for MACPF/CDC proteins. By tagging proteins of interest with a fluorescent probe, single proteins can be tracked both in vivo and in vitro using optical microscopy [44]. Single-molecule fluorescence methods typically exploit detection of signal from only a small volume in order to reduce background noise to a level sufficient for single-molecule detection. Several detection methods are common, including confocal imaging [81], and total internal reflection fluorescence (TIRF) microscopy [1, 2]. To achieve single-molecule detection, confocal microscopy uses a pinhole at a conjugate focal plane to reject out-of-focus light, whereas TIRF microscopy relies on the limited ($\sim$200 nm) penetration depth of the evanescent field created by the total internal reflection of light at an interface between two materials with differing refractive indices.

Imaging fluorescently-labelled α-hemolysin oligomerization using single-molecule TIRF microscopy revealed that pore assembly is significantly cooperative, with the full heptameric pore forming in less than 5 ms (Fig. 15.2a) [92]. Importantly, no assembly intermediates were present on a timescale longer than 5 ms (Fig. 15.2b). Similarly, γ-hemolysin monomers and dimers were found to oligomerize cooperatively by Nguyen et al. [58]. These results place significant limitations on the possible mechanisms for β-barrel pore-forming toxin assembly, and predict [92] that either a single monomer activation step is required, or perhaps more likely, that the individual steps in pore assembly are reversible on a sub-millisecond timescale, with final prepore or pore formation being irreversible (Fig. 15.2b, inset). The results are reconcilable with a recent theory of CDC assembly, where monomers are activated by the rearrangement of domain 3 β-strands, by thermal fluctuations at first and then cooperatively between monomers, promoting oligomerization [40].

In addition to tracking the position of individual proteins, photobleaching step-counting can be used to determine the stoichiometries of membrane protein complexes [50, 56, 67, 92, 94]. Photobleaching is the irreversible photochemical modification of a fluorescent probe that results in a loss of fluorescence, typically through singlet oxygen production via the triplet state of the probe followed by reaction of the probe with singlet oxygen [86, 87]. For an ensemble measurement, this process is observed as a gradual fading of fluorescence intensity. For a single molecule this process is observed as an essentially instantaneous step-wise drop in intensity. If care is taken to ensure unitary labelling of monomers that form part of a protein complex, counting of the number of steps in a photobleaching trace from a single complex will reveal its stoichiometry.

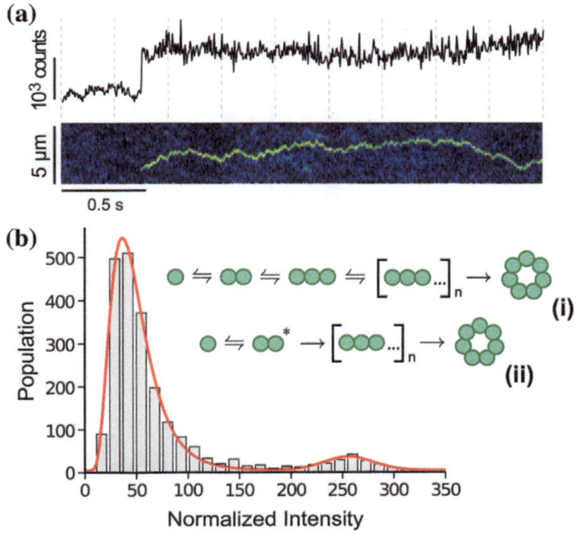

Fig. 15.2 Single-molecule imaging of pore-assembly. TIRF microscopy on droplet interface bilayers presents one way to observe assembly of pore-forming toxins [92]. **a** *upper*, intensity trace of a single pore at 5 ms temporal resolution, and *lower*, kymograph of the same data, the vertical axis represents lateral position in one dimension on the bilayer. **b** normalised fluorescent spot intensity histogram, showing a binary population of spots corresponding to monomers and heptamers (the heptamer peak is at 7.0 ± 1.2 multiples of the intensity of the monomer peak). Inset, these results suggest two likely mechanisms for α-hemolysin oligomerization: (*i*) reversible addition of monomers, or (*ii*), activation of a monomer causing sequential addition of other monomers. Adapted from [92] with permission from Elsevier © 2011

Such methods have been used to measure the stoichiometry of a variety of different multimeric protein complexes, including pore-forming toxins [57, 92], ion channels [56, 94], and biological motors [50]. For example, Fig. 15.3 highlights this concept using the work by Isacoff and colleagues [94], with the homotetrameric CNG channel as the protein of interest. In the original work the aim was to investigate the stoichiometry of NR1:NR3 NMDA receptors, with CNG used as a control of known stoichiometry, but the CNG channel data provides a simple example of the principle of stepwise photobleaching.

What might we expect if such single-molecule experiments were extended to study MACPF/CDC proteins? The basic steps of pore formation (monomer → prepore → pore) are common across different classes of β-barrel pore-forming toxin, so it would be reasonable to use MACPF/CDC proteins to test the prediction of pore assembly with rapid reversible intermediates followed by ring closure as a final step. Photobleaching would allow us to determine stoichiometry both during and after pore formation. The larger size of the MACPF/CDC pores could open up the opportunity to observe these intermediates, as the timescale for an assembly event is likely to be within the observation limits of current single-molecule methods ($\sim$2 ms).

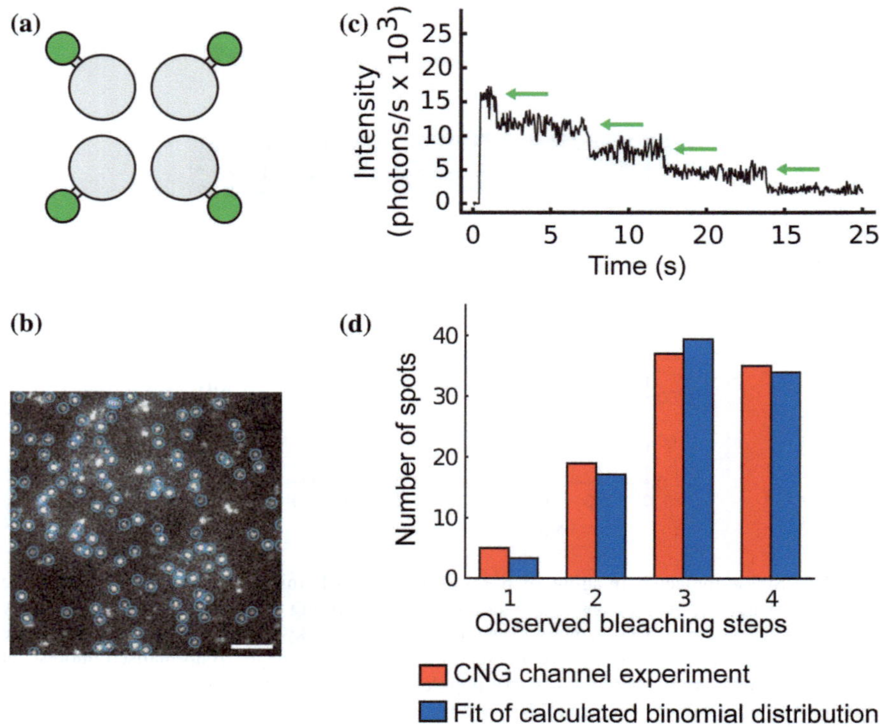

Fig. 15.3 Counting subunits by stepwise decrements in intensity as single fluorophores photobleach. Isacoff and co-workers used this technique to confirm the stoichiometry of the homotetrameric CNG channel, while investigating the stoichiometry of NR3 subunits in NMDA receptors [94]. They genetically engineered subunits with a green fluorescent protein (GFP) label, and used TIRF microscopy to observe proteins of interest in live cell membranes. **a** labelling of the tetrameric CNG channel. **b** an example frame showing fluorescent spots that identify labelled protein complexes in the cell membrane (*blue circles* indicate spots analyzed by the authors). Scale bar, 2 μm. **c** an intensity trajectory for one fluorescent spot corresponding to a CNG channel. 4 decremental steps are highlighted by the *green arrows*, one for each of the subunits as their attached GFP molecule photobleaches. **d** the distribution of photobleaching steps for fluorescent CNG complexes was found to follow a binomial distribution, with the majority of spots exhibiting 3 photobleaching steps. This distribution is due to a particular probability of successfully detecting a step. This could be due to limited signal to noise or GFP molecules being non-fluorescent, due to misfolding or incomplete maturation. The distribution of steps closely matched a binomial when the probability of a step being detected was 77.5 %. Adapted by permission from Macmillan Publishers Ltd: Nature Methods [94], © 2007

These time limits have only been exceeded in a number of limited cases. For example, Semrau et al. achieved microsecond temporal resolution in a single-molecule fluorescence study, tracking individual DNA oligonucleotides in solution (microsecond single-molecule tracking, μsSMT [74]). They labelled each oligonucleotide with two different fluorophores, spatially far enough apart that no energy transfer took place between the two, and introduced a time lag between pulses of excitation at the two distinct excitation wavelengths [74]. The time lag allowed localization of the molecule at two different time points that could in principle be made arbitrarily close together, so temporal resolution was limited by the rate of diffusion, number of localizations and desired localization accuracy. Mean-squared-displacement values for a time-step of 100 μs were extracted (found to be $<0.005\ \mu m^2$). The authors suggest that with a bright enough fluorophore, the two detection channels could both originate from the same fluorophore; imaging the fluorophore onto two different regions of the camera allows the time lag to be introduced in the exposure of these distinct regions rather than in the excitation pulses. This method could be applied to MACPF/CDC proteins to improve the temporal resolution with which they can be dynamically observed during pore assembly. Providing that monomers could be labelled with two fluorophores, far enough apart to avoid energy transfer and not inhibiting the function of the protein, then microsecond tracking may be able to resolve discrete increments in brightness as monomers add to an oligomer.

How and When do β-Barrels Insert?

Structural studies have revealed a great deal about the mechanism behind *β*-barrel insertion [38, 40, 41, 78, 93]. Two fluorescent approaches that could contribute to the temporal resolution of *β*-barrel insertion are the use of Förster resonance energy transfer (FRET), and environment-sensitive fluorophores.

FRET is a strongly distance-dependent non-radiative transfer of energy between two fluorophores (the donor and the acceptor) that acts on the sub-10 nm range [9, 21, 45]. It can measure structural transitions in proteins, which often occur on this length scale. The *β*-barrel insertion of MACPF/CDC proteins could be observed using FRET providing that two fluorophores could be attached to the protein monomers in spatial locations that allowed monitoring of the transition without inhibiting function. Ensemble FRET has been used previously with significant success to study the geometry of PFO in the membrane-bound monomer, prepore and pore states [66]: a donor BODIPY molecule was attached to domain 1 and also domain 3 in distinct mutants of PFO, and lipids labelled with an acceptor rhodamine molecule were used in vesicle preparation. The assembly of the PFO mutants was halted at different discrete stages of pore formation. The height of the donor above the membrane was reported by the FRET efficiency between the donor and acceptor. The results suggested that the long axis of the monomer stands perpendicular to the membrane upon binding, with a minor reduction of the height of domains 1 and 3 in the prepore state, and a major reduction in their height upon *β*-barrel insertion; domains 1 and 3 did not significantly sink towards the membrane before *β*-barrel insertion.

A second example examined the dependence of PFO's lipid domain affinity on hydrophobic mismatch [51]. Here, rather than using two fluorophores for the donor and acceptor, the authors used the tryptophan residues of the protein as the donor and the fluorophore NBD (7-nitro-2-1,3-benzoxadiazol-4-yl), attached to a lipid marker for disordered domains, as the acceptor. The relative concentration of the acceptor around the donor in unilamellar vesicles was measured, revealing the partitioning of the protein into lipid domains; partitioning was found to depend on matching of the protein's transmembrane segment length with the thickness of the bilayer in the domains.

When measuring conformational changes in bulk, only the average state of the molecules over time can be extracted; these studies are therefore not able to temporally resolve the insertion of the β-barrel with respect to oligomerization unless precise synchronization of the protein population can be achieved. Such synchronization is often impossible for many multi-step complex biological reactions. For resolving β-barrel insertion, single-molecule methods provide the advantage of removing this need for synchronization. Resolving individual pores allows observation of when insertion occurs relative to the assembly of the monomers.

Nguyen et al. applied single-molecule FRET (smFRET) to γ-hemolysin to observe assembly of this bicomponent toxin labelled heterogeneously with donor and acceptor fluorophores [58]. However, while applicable to the MACPF/CDC proteins in that it reports on assembly of monomers into a pore, the work by Nguyen et al. does not help with measuring the conformational change associated with β-barrel insertion using smFRET. For this experiment, the donor and acceptor must be attached to the same molecule and must change their spatial separation during β-barrel insertion. With sufficient temporal resolution, such an arrangement would allow tracking of the monomers as they oligomerize, and detection of the β-barrel insertion by the corresponding changes in donor and acceptor emission intensities.

One major problem with such FRET measurements is the need to selectively label a protein at two distinct sites of interest. The simplest method for dual labelling of a protein with both donor and acceptor dyes is double cysteine substitution; however, such modification will result in four permutations of labelling at the two residues. Although donor-donor and acceptor-acceptor labelled molecules can be excluded in a single-molecule measurement, care must be taken in treating donor-acceptor and acceptor-donor labelled constructs as identical [42].

A second approach to detecting the conformational change as the β-barrel inserts is to exploit environmental changes to the fluorophore, and use specifically environment-sensitive probes. During β-barrel insertion, certain residues of the MACPF/CDC protein that have been exposed to a hydrophilic aqueous solution become buried in the hydrophobic core of the membrane [36]. Fluorophores that exhibit changes in their properties depending on the hydrophobicity of their local environment [11] can therefore be used to report on β-barrel insertion. Environment-sensitive fluorophores were used in bulk to identify the membrane-spanning region of PFO and the secondary structure of this region [35, 76, 77] (summarized in a review by Heuck and Johnson [36]). As with FRET measurements, however, bulk changes in fluorescence based on local hydrophobicity are not sufficient to resolve insertion of the β-barrel with respect to oligomerization without synchronization.

Nguyen et al. report the use of the environment-sensitive fluorophore Badan (6-bromoacetyl-2-dimethylaminonaphthalene) in a single-molecule investigation of the membrane insertion of γ-hemolysin [57]. They attached Badan to leukocidin fast fraction, one of the two components comprising the γ-hemolysin pore, at a residue that changes environment during insertion. Using TIRF microscopy, they identified membrane insertion by the increase in emission intensity and shift in emission spectra of Badan as it moved into the hydrophobic core of the membrane.

A combination of the single-molecule experiments presented here would enable us to resolve oligomerization and β-barrel insertion of the MACPF/CDC proteins, and thereby develop a better understanding of how these processes are regulated by the protein itself or the local environment. The biggest challenge will be sufficient temporal resolution to clearly detect intermediate stages of assembly.

Are Arc-Pores Functional?

Cryo-electron microscopy and atomic force microscopy have revealed the existence of incomplete oligomeric arc structures on membranes exposed to CDCs [5, 28, 55, 60, 100]. It is unresolved as to whether these objects are functional as pores, artefacts of the experimental procedure, or simply by-products of the oligomerization process [39]. However, their height above the membrane is equivalent to the pore rather than the prepore height [10], and a wide distribution of pore conductivities is also observed [16, 28, 63]. Taken together, these results suggest that the arcs may be functional as pores [26]. If arc-shaped oligomers do form functional pores then questions immediately arise as to the membrane effects of these structures [27]. How are the hydrophobic interactions at either end of the arc satisfied, and how are the lipids arranged on the other side of the pore? Determining the nature of these incomplete structures may have implications for other pore-forming proteins, regarding their ability to form pores seemingly without a full set of components.

Although cryo-electron microscopy and atomic force microscopy give us a glimpse of the structure of arcs and rings, they are not able to confirm that such structures form functional pores [26, 28, 39]. We need a method capable of simultaneously confirming the size and structure of a pore, and the flow of ions across the membrane. Fluorescence can be used to measure ionic flux (i.e. function) through MACPF/CDC pores using fluorogenic dyes that respond to the presence of particular ions. For example, calcium flux through individual ion channels can be measured both in vivo and in vitro [12, 34, 102]. We exploited these measurements to image calcium flux through α-hemolysin [34].

By itself, fluorescence imaging of ionic flux is insufficient to discriminate arc--pore function. We must also determine the nature, and ideally the shape, of the oligomer which has formed a pore. Super-resolution methods are capable of overcoming the diffraction limit that affects conventional optical microscopy [19, 20]; they can be used to resolve single molecules that are within a diffraction-limit of

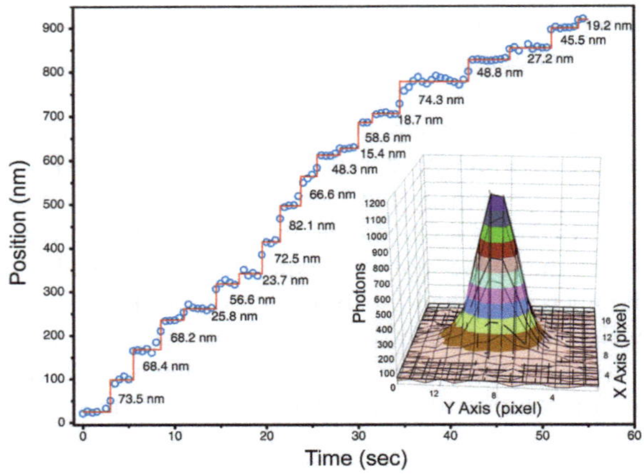

Fig. 15.4 FIONA resolving the hand-over-hand walking mechanism of Myosin V [97]. Myosin V was labelled with the fluorophore bisiodoacetamidorhodamine on the light chain domain of the head, then added to immobilised F-actin filaments on a coverslip. Adding ATP allowed the Myosin V to walk along the actin in steps; a typical fluorophore localization precision of ±3 nm was achieved. In the example shown, alternating 52 and 23 nm steps are observed. The integration time was 0.5 s, so some steps are not fully resolved, resulting in an apparent step-size of approximately 74 nm. Other observed step sizes included 74 nm steps, and alternating 42 and 33 nm steps. The distribution of step sizes fits the hand-over-hand walking mechanism of Myosin V, for which alternating step sizes of 37 + 2x and 37 – 2x nm are expected. Inset, a control to demonstrate localization of a stationary fluorophore. A Gaussian function (*solid lines*) was fitted to the PSF of a single Cy3 fluorophore attached to a coverslip. The close fit between the Gaussian and the PSF allows the centre of the PSF to be localized with 1.3 nm precision. PSF width: 287 nm. Adapted from [97] with permission from American Association for the Advancement of Science

each other, and to determine the position of single molecules to within a few nanometres. In single-molecule imaging, individual molecules are resolved as separate point sources of light; this distribution of fluorescence can be fitted with a function that approximates the point spread function (PSF) of the microscope, typically a Gaussian. Given sufficient photons, the location of this Gaussian peak can be determined with arbitrary precision of around 1–10 nm is possible (e.g. [97]).

Fluorescence imaging with one-nanometre accuracy (FIONA) provided the first imaging of biological samples at a truly nanometre scale [49, 97–99]. This technique allows individual fluorophores to be tracked with millisecond temporal resolution [49]. FIONA was first employed to observe the hand-over-hand walking mechanism of myosin V (Fig. 15.4) [97]. Using TIRF microscopy in vitro, with a 0.5 s integration time, Selvin and colleagues excited fluorophores attached to the light chain domain of myosin V molecules, and fitted the PSF of each fluorophore with a Gaussian function. This enabled them to locate the centre of each PSF and hence localize the fluorophores with a typical accuracy on the order of single nanometres. Adding ATP to a solution of myosin V molecules attached to immobilised F-actin filaments allowed the myosin V to walk along the actin in steps, as observed by the step-wise movement of the fluorophore's PSF. From the distribution of step-sizes they were able to validate the hand-over-hand walking mechanism and reject the inchworm theory. They went on to apply this technique in vivo with millisecond temporal resolution to look at the efforts of kinesin and dynein in transporting the peroxisome [49].

As presented, this method can only resolve the position of a single well-separated fluorescent probe. It would therefore be sufficient to track the location of individual MACPF/CDC monomers or pores, but not to resolve their structure. However, if we can turn nearby fluorophores on or off, we can exploit this super-resolution fitting to determine the positions of fluorophores in near proximity, thereby revealing the shape of the protein complex. The simplest method to do this is to wait for photobleaching during excitation of the multi-labelled complex. When one fluorophore is left active, its position can be localized as in FIONA; following the recorded wide-field image sequence backwards, the Gaussian corresponding to the last remaining fluorophore can be subtracted from the intensity profile of the last two remaining fluorophores to determine the position of the second to last remaining fluorophore. The intensity profile of the last two can be subtracted from that of the last three to localize the third from last fluorophore to photobleach, and so on. In this manner, in principle, the position of all the fluorophores present in the complex could be determined. Continuing the penchant for colourful acronyms in this field, this technique is called single-molecule high-resolution imaging with photobleaching (SHRImP) [29].

Going beyond simple photobleaching, stochastically or actively modulating the state of a fluorophore from 'on' to 'off' allows us to switch between different subsets of the total fluorophore population; within each subset, the fluorophores are further than a diffraction limit apart and can therefore be localized to within nanometers by fitting their PSF [19, 20, 72]. This principle is the basis for stochastic optical reconstruction microscopy (STORM) [72] and photoactivated localization microscopy (PALM) [4]. These techniques are discussed further in the section "Cellular Function". Such super-resolution schemes, including FIONA and SHRImP, would in theory allow us to discriminate between complete rings and arc-shaped structures. However the small size of even 'large' MACPF/CDC pores (30–35 nm diameter [26]) is not trivial.

What Controls the Diameter of the Pore?

We do not know how the diameter of MACPF/CDC pores is determined. Is it by the shape of the molecules, so that the effective pore size depends on the completeness of the arc- or ring-shaped structure? Does it depend on local membrane properties like curvature or composition? The consistent level of observed oligomeric curvature, combined with varying pore conductance, seems to imply that the former, at least, is true [16, 28, 63]. Structural changes in the monomers induced by membrane binding [14, 40, 65] will also affect the size of the monomer footprint in an oligomer and therefore the pore diameter.

If our goal is simply to measure the distribution in pore diameters, then high-resolution structural methods such as atomic force microscopy and cryo-electron microscopy appear best placed to pursue this question. However, if dynamic factors that affect pore diameter are of interest, then single-molecule, and in particular super-resolution methods may have a role to play. The techniques discussed in the section “Are Arc-pores Functional?” on arc-pores are also directly applicable to studying pore diameter.

What Happens to the Lipids When a Pore Inserts?

When a full >30 mer prepore punctures a hole through a membrane, the lipids and other membrane components from the lumen must be redistributed to another location; what happens to them following β-barrel insertion is unknown [39]. Do the lipids reincorporate into the surrounding membrane or are they lost from the bilayer? This is a difficult question to answer, because the membrane components affected by the MACPF/CDC proteins are few in relation to the whole bilayer. Worse still, they cannot be identified until the β-barrel has inserted, because until that point the final site of the pore is not definite. As a result, the desired lipids could not be labelled specifically, and if all of the lipids were labelled then any change in fluorescence corresponding to pore formation would be masked by the signal from the other lipids in the bilayer.

Other Than Cholesterol Binding, What Role do Lipid Properties Have in Pore Formation?

Pilzer et al. review the role of complement proteins in inducing vesicle emission from cell membranes [62]. Furthermore, a report by Gilbert summarizes structural investigations into the effects of the CDC pneumolysin (PLY) on liposomal bilayer structure [25], including depth of the bilayer, liposome aggregation and the formation of vesicles from the liposomal membrane. Given the ability of MACPF/CDC proteins to alter the lipid arrangement in membranes, it follows that the membrane structure has a significant effect on pore formation. Further evidence that pore formation and membrane binding are influenced by the structure of the target bilayer comes from the example discussed previously in relation to ensemble FRET measurements [51]. Lin and London used ensemble FRET to demonstrate that PFO affinity for lipid domains depends on matching of the lipid tail length with the transmembrane segment of the protein. This example highlights the relevance of lipid domains for pore formation; there is much evidence for a non-trivial relationship between lipid rafts and domains and pore formation by the MACPF/CDC proteins [22, 51, 90].

The dependence of oligomerization and β-barrel insertion on membrane properties can be studied using the fluorescence techniques already described, if a method for manipulating the membrane properties can be devised. The type and ratio of lipids comprising the membrane in vitro can usually be tuned using the starting concentrations, and in a similar way the lipids can be chosen to have a specific length of tail to alter bilayer thickness [51]. Creating membrane curvature in a controlled manner is harder to implement.

In a review by Oliver and Parikh [59], methods for manipulating a supported lipid bilayer are discussed, with a particular focus on membrane curvature. In one technique, artificial curvature was created by rupture of small phospholipid unilamellar vesicles on a substrate formed of hydrophilic colloidal crystals. A bilayer formed on the substrate by hydrophobic interaction, with curvature matching the surface of the colloids [8]. A second technique induced curvature by stretching a poly(dimethyl) siloxane (PDMS) surface and exposing it to ultraviolet radiation to make it stiff by oxidation. When the strain was removed, the PDMS became corrugated along the stretched axis, and rupture of vesicles on this surface again created a curved bilayer matching the surface profile. By adding the vesicles and forming the bilayer first, then releasing the tension in the PDMS, dynamic curvature of the bilayer was also induced [73].

Methods for manipulating membrane curvature and altering lipid composition could be combined with the fluorescence imaging techniques discussed to observe the effects of varying membrane parameters on the whole MACPF/CDC assembly process, from membrane binding to oligomerization and β-barrel insertion. Thus, the role of lipid properties in pore formation could be investigated.

Why Don't Monomers Oligomerize in Solution?

Finally, for the formation of pores it is necessary for monomers to diffuse laterally across the membrane. This requirement means that the protein must somehow avoid oligomerization in solution before binding to the membrane. There are different theories that could explain how premature oligomerization is avoided. The first is by dimerization: Solovyova et al. used analytical ultracentrifuge and small-angle x-ray scattering to show that PFO forms antiparallel dimers in solution even at low concentrations [84]; antiparallel dimers were also observed in the crystal forms of PFO [71]. PLY, on the other hand, does not seem to form an antiparallel dimer in solution [84], and therefore must avoid oligomerization by alternative means. One theory is that becoming constrained on a 2D membrane increases the frequency of interaction between monomers, therefore leading to a higher rate of oligomerization and pore formation than in solution, where the concentration is mainly insufficient for oligomerization [26]. The final argument is that membrane binding induces a structural transition in the monomers that is necessary to allow monomer-monomer interaction [65]. This topic is discussed in detail in Gilbert's review [26].

The formation of dimers is a property suitable for investigation by smFRET: labelling a subset of monomers with a donor and the rest with an acceptor fluorophore, at appropriate residues, would allow detection of the dimer by high FRET efficiency between the donor and acceptor. Combined with imaging experiments, a change in FRET efficiency either in solution or upon contact with the membrane could be discriminated. smFRET or environment-sensitive fluorophores could also be used to investigate the structural transitions induced by membrane binding that are reportedly required for oligomerization. In the original study by Ramachandran et al. [65], ensemble fluorescence techniques were used to characterise the exposure of a β-strand upon membrane binding to allow monomer-monomer interaction. Subsequent investigations into the nature of this structural rearrangement again utilised bulk fluorescence of the environment-sensitive fluorophore NBD [14, 40]. However, single-molecule methods could be used to obtain dynamic information about the transition relative to assembly of the pore. We could thereby verify that the exposure of the β-strand is necessary for oligomerization. The less dramatic conformational change in this process would make the signal change harder to detect than in the previous examples relating to β-barrel insertion.

Cellular Function

Do MACPF/CDC Proteins Have Roles Other Than Simple Pore Formation?

In addition to the conventional roles in immune surveillance and bacterial pathogenesis [17, 48, 95], biochemical studies have revealed that proteins with MACPF domains also play a part in neural development [101] and embryonic structure [54, 89]. Furthermore, there is evidence that certain major roles of the MACPF/CDC proteins go beyond just simple pore formation and target-cell lysis. For example, perforin appears to be involved in the delivery of proteases to target cells [91], and the CDC streptolysin O translocates *Streptococcus pyogenes* NAD-glycohydrolase into keratinocytes with cytotoxic effects [52], but seemingly without the necessity of a pore [53]. The CDCs listeriolysin O (LLO), PLY and PFO appear to control host transcriptional activity by histone-modification [32], and the observation that various CDCs have an affinity for rafts or domains [22, 51, 80, 90] hints at membrane remodelling and signalling roles for the MACPF/CDC proteins. These roles clearly illustrate that MACPF/CDC proteins do more than simply make holes in membranes, and suggest that we may be ignorant of other important functions.

Each of these roles is suitable for investigation with fluorescence imaging, reporting on location of the protein population at a given time and their colocalization with other molecules. In contrast to the single-molecule investigations discussed in the previous sections, our focus here is on the larger scale cellular effects of the proteins and not individual monomers and pores. The value of fluorescence imaging in this setting has already been demonstrated with conventional methods: confocal microscopy was used, for example, to study the functional role of perforin [91]. Thiery et al. observed endocytosis of perforin and the protease granzyme B due to a perforin-induced calcium gradient across the cell membrane, and subsequent perforin-induced rupture of endosomes to release the protease inside the target cell.

However, despite the useful information that can be extracted from conventional fluorescence imaging methods, they are all limited to a spatial resolution governed by the wavelength of light ($\sim$250 nm). To describe the cellular processes influenced by the MACPF/CDC proteins with greater accuracy, we must achieve resolution beyond this limit; we again look to the use of super-resolution fluorescence imaging techniques.

STORM [72] and PALM [4] imaging follow a similar principle to that introduced above for FIONA and SHRImP, in that they involve localizing individual fluorophores. Subsets of fluorophores are sequentially activated by excitation with one laser, then localized and deactivated by excitation with a different laser (Fig. 15.5a (i)) [19, 20, 72]. The power of the activation laser is used to control how many fluorophores are active in each subset; the fluorophores in a subset must be at least a diffraction limit apart on average so that their PSF can be fitted and they can be individually localized [20]. The iterations continue until the labelled cellular feature is resolved to the desired accuracy. Fernández-Suárez and Ting provide an excellent review of super-resolution techniques and the fluorophores best suited to this type of imaging [19].

In addition to super-resolution methods based on the localization of individual molecules, sub-diffraction imaging can be achieved via manipulation of the ensemble excitation state of the fluorophores (stimulated emission depletion, STED) [13, 33], or through the use of structured illumination to excite the sample (structured illumination microscopy, SIM) [30, 31, 75]. In STED (Fig. 15.5c), the sample is illuminated with an excitation beam superposed with a donut-shaped STED beam [33]. The fluorophores that are excited around the periphery of the diffraction-limited excitation spot are then depleted by stimulated emission, by the STED beam, rather than fluorescence. This effect can reduce the size of the effective illumination PSF to around 20 nm full width at half maximum (FWHM) in the lateral direction (compared to nearer 200 nm in the diffraction-limited case) [13]. In SIM (Fig. 15.5b (i)), a structured excitation beam generates an interference pattern within the sample, so that high frequency information from the sample can be detected by conventional methods and decoded later using software [20, 31].

To illustrate the principle of super-resolution imaging of 3D cellular features, both by the localization of individual fluorophores (STORM/PALM) and by sub-diffraction illumination (STED/SIM), two relevant case studies are presented. The use of STORM by Jones et al. [43], and 3D-SIM by Shao et al. [75], are chosen

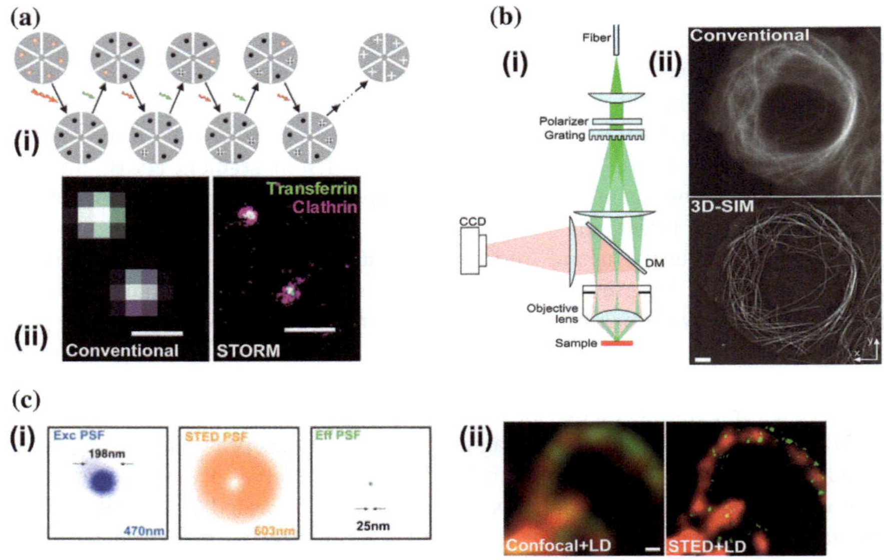

Fig. 15.5 Super-resolution microscopy. **a** STORM/PALM. (*i*) The concept for STORM and PALM is the same. Here we illustrate it for STORM [72]. To resolve a feature, all the fluorophores are first deactivated (*black dots*) with one (*red*) laser; a second (*green*) laser then activates (*orange dots*) a subset of the fluorophores that are far enough apart to be resolved: further than a diffraction limit. The *red* laser then excites these active fluorophores, whose position (*white crosses*) is located to within a few nanometres using their fluorescence, before being again deactivated by the red laser. The fluorophores are iteratively activated, localized and deactivated until a full image can be constructed from all of the iterations. (*ii*) The advantage of two-colour 3D-STORM (*right panel*) over conventional imaging (*left panel*) demonstrated with transferrin encapsulated in clathrin-coated pits [43]. The STORM image, an x–y projection of a 3D volume, was taken in 30 s and displays the same area as the conventional image. Clathrin-coated pits (*magenta*) were labelled with Alexa647 via a SNAP tag [46], transferrin (*green*) was directly labelled with Alexa568. Scale bar, 500 nm. **b** 3D-SIM. (*i*) An implementation of 3D-SIM where a phase grating is mechanically adjusted to illuminate the sample with patterned light in various orientations [31]. Spatially incoherent light exits an optical fibre and is collimated onto the phase grating. The diffraction orders generated by the grating are blocked save for orders −1, 0 and +1. These orders are focused onto the back focal plane of the objective, which collimates the beams such that they intersect at the focal point, in the sample. The interference pattern generated between the diffraction orders excites the sample; the resulting fluorescence intensity pattern is directed by a dichroic mirror (DM) onto a charge-coupled device (CCD) camera and can be decoded computationally. (*ii*) Conventional wide-field microscopy (*upper panel*) and 3D-SIM (*lower panel*) images of a Drosophila S2 cell expressing α-tubulin genetically labelled with enhanced GFP [75]. These images were taken using an implementation of 3D-SIM where the mechanical phase grating was replaced with a spatial light modulator allowing faster image acquisition and application of 3D-SIM to live cells. Scale bar, 2 μm. **c** STED [13]. (*i*) The concept for STED. A diffraction-limited confocal excitation PSF (198 nm FWHM) is superposed with a donut-shaped STED PSF. The fluorophores excited at the periphery of the diffraction-limited confocal spot are depleted by stimulated emission, reducing the effective excitation PSF to around 25 nm FWHM. The wavelengths displayed are those used for the *green* fluorescent

channel in (ii). (ii) Two-colour conventional confocal (*left panel*) and STED (*right panel*) microscopy applied to mitochondria labelled with antibodies specific for the F_1F_0-ATP synthase (*red*) and the translocase of the outer membrane (TOM) complex (*green*). Both images were subsequently enhanced by a linear deconvolution (LD). In the STED image, the TOM complex seems to cluster at the mitochondrial surface while F_1F_0-ATP synthase is more evenly distributed; this pattern cannot be resolved in the confocal image. Scale bar, 200 nm. **a** (i), reprinted by permission from Macmillan Publishers Ltd: Nature Methods [72], © 2006. **a** (*ii*), adapted by permission from Macmillan Publishers Ltd: Nature Methods [43], © 2011. **b** (i), reprinted from [31] with permission from Elsevier © 2008. **b** (*ii*), adapted by permission from Macmillan Publishers Ltd: Nature Methods [75], © 2011. **C**, adapted from [13] with permission from Elsevier © 2007

because they exemplify super-resolution techniques that could be used to observe novel MACPF/CDC cellular functions.

In the work of Jones et al. STORM was used to image clathrin-coated pits and transferrin cargo in live cells in 3D, with resolution reaching 30 nm in the lateral and 50 nm in the axial directions [43]. Traditionally, because of the large number of iterations and many photons required to obtain a super-resolution image, the technique is relatively slow, with temporal resolution on the order of tens of seconds [6, 20, 82]. In this study, temporal resolution of 0.5 s in 2D and 2 s in 3D was achieved by using a high excitation-beam intensity (15 kW cm^{-2}). However, only six consecutive 0.5- or 2-second images could be recorded at this intensity before significant photophysical damage to the sample. Longer imaging was possible after reducing the photobleaching rate by using a lower activation-beam intensity (although fewer active fluorophores necessitates longer frames—meaning lower temporal resolution—to maintain spatial resolution). 3D information was obtained using a cylindrical lens in the imaging path so that each fluorescent molecule appeared elliptical. The centroid was then used for lateral localization, the ellipticity for axial localization. Finally, using spectrally distinct fluorophores, they also resolved transferrin clusters encapsulated in cup-shaped clathrin coats by two-colour fast 3D-STORM (Fig. 15.5a (ii)).

Shao et al. used 3D-SIM to obtain images of α-tubulin labelled with enhanced GFP (Fig. 15.5b (ii)) and mitochondria labelled with MitoTracker Green [75]. They achieved 120 and 360 nm resolution in the lateral and axial directions respectively. Whole live cells were imaged as stacks of 2D axial planes, with the significant improvement of tens to hundreds of time-points acquired (compared to six in the previously discussed STORM experiment). The full volume took in the region of 5 s to acquire for the α-tubulin images, 20 s for the mitochondria. High imaging speed, relative to previous SIM, was achieved by replacing a mechanical transmission phase grating with a ferroelectric liquid crystal spatial light modulator, allowing 3D-SIM on live cells.

These case studies provide examples of super-resolution methods that could feasibly be applied to studying the MACPF/CDC proteins, enabling observation of their functions in previously unseen detail. By labelling molecules of interest such as the MACPF/CDC proteins themselves, or interacting molecules, we could study their location and long-term movement in cells with resolution reaching tens of

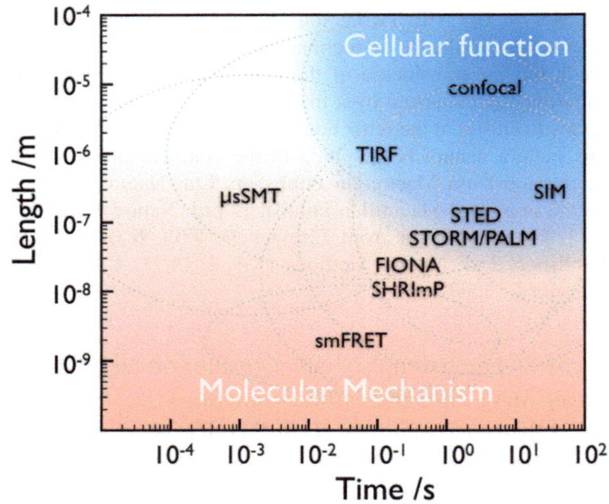

Fig. 15.6 An overview of the biological processes and the temporal and spatial limits of the techniques that we have discussed in relation to the MACPF/CDC proteins. See [19, 20] and references in the text

nanometres. The disadvantages of these techniques include the long image acquisition times, possible photodamage of the relevant cellular components under intense illumination and the high density of fluorophores necessary for STORM and PALM. These factors will make resolution of individual pores in live cells difficult. However, as discussed, the roles that we might expect for the MACPF/CDC proteins include manipulating cellular structure, delivering molecules across target-cell membranes, signalling and membrane remodelling. For understanding these roles, accurate localization of the population of MACPF/CDC proteins is required, and this can be provided by 3D super-resolution techniques.

What Membrane Features are Targeted by the MACPF/CDC Proteins?

Canonically, CDCs depend on cholesterol for binding to the membrane. A recent study demonstrated that the primary cholesterol recognition motif in a number of CDCs consists of just two residues in the L1 loop [18]. Despite the apparent simplicity of this conventional cholesterol binding process, other important features of membrane targeting complicate the issue in cells. Intermedilysin (ILY), vaginolysin and lectinolysin are known to bind the human CD59 protein receptor [23, 24, 96]. In humans CD59 primarily inhibits the pore-forming function of the MAC, and yet it is necessary for pore formation by ILY for example [39]. Therefore, the effect of ILY binding CD59 on the membrane is two-fold: allowing MAC pore formation in addition to ILY pore formation. As demonstrated by the CD59-binding proteins, the overall effect of a MACPF/CDC protein on the target cell can depend not only on its action subsequent to membrane binding, but also on the binding itself. It is important to determine whether other MACPF/CDC proteins have novel receptors and if they exhibit the two-fold effect of those that bind CD59.

Membrane targeting by these proteins is further complicated by the reported ability of LLO to aggregate lipid rafts and their associated molecules on cell membranes [22]. Pre-treatment of LLO with cholesterol to prevent lytic activity, but not binding or oligomerization, did not stop the protein from aggregating lipid raft markers. However, treatment with an antibody to prevent oligomerization did inhibit this function, suggesting that the aggregation of rafts caused by LLO was dependent on its oligomerization rather than its pore-forming capability [22]. Another study found that mutants of PLY that were able to bind to the membrane, but were not able to oligomerize, did not associate with cholesterol-rich rafts, whereas those that were fully active did associate with rafts [90]; this study, however, did not discriminate between the ability to oligomerize and the ability to form pores. More experiments are required to consolidate these ideas to describe exactly how the MACPF/CDC proteins interact with lipid rafts: do rafts affect protein binding affinity, and can they be induced to form and aggregate directly by MACPF/CDC proteins?

Attaching fluorophores to both the MACPF/CDC protein and different membrane components will allow observation of the interaction of the proteins with membrane receptors or molecules involved in signalling. One example would be to label the CD59 receptor and ILY to explore how pore formation depends on their interaction, providing this is possible without affecting their natural function. Labelled proteins and fluorescent markers for lipid rafts could be exploited in a similar way [22]. The techniques that are relevant for answering questions about membrane binding will inevitably also apply to revealing novel functions that involve the cell membrane: those that involve manipulating the 3D membrane structure and the remodelling of membrane domains, for example.

Super-resolution tracking such as FIONA is capable of capturing MACPF/CDC proteins interacting dynamically with membrane receptors, particularly if different fluorophores are used to label the interacting molecules. However, in order to observe the 3D structural effects of the MACPF/CDC proteins on the target-cell membrane, and the location of the proteins on the membrane relative to its composition (such as in rafts), STORM and PALM imaging seem preferable. The following case studies are examples of how cell membranes can be observed with these super-resolution imaging techniques.

Shim et al. used the super-resolution technique STORM, with photoswitchable membrane-specific fluorophores (that can reversibly be made active or inactive by illumination), to label and observe the dynamics of the plasma membrane, mitochondria, the endoplasmic reticulum (ER) and lysosomes in live cells [79]. The fluorophores were selected from existing membrane-specific fluorophores, and their photoswitching capabilities were demonstrated. The experiment resolved tubular intermediates connecting neighbouring mitochondria before fusion and fission events, as well as the dynamic growth of new tubules in the ER. As an example, the authors obtained a spatial resolution of 30–60 nm at a temporal resolution of 2 s, and hence were able to capture speeds of up to 15–30 nm s^{-1} in mitochondria fission and fusion events without substantial blur due to motion.

The results from Shim et al. show that super-resolution monitoring of membrane structure is possible with specificity down to the individual organelle level. An example where this capability is relevant is the targeting of mitochondria in neurons by PLY [7]: PLY was reported to damage mitochondrial membranes and induce native cell death pathways by increasing intracellular levels of calcium and reactive oxygen species. The tracking of mitochondrial membranes in the Shim study suggests an experiment where the effects of PLY on the target mitochondrial membrane could be observed. Moreover, STORM could be used to improve the resolution that was seen in the confocal experiments by Gekara et al. where aggregation of lipid rafts by LLO was observed [22], providing further detail on a non-pore-forming role of the MACPF/CDC proteins. In their study, Shim et al. demonstrated two-colour STORM imaging of the ER and mitochondria using spectrally distinct fluorophores. Two-colour STORM could be applied to resolve PLY and the mitochondrial membrane, or LLO and lipid raft markers, simultaneously.

A study by Shtengel et al. describes the interference of emitted fluorescent light as a means to achieve exceptional axial resolution, in a technique called interferometric PALM (iPALM) [83]. As discussed by Shtengel et al. there are conflicting obstacles to using conventional interference methods in the study of biological samples. Cells do not interact strongly with visible light, they lack reflective surfaces, and molecular specificity is a problem. Conversely, labelling features with fluorophores has excellent specificity, but problems arise in that the brightness of fluorophores is variable due to their photophysical properties, and most importantly it is difficult to form two phase-coherent beams of fluorescence emission that will interfere depending on their path difference [83].

In the Shtengel study, fixed cells were observed with photoswitchable genetically encoded fluorescent proteins (FPs). Two objective lenses were placed opposing each other across the sample, and the fluorescence light emitted by individual fluorophores was collected by the objective lenses into two beams that were directed onto a beam-splitter. Light from the two beams (from the same source fluorophore) interfered in the beam-splitter and exited in three channels. The intensity in each channel depended on the relative phase of the incoming beams, which depended on the distance of the source fluorophore along the optical axis between the two objectives. The axial position of the source fluorophores was extracted from the relative intensity of their signal in the three imaging channels exiting the beam splitter, at a resolution of 9.8 nm FWHM for the PSF of an individual fluorophore. Lateral localization was according to PALM but with resolution improved by a factor of two because the dual objectives collect twice as much light as a single objective, reaching 22.8 nm FWHM. The technique was demonstrated by measuring the 25 nm diameter of microtubules, observing the arrangement of integrin receptors in the ER, and resolving the dorsal and ventral plasma membrane.

The improved photon collection efficiency of iPALM meant that FPs could be used in place of synthetic dyes (which have lower specificity despite their added brightness); using FPs is beneficial if possible because it improves the efficiency of labelling. However, the experimental set-up for iPALM only allows imaging of thin samples (15–20 μm), an imaging depth of a few hundred nanometres, and is ideally suited to studying fixed cells [20]. It remains to be seen whether or not these limitations would prevent the technique being useful for MACPF/CDC proteins, or if, for example, the plasma membrane could be observed using iPALM during exposure to a MACPF/CDC protein, to reveal any structural changes.

How are MACPF/CDC Proteins Released?

Even the initial release of MACPF/CDC proteins is unclear in some cases [39, 69]. Given that it lacks a typical secretion-signalling peptide sequence, the process by which PLY is released into the environment was assumed to be autolysis of the bacteria *Streptococcus pneumoniae* [61], but secretion from the WU2 strain in the absence of autolysis has been demonstrated [3]. There is therefore some ambiguity in the secretion pathways available to PLY, and potentially other MACPF/CDC proteins. Revealing the secretion pathway for PLY could have important consequences for studies of other pore-forming toxins.

In order to observe secretion of the proteins, it may be necessary to track them immediately following their expression. The most tractable way to achieve this is to use genetically encoded FPs. Moreover, for the most detail about the secretion process we may again look to super-resolution techniques. However, choosing the right FP for super-resolution studies is not trivial. For super-resolution of the STORM and PALM kind, the fluorophore must have special properties that enable it to be rendered fluorescently active upon exposure to light of a specific wavelength; both synthetic fluorophores and FPs with such properties exist [19]. Synthetic fluorophores are generally brighter than FPs [19], but their conjugation to proteins in live cells is difficult and may prevent their use in observing the secretion process.

The creation of a suitable FP-MACPF/CDC fusion construct could allow 3D super-resolution imaging of the protein as it is produced and secreted. Secretion may coincide with autolysis of the cell, an event that could be observed sufficiently with conventional techniques; alternatively, secretion may involve multiple molecules of interest. To resolve multiple molecules, two- or multi-colour fluorescence experiments are necessary, where each type of molecule has a spectrally distinct fluorophore attached to it and corresponding excitation wavelengths and emission filters are used. Such an experiment would allow us to resolve, for example, cell-machinery transporting the MACPF/CDC protein across the cell and into the extracellular environment, if both components were labelled with fluorophores.

Conclusion and Outlook

In this chapter we have briefly covered some of the remaining questions about MACPF/CDC proteins, and have presented a selection of novel fluorescence imaging methods based on their particular capabilities to aid our understanding of these proteins (Fig. 15.6). Recent advances in fluorescence imaging have extended both temporal and spatial resolution limits into regions that are pertinent to a range of questions regarding MACPF/CDC function. For these reasons, we believe high-resolution fluorescence imaging will play a significant role in future investigations of MACPF/CDC proteins.

References

1. Axelrod D (1981) Cell-substrate contacts illuminated by total internal reflection fluorescence. J Cell Biol 89:141–145
2. Axelrod D (2001) Total internal reflection fluorescence microscopy in cell biology. Traffic 2:764–774
3. Balachandran P, Hollingshead SK, Paton JC, Briles DE (2001) The autolytic enzyme LytA of Streptococcus pneumoniae is not responsible for releasing pneumolysin. J Bacteriol 183:3108–3116
4. Betzig E, Patterson GH, Sougrat R, Lindwasser OW, Olenych S, Bonifacino JS, Davidson MW, Lippincott-Schwartz J, Hess HF (2006) Imaging intracellular fluorescent proteins at nanometer resolution. Science 313:1642–1645
5. Bhakdi S, Tranum-Jensen J, Sziegoleit A (1985) Mechanism of membrane damage by streptolysin-O. Infect Immun 47:52–60
6. Biteen JS, Thompson MA, Tselentis NK, Bowman GR, Shapiro L, Moerner WE (2008) Super-resolution imaging in live Caulobacter crescentus cells using photoswitchable EYFP. Nat Methods 5:947–949
7. Braun JS, Hoffmann O, Schickhaus M, Freyer D, Dagand E, Bermpohl D, Mitchell TJ, Bechmann I, Weber JR (2007) Pneumolysin causes neuronal cell death through mitochondrial damage. Infect Immun 75:4245–4254
8. Brozell AM, Muha MA, Sanii B, Parikh AN (2005) A class of supported membranes: formation of fluid phospholipid bilayers on photonic band gap colloidal crystals. J Am Chem Soc 128:62–63
9. Clegg RM (1992) Fluorescence resonance energy transfer and nucleic acids. Meth Enzymol 211:353–388
10. Czajkowsky DM, Hotze EM, Shao Z, Tweten RK (2004) Vertical collapse of a cytolysin prepore moves its transmembrane β-hairpins to the membrane. EMBO J 23:3206–3215
11. Demchenko AP, Mély Y, Duportail G, Klymchenko AS (2009) Monitoring biophysical properties of lipid membranes by environment-sensitive fluorescent probes. Biophys J 96:3461–3470
12. Demuro A, Parker I (2005) "Optical Patch-clamping" single-channel recording by imaging Ca2+ flux through individual muscle acetylcholine receptor channels. J Gen Physiol 126:179–192
13. Donnert G, Keller J, Wurm CA, Rizzoli SO, Westphal V, Schönle A, Jahn R, Jakobs S, Eggeling C, Hell SW (2007) Two-color far-field fluorescence nanoscopy. Biophys J 92:L67–L69

14. Dowd KJ, Farrand AJ, Tweten RK (2012) The cholesterol-dependent cytolysin signature motif: a critical element in the allosteric pathway that couples membrane binding to pore assembly. PLoS Pathog 8:e1002787
15. Dunstone MA, Tweten RK (2012) Packing a punch: the mechanism of pore formation by cholesterol dependent cytolysins and membrane attack complex/perforin-like proteins. Curr Opin Struct Biol 22:342–349
16. El-Rachkidy RG, Davies NW, Andrew PW (2008) Pneumolysin generates multiple conductance pores in the membrane of nucleated cells. Biochem Biophys Res Comm 368:786–792
17. Epstein FH, Tuomanen EI, Austrian R, Masure HR (1995) Pathogenesis of pneumococcal infection. New Engl J Med 332:1280–1284
18. Farrand AJ, LaChapelle S, Hotze EM, Johnson AE, Tweten RK (2010) Only two amino acids are essential for cytolytic toxin recognition of cholesterol at the membrane surface. Proc Natl Acad Sci USA 107:4341–4346
19. Fernández-Suárez M, Ting AY (2008) Fluorescent probes for super-resolution imaging in living cells. Nature Rev Mol Cell Biol 9:929–943
20. Fischer RS, Wu Y, Kanchanawong P, Shroff H, Waterman CM (2011) Microscopy in 3D: a biologist's toolbox. Trends Cell Biol 21:682–691
21. Förster T (1948) Zwischenmolekulare energiewanderung und fluoreszenz. Annal Physik 437:55–75
22. Gekara NO, Jacobs T, Chakraborty T, Weiss S (2005) The cholesterol-dependent cytolysin listeriolysin O aggregates rafts via oligomerization. Cell Microbiol 7:1345–1356
23. Gelber SE, Aguilar JL, Lewis KLT, Ratner AJ (2008) Functional and phylogenetic characterization of vaginolysin, the human-specific cytolysin from gardnerella vaginalis. J Bacteriol 190:3896–3903
24. Giddings KS, Zhao J, Sims PJ, Tweten RK (2004) Human CD59 is a receptor for the cholesterol-dependent cytolysin intermedilysin. Nat Struct Mol Biol 11:1173–1178
25. Gilbert RJ (2002) Pore-forming toxins. Cell Mol Life Sci 59:832–844
26. Gilbert RJ (2005) Inactivation and activity of cholesterol-dependent cytolysins: what structural studies tell us. Structure 13:1097–1106
27. Gilbert RJ (2010) Cholesterol-dependent cytolysins. Adv Exp Med Biol 677:56–66
28. Gilbert RJ, Mikelj M, Dalla Serra M, Froelich CJ, Anderluh G (2013) Effects of MACPF/CDC proteins on lipid membranes. Cell Mol Life Sci 70:2083–2098
29. Gordon MP, Ha T, Selvin PR (2004) Single-molecule high-resolution imaging with photobleaching. Proc Natl Acad Sci USA 101:6462–6465
30. Gustafsson MGL (2000) Surpassing the lateral resolution limit by a factor of two using structured illumination microscopy. J Microscopy 198:82–87
31. Gustafsson MGL, Shao L, Carlton PM, Wang CR, Golubovskaya IN, Cande WZ, Agard DA, Sedat JW (2008) Three-dimensional resolution doubling in wide-field fluorescence microscopy by structured illumination. Biophys J 94:4957–4970
32. Hamon MA, Batsché E, Régnault B, Tham TN, Seveau S, Muchardt C, Cossart P (2007) Histone modifications induced by a family of bacterial toxins. Proc Natl Acad Sci USA 104:13467–13472
33. Hell SW, Wichmann J (1994) Breaking the diffraction resolution limit by stimulated emission: stimulated-emission-depletion fluorescence microscopy. Optics Lett 19:780–782
34. Heron AJ, Thompson JR, Cronin B, Bayley H, Wallace MI (2009) Simultaneous measurement of ionic current and fluorescence from single protein pores. J Am Chem Soc 131:1652–1653
35. Heuck AP, Hotze EM, Tweten RK, Johnson AE (2000) Mechanism of membrane insertion of a multimeric β-barrel protein: perfringolysin O creates a pore using ordered and coupled conformational changes. Mol Cell 6:1233–1242
36. Heuck AP, Johnson AE (2002) Pore-forming protein structure analysis in membranes using multiple independent fluorescence techniques. Cell Biochem Biophys 36:89–101

37. Heuck AP, Tweten RK, Johnson AE (2003) Assembly and topography of the prepore complex in cholesterol-dependent cytolysins. J Biol Chem 278:31218–31225
38. Hotze EM, Heuck AP, Czajkowsky DM, Shao Z, Johnson AE, Tweten RK (2002) Monomer-monomer interactions drive the prepore to pore conversion of a β-barrel-forming cholesterol-dependent cytolysin. J Biol Chem 277:11597–11605
39. Hotze EM, Tweten RK (2012) Membrane assembly of the cholesterol-dependent cytolysin pore complex. Biochim Biophys Acta 1818:1028–1038
40. Hotze EM, Wilson-Kubalek E, Farrand AJ, Bentsen L, Parker MW, Johnson AE, Tweten RK (2012) Monomer-Monomer interactions propagate structural transitions necessary for pore formation by the cholesterol-dependent cytolysins. J Biol Chem 287:24534–24543
41. Hotze EM, Wilson-Kubalek EM, Rossjohn J, Parker MW, Johnson AE, Tweten RK (2001) Arresting pore formation of a cholesterol-dependent cytolysin by disulfide trapping synchronizes the insertion of the transmembrane β-sheet from a prepore intermediate. J Biol Chem 276:8261–8268
42. Jäger M, Nir E, Weiss S (2006) Site-specific labeling of proteins for single-molecule FRET by combining chemical and enzymatic modification. Prot Sci 15:640–646
43. Jones SA, Shim S-H, He J, Zhuang X (2011) Fast, three-dimensional super-resolution imaging of live cells. Nat Methods 8:499–505
44. Joo C, Balci H, Ishitsuka Y, Buranachai C, Ha T (2008) Advances in single-molecule fluorescence methods for molecular biology. Ann Rev Biochem 77:51–76
45. Kapanidis AN, Lee NK, Laurence TA, Doose Sr, Margeat E, Weiss S (2004) Fluorescence-aided molecule sorting: analysis of structure and interactions by alternating-laser excitation of single molecules. Proc Natl Acad Sci USA 101:8936–8941
46. Keppler A, Gendreizig S, Gronemeyer T, Pick H, Vogel H, Johnsson K (2003) A general method for the covalent labeling of fusion proteins with small molecules in vivo. Nature Biotechnol 21:86–89
47. Keyel PA, Loultcheva L, Roth R, Salter RD, Watkins SC, Yokoyama WM, Heuser JE (2011) Streptolysin O clearance through sequestration into blebs that bud passively from the plasma membrane. J Cell Sci 124:2414–2423
48. Kondos SC, Hatfaludi T, Voskoboinik I, Trapani JA, Law RHP, Whisstock JC, Dunstone MA (2010) The structure and function of mammalian membrane-attack complex/perforin-like proteins. Tissue Antigens 76:341–351
49. Kural C, Kim H, Syed S, Goshima G, Gelfand VI, Selvin PR (2005) Kinesin and dynein move a peroxisome in vivo: a tug-of-war or coordinated movement? Science 308:1469–1472
50. Leake MC, Chandler JH, Wadhams GH, Bai F, Berry RM, Armitage JP (2006) Stoichiometry and turnover in single, functioning membrane protein complexes. Nature 443:355–358
51. Lin Q, London E (2013) Altering hydrophobic sequence lengths shows that hydrophobic mismatch controls affinity for ordered lipid domains (rafts) in the multitransmembrane strand protein perfringolysin O. J Biol Chem 288:1340–1352
52. Madden JC, Ruiz N, Caparon M (2001) Cytolysin-mediated translocation (CMT): a functional equivalent of type III secretion in gram-positive bacteria. Cell 104:143–152
53. Magassa NG, Chandrasekaran S, Caparon MG (2010) Streptococcus pyogenes cytolysin-mediated translocation does not require pore formation by streptolysin O. EMBO Rep 11:400–405
54. Martin JR, Raibaud A, Ollo R (1994) Terminal pattern elements in Drosophila embryo induced by the torso-like protein. Nature 367:741–745
55. Morgan PJ, Hyman SC, Byron O, Andrew PW, Mitchell TJ, Rowe AJ (1994) Modeling the bacterial protein toxin, pneumolysin, in its monomeric and oligomeric form. J Biol Chem 269:25315–25320
56. Nakajo K, Ulbrich MH, Kubo Y, Isacoff EY (2010) Stoichiometry of the KCNQ1-KCNE1 ion channel complex. Proc Natl Acad Sci USA 107:18862–18867

57. Nguyen AH, Nguyen VT, Kamio Y, Higuchi H (2006) Single-molecule visualization of environment-sensitive fluorophores inserted into cell membranes by staphylococcal γ-hemolysin. Biochemistry 45:2570–2576
58. Nguyen VT, Kamio Y, Higuchi H (2003) Single-molecule imaging of cooperative assembly of γ-hemolysin on erythrocyte membranes. EMBO J 22:4968–4979
59. Oliver AE, Parikh AN (2010) Templating membrane assembly, structure, and dynamics using engineered interfaces. Biochim Biophys Acta 1798:839–850
60. Olofsson A, Hebert H, Thelestam M (1993) The projection structure of perfringolysin O (Clostridium perfringens θ-toxin). FEBS Lett 319:125–127
61. Paton JC, Andrew PW, Boulnois GJ, Mitchell TJ (1993) Molecular analysis of the pathogenicity of Streptococcus pneumoniae: the role of pneumococcal proteins. Ann Rev Microbiol 47:89–115
62. Pilzer D, Gasser O, Moskovich O, Schifferli J, Fishelson Z (2005) Emission of membrane vesicles: roles in complement resistance, immunity and cancer. Springer Semin Immun 27:375–387
63. Praper T, Sonnen A, Viero G, Kladnik A, Froelich CJ, Anderluh G, Dalla Serra M, Gilbert RJ (2011a) Human perforin employs different avenues to damage membranes. J Biol Chem 286:2946–2955
64. Praper T, Sonnen AF, Kladnik A, Andrighetti AO, Viero G, Morris KJ, Volpi E, Lunelli L, Dalla Serra M, Froelich CJ, Gilbert RJ, Anderluh G (2011b) Perforin activity at membranes leads to invaginations and vesicle formation. Proc Natl Acad Sci USA 108:21016–21021
65. Ramachandran R, Tweten RK, Johnson AE (2004) Membrane-dependent conformational changes initiate cholesterol-dependent cytolysin oligomerization and intersubunit β-strand alignment. Nat Struct Mol Biol 11:697–705
66. Ramachandran R, Tweten RK, Johnson AE (2005) The domains of a cholesterol-dependent cytolysin undergo a major FRET-detected rearrangement during pore formation. Proc Natl Acad Sci USA 102:7139–7144
67. Reyes-Lamothe R, Sherratt DJ, Leake MC (2010) Stoichiometry and architecture of active DNA replication machinery in Escherichia coli. Science 328:498–501
68. Rosado CJ, Buckle AM, Law RH, Butcher RE, Kan WT, Bird CH, Ung K, Browne KA, Baran K, Bashtannyk-Puhalovich TA, Faux NG, Wong W, Porter CJ, Pike RN, Ellisdon AM, Pearce MC, Bottomley SP, Emsley J, Smith AI, Rossjohn J, Hartland EL, Voskoboinik I, Trapani JA, Bird PI, Dunstone MA, Whisstock JC (2007) A common fold mediates vertebrate defense and bacterial attack. Science 317:1548–1551
69. Rosado CJ, Kondos S, Bull TE, Kuiper MJ, Law RH, Buckle AM, Voskoboinik I, Bird PI, Trapani JA, Whisstock JC, others (2008) The MACPF/CDC family of pore-forming toxins. Cell Microbiol 10:1765–1774
70. Rossjohn J, Feil SC, McKinstry WJ, Tweten RK, Parker MW (1997) Structure of a cholesterol-binding, thiol-activated cytolysin and a model of its membrane form. Cell 89:685–692
71. Rossjohn J, Polekhina G, Feil SC, Morton CJ, Tweten RK, Parker MW (2007) Structures of perfringolysin O suggest a pathway for activation of cholesterol-dependent cytolysins. J Mol Biol 367:1227–1236
72. Rust MJ, Bates M, Zhuang X (2006) Sub-diffraction-limit imaging by stochastic optical reconstruction microscopy (STORM). Nat Methods 3:793–796
73. Sanii B, Smith AM, Butti R, Brozell AM, Parikh AN (2008) Bending membranes on demand: fluid Phospholipid Bilayers on Topographically deformable substrates. Nano Lett 8:866–871
74. Semrau S, Pezzarossa A, Schmidt T (2011) Microsecond single-molecule tracking (μsSMT). Biophys J 100(4):L19–L21
75. Shao L, Kner P, Rego EH, Gustafsson MG (2011) Super-resolution 3D microscopy of live whole cells using structured illumination. Nat Methods 8:1044–1046

76. Shatursky O, Heuck AP, Shepard LA, Rossjohn J, Parker MW, Johnson AE, Tweten RK (1999) The mechanism of membrane insertion for a cholesterol-dependent cytolysin: a novel paradigm for pore-forming toxins. Cell 99:293–299
77. Shepard LA, Heuck AP, Hamman BD, Rossjohn J, Parker MW, Ryan KR, Johnson AE, Tweten RK (1998) Identification of a membrane-spanning domain of the thiol-activated pore-forming toxin Clostridium perfringens perfringolysin O: an α-helical to β-sheet transition identified by fluorescence spectroscopy. Biochemistry 37:14563–14574
78. Shepard LA, Shatursky O, Johnson AE, Tweten RK (2000) The mechanism of pore assembly for a cholesterol-dependent cytolysin: formation of a large prepore complex precedes the insertion of the transmembrane β-hairpins. Biochemistry 39:10284–10293
79. Shim S-H, Xia C, Zhong G, Babcock HP, Vaughan JC, Huang B, Wang X, Xu C, Bi G-Q, Zhuang X (2012) Super-resolution fluorescence imaging of organelles in live cells with photoswitchable membrane probes. Proc Natl Acad Sci USA 109:13978–13983
80. Shimada Y, Maruya M, Iwashita S, Ohno-Iwashita Y (2002) The C-terminal domain of perfringolysin O is an essential cholesterol-binding unit targeting to cholesterol-rich microdomains. Eur J Biochem 269:6195–6203
81. Shotton DM (1989) Confocal scanning optical microscopy and its applications for biological specimens. J Cell Sci 94:175–206
82. Shroff H, Galbraith CG, Galbraith JA, Betzig E (2008) Live-cell photoactivated localization microscopy of nanoscale adhesion dynamics. Nat Methods 5:417–423
83. Shtengel G, Galbraith JA, Galbraith CG, Lippincott-Schwartz J, Gillette JM, Manley S, Sougrat R, Waterman CM, Kanchanawong P, Davidson MW, Fetter RD, Hess HF (2009) Interferometric fluorescent super-resolution microscopy resolves 3D cellular ultrastructure. Proc Natl Acad Sci USA 106:3125–3130
84. Solovyova AS, Nöllmann M, Mitchell TJ, Byron O (2004) The solution structure and oligomerization behavior of two bacterial toxins: pneumolysin and perfringolysin O. Biophys J 87:540–552
85. Soltani CE, Hotze EM, Johnson AE, Tweten RK (2007) Structural elements of the cholesterol-dependent cytolysins that are responsible for their cholesterol-sensitive membrane interactions. Proc Natl Acad Sci USA 104:20226–20231
86. Song L, Hennink EJ, Young IT, Tanke HJ (1995) Photobleaching kinetics of fluorescein in quantitative fluorescence microscopy. Biophys J 68:2588–2600
87. Song L, Varma CA, Verhoeven JW, Tanke HJ (1996) Influence of the triplet excited state on the photobleaching kinetics of fluorescein in microscopy. Biophys J 70:2959–2968
88. Steinfort C, Wilson R, Mitchell T, Feldman C, Rutman A, Todd H, Sykes D, Walker J, Saunders K, Andrew PW, Boulnois GJ, Cole PJ (1989) Effect of Streptococcus pneumoniae on human respiratory epithelium in vitro. Infect Immun 57:2006–2013
89. Stevens LM, Frohnhofer HG, Klingler M, Nusslein-Volhard C (1990) Localized requirement for torso-like expression in follicle cells for development of terminal anlagen of the Drosophila embryo. Nature 346:660–663
90. Taylor SD, Sanders ME, Tullos NA, Stray SJ, Norcross EW, McDaniel LS, Marquart ME (2013) The cholesterol-dependent cytolysin pneumolysin from Streptococcus pneumoniae binds to lipid raft microdomains in human corneal epithelial cells. PLoS ONE 8:e61300
91. Thiery J, Keefe D, Boulant S, Boucrot E, Walch M, Martinvalet D, Goping S, Bleackley RC, Kirchhausen T, Lieberman J (2011) Perforin pores in the endosomal membrane trigger the release of endocytosed granzyme B into the cytosol of target cells. Nat Immunol 12:770–777
92. Thompson JR, Cronin B, Bayley H, Wallace MI (2011) Rapid assembly of a multimeric membrane protein pore. Biophys J 101:2679–2683
93. Tilley SJ, Orlova EV, Gilbert RJ, Andrew PW, Saibil HR (2005) Structural basis of pore formation by the bacterial toxin pneumolysin. Cell 121:247–256
94. Ulbrich MH, Isacoff EY (2007) Subunit counting in membrane-bound proteins. Nat Methods 4:319–321

95. Voskoboinik I, Smyth MJ, Trapani JA (2006) Perforin-mediated target-cell death and immune homeostasis. Nat Rev Immunol 6:940–952
96. Wickham SE, Hotze EM, Farrand AJ, Polekhina G, Nero TL, Tomlinson S, Parker MW, Tweten RK (2011) Mapping the intermedilysin-human CD59 receptor interface reveals a deep correspondence with the binding site on CD59 for complement binding proteins C8α and C9. J Biol Chem 286:20952–20962
97. Yildiz A, Forkey JN, McKinney SA, Ha T, Goldman YE, Selvin PR (2003) Myosin V walks hand-over-hand: single fluorophore imaging with 1.5 nm localization. Science 300:2061–2065
98. Yildiz A, Selvin PR (2005) Fluorescence imaging with one nanometer accuracy: application to molecular motors. Accounts Chem Res 38:574–582
99. Yildiz A, Tomishige M, Vale RD, Selvin PR (2004) Kinesin walks hand-over-hand. Science 303:676–678
100. Young JD-E, Hengartner H, Podack ER, Cohn ZA (1986) Purification and characterization of a cytolytic pore-forming protein from granules of cloned lymphocytes with natural killer activity. Cell 44:849–859
101. Zheng C, Heintz N, Hatten ME (1996) CNS gene encoding astrotactin, which supports neuronal migration along glial fibers. Science 272:417–419
102. Zou H, Lifshitz LM, Tuft RA, Fogarty KE, Singer JJ (1999) Imaging Ca2+ entering the cytoplasm through a single opening of a plasma membrane cation channel. J Gen Physiol 114:575–588

About the Editors

Gregor Anderluh is head of the Laboratory for Molecular Biology and Nanobiotechnology at the National Institute of Chemistry, Slovenia, and Professor of Biochemistry at the University of Ljubljana, Slovenia. He received his Ph.D in Biology from University of Ljubljana, where he worked on molecular mechanisms of action of pore forming proteins from sea anemones. He did his postdoctoral studies at the University of Newcastle, United Kingdom. He and his coworkers are studying protein-membrane interactions and how cellular membranes are damaged by proteins in bacterial pathogenesis and immune system. He is also head of the Infrastuctural Centre for Molecular Interaction Analysis at the University of Ljubljana, where they study molecular interactions and are developing novel approaches on how to study protein binding to membranes.

Robert Gilbert holds a Senior Departmental Research Fellowship in Structural Biology within the Nuffield Department of Medicine at the University of Oxford. He is also fellow and tutor in Biochemistry at Magdalen College, Oxford. After a B.Sc degree at Durham University he received his Ph.D from the University of Leicester, where he worked on the cholesterol-dependent cytolysin pneumolysin. He undertook his initial post-doctoral studies in the Division of Structural Biology at Oxford and held a Royal Society University Research Fellowship there from 2004–2012. He and his co-workers are studying

G. Anderluh and R. Gilbert (eds.), *MACPF/CDC Proteins - Agents of Defence, Attack and Invasion*, Subcellular Biochemistry 80, DOI: 10.1007/978-94-017-8881-6,

protein interactions using biophysical techniques and also making use of cryo-electron microscopy and X-ray crystallography to study pore formation processes and also the RNA-directed control of gene expression, especially ribosomal frame shifting and RNA interference.

Index

Symbols
α-hemolysin, 134, 296, 301
α-PFP, 127, 134
β-barrel, 4, 9, 63, 64, 66, 72–76, 93, 95, 100, 101, 103, 106, 127, 168, 208, 296, 297, 299–301, 304–307
β-hairpin, 63–65, 72, 74, 75, 95, 99, 100, 105, 147, 168, 245
β-PFP, 127
β-sandwich domain, 55, 245
γ-hemolysin, 127, 196, 300, 301
μsSMT (microsecond single-molecule tracking), 294, 298
π-stacking interaction, 73, 75

A
AA (amino acid), 88, 91, 92, 97–99, 100, 102
AD (atopic dermatitis), 229
Adaptive immunity, 183
ADCC (antibody-dependent cell-mediated cytotoxicity), 205
Aegerolysin-like proteins, 274, 279, 285
Aegerolysins, 274–278, 285
Agrocybeaegerita, 274
AIF (apoptosis-inducing factor), 152
Allografts, 221, 232
ALO (anthrolysin), 49–51, 56
ANATO (anaphylatoxin), 31
Anopheles, 243, 245
APC (antigen presenting cells), 203
APC-β (Apicomplexan perforin-like proteins C-terminal β-pleated domain), 8, 15
APC-β (Apicomplexan perforin-like protein C-terminal β-sheet), 241, 245
Apicomplexa, 241– 244, 250, 251
Apoptosis, 121, 130, 133, 145, 152, 200, 203, 211
Arcs, 128, 129
ASM (acid sphingomyelinase), 177, 212
Aspergillus fumigatus, 274
Asp-haemolysin, 274, 278, 279
Asthma, 232
ASTN (astrotactin), 10, 11
AUC (analytical ultracentrifugation), 56
Autoimmunity, 228, 233
Autologous cells, 221, 227

B
Bacillus thuringiensis, 277–279
Bacterial toxin, 165, 183
Badan (6-bromoacetyl-2-dimethylaminonaphthalene), 301
Bf (Factor B), 32
Bilateria, 33, 36
BRINP (bone morphogenic protein/retinoic acid inducible neural-specific protein), 10

C
C2 domain, 9, 15, 19, 47, 53–55, 122, 207, 208
C3 convertase, 32
C3 (the third component of complement), 32
C3a, 32, 38
C3b, 32, 38
C4 (the fourth component of complement), 32, 33
C5 (the fifth component of complement), 32, 33, 35, 38, 40, 83, 85–89, 90, 93, 106
C5b-6, 87, 91–93, 101, 102
C5b-7, 85, 93–95, 101–104
C5b-8, 85, 95–97, 100, 102, 103, 105
C5b-9, 89, 95–97, 102, 103, 105
C6 (the sixth component of complement), 32, 37, 39, 83, 85, 87, 90–94, 97, 99–102

G. Anderluh and R. Gilbert (eds.), *MACPF/CDC Proteins - Agents of Defence, Attack and Invasion*, Subcellular Biochemistry 80, DOI: 10.1007/978-94-017-8881-6,

C7 (the seventh component of complement), 33, 37, 91–94, 100, 102
C8 (the eighth component of complement), 83, 85, 87, 93–97, 99, 100, 102–106
C9 (the ninth component of complement), 32, 37, 39, 40, 85, 88, 90, 91, 93, 95–97, 100, 103–105
CAD1 (constitutively activated cell death 1), 16
Cathepsin B, 199, 201, 209
CCP (complement control protein), 36
CCP (complement control protein repeat), 91
$CD4^+$, 183, 198, 205, 213
CD4 (cluster of differentiation 4), 227, 229–231, 233
$CD8^+$, 183, 198, 203, 205–207, 213
CD8 (cluster of differentiation 8), 222, 227–233
CD59, 49, 51
CD107, 209
CD109, 33, 35, 40
CDC (cholesterol-dependent cytolysin), 7–9, 11, 13, 15, 18, 20–25, 85, 86, 91, 96–99, 100, 102–106, 259
Cell cytoskeleton, 151
Cell death, 145, 157, 150, 152
Cell lysis, 121
Cell signalling, 120
Cell traversal, 243, 248–250
Cellular effects, 132, 134, 135
Cerebral inflammation, 229
Cerebral malaria, 229
CFTR (cystic fibrosis transmembrane conductance regulator), 171
CFU (colony forming units), 154
CH1 (helical cluster 1), 85, 93–95, 99, 100
CH2 (helical cluster 2), 85, 92, 93, 99, 100, 105
Chlamydia, 123, 242, 255–260, 262, 265
Chlamydia abortus, 256, 259
Chlamydia caviae, 256, 259, 261
Chlamydia felis, 256, 259
Chlamydia muridarum, 256, 259
Chlamydia pecorum, 256
Chlamydia pneumonia, 256, 259
Chlamydia psittaci, 256, 259
Chlamydia suis, 256
Chlamydia trachomatis, 256, 258–263
Chlamydiaceae, 256, 258, 265
Chlamydial developmental cycle, 256, 257, 264
Chlamydial plasticity zone, 255, 258–260, 262
Cholesterol, 64, 65, 67–72, 76, 122, 123, 279, 280, 281, 286
Cholesterol accessibility, 67–71, 76
Cholesterol recognition, 63, 65, 66, 69, 70, 71
Cholesterol threshold, 67, 71
Cholesterol-binding, 167
Cholesterol-dependent cytolysins, 47, 48, 63–66, 72, 74, 119, 120, 122, 123, 127, 129, 132, 134, 135, 161
CIA (collagen-induced arthritis), 228
Cilial beat, 152, 153
CI-MPR (cation-independent mannose-6-phosphate receptor), 211
CL (cytotoxic lymphocytes), 222
Clostridium perfringens, 63, 64
Cnidaria, 4
Complement, 31–33, 35–41
Complement activation, 148
Complement components, 83, 85, 86, 97
Complement system, 8, 83–85, 88
Confocal microscopy, 296, 307
CPAMD8 (PZP-like A2M domain-containing 8), 33, 35
CRP (C-reactive protein), 148
Cryo-electron microscopy, 47, 56, 86, 90, 95, 96, 99, 101, 103, 104
c-SMAC (the central region of the immune synapse), 210
CT_153, 124, 125
CT153, 259
CTL (cytotoxic T lymphocytes), 198, 206, 209, 230–232
CUB (C1r, C1s, uEGF and bone morphogenic protein), 33, 36
CUB domain, 89
Cutaneous inflammation, 221, 229
CVF (cobra venom factor), 89
CYLD (deubiquitinating enzyme cylindromatosis), 149, 151
Cytotoxic granules, 197, 198, 208–210

D

D1 (domain 1), 64, 65, 73, 96–99
D1-3 (domains 1-3), 146, 151
D2, 64, 65, 72–75
D3, 64, 65, 69, 72–75
D3 (domain 3), 96, 97, 99
D4 (domain 4), 64–66, 69–71, 73, 76, 98, 105
Destruction of allogeneic cells, 232
Deuterostome, 31, 33, 36–40
DiΦPC (diphytanoylphosphatidylcholine), 280, 283
Diabetes, 221, 226, 231, 231
Diatom, 8, 21, 22, 24

DM (dermatomyositis), 227, 228
DMD (duchenne muscular dystrophy), 230
Domain, 32, 33, 35–40, 64, 65, 71

E

Early endosome, 259
EB (elementary body), 257, 260, 261, 263
Echinoderm, 33, 36, 38, 39
Effect on gene expression, 148
EGF (epidermal growth factor like repeat), 91–94, 96, 101, 103
EGF (epidermal growth factor regulatory domain), 224, 225
EGFCA (calcium-binding EGF-like), 36
EGF-like domain, 207, 208
Egress, 243, 246–248, 250
Endocytosis, 129, 132–134, 177
Enterolobin, 21
Enterolysin, 20
Environment-sensitive flourophores, 299–301, 306
EOMES (eomesodermin), 206, 207
EPCS50, 17
Epigenetic regulation, 149
EPM (erythrocyte plasma membrane), 247, 250
Equinatoxin II, 122, 127
ER (endoplasmic reticulum), 178, 181, 182, 208, 209
ERK (extracellular-signal-regulated kinase), 178
Erylysin B, 14, 278
Erythrocyte, 241, 247, 250
Eumetazoa, 31, 33, 38
Evolution, 33, 35–37, 40
Exocytic vesicle, 259, 262
Extrusion, 257, 258, 263, 265

F

FasL (Fas ligand, a type-II transmembrane protein that belongs to the TNF family), 228, 232
FHL (familial hemophagocytic lymphohistiocytosis), 224, 226
FHL-2 (familial hemophagocytic lymphohistiocytosis type 2), 222, 224
FIM (Factor I/membrane attack complex 6/7 module), 90–93, 101
FIMAC (factor I membrane attack complex), 36, 241, 245
FIONA (fluorescence imaging with one-nanometre accuracy), 302, 303, 311
Fluorescence, 294, 296, 298, 301, 305–308, 312
Fluorescent protein, 312, 313
Fluorophore, 299, 300, 302, 303
FRET (Förster resonance energy transfer), 299, 300
Fungi, 11, 13, 14, 24

G

Gametocytes, 246, 247
Gene duplication, 35, 37, 38, 40
GILT (gamma-interferon inducible lysosomalthiolreductase), 169, 171
GNLY (granulysin), 209
Gram-negative bacteria, 8, 18, 121
Gram-positive bacteria, 8, 11, 19, 63, 120
Granzymes, 8
GvHD (graft-versus-host disease), 213
Gzm (granzymes), 198, 201, 207, 212

H

HBMEC (Human Brain Microvascular Endothelial Cells), 153
Hearing loss, 153–155
Hemibiotrophic pathogen, 285
Heterobasidion irregulare, 276
HGF (hepatocyte growth factor receptor), 164
HGH (hemophagocyticlymphohistiocytosis), 164
HGMD (human gene mutation database), 224
Histone modifications, 130, 132, 134
HNP1 (human neutrophil peptide), 166
Host cell invasion, 163, 165, 172, 185
HPM (hepatocyte plasma membrane), 250
Human, 32, 33, 35–38
Human respiratory epithelium, 152
Hyperkalaemia, 284

I

ICAM-1 (intracellular adhesion molecule-1), 148
If (Factor I), 33
IL2 (interleukin-2), 207
ILY (intermedilysin O), 173
ILY (intermedilysin), 50–54
Immune signalling, 135
Inclusion, 257, 259, 261–265
Infectious diseases, 163, 174
Inflammation, 31, 32, 38, 224, 227, 229, 232
Inflammatory lesions, 228
InlA (internalin A), 164, 179

InlB (internalin B), 164, 179
Innate immunity, 183–185
Interleukin-8, 148, 150
Intermedilysin, 21, 22, 64, 72, 75, 122
Intracellular delivery, 154
Intracellular parasite, 256, 263
Intracellular pathogen, 161, 163, 180
Invaginations, 129
Ion conductivity, 120
Ion fluxes, 180
Ions, 122, 129, 130, 134
IPALM (interferometric PALM), 312, 313
IPTG (isopropyl β-D-1-thiogalactopyranoside), 172
IS (immune synapse), 198, 209, 210
iTEP (insect TEP), 33

J
Jacalin domain, 15
JNK (c-Jun N-terminal kinase), 178

K
Killer cell, 198, 200, 201, 207–210, 213
Kupffer cells, 246, 248, 249

L
LCMV (lymphochoriomeningitis virus), 231
LCMV-GP (glycoprotein of lymphocytic choriomeningitis virus), 231
LCR (locus control region), 207
LDLa (low-density lipoprotein receptor domain class a), 36, 244
LGV (lymphogranuloma venereum), 256
LIPI-1 (*Listeria* pathogenicity island), 163, 165, 170
Lipid bilayer, 84, 99, 100, 102, 103
Lipid droplets, 258, 259, 262
Lipid rafts, 279
Liquid-ordered lipid phase, 281
Listeria monocytogenes, 161, 162, 185
Listeriolysin O, 19
Listeriosis, 162, 163, 171
Liver, 246, 248–250
LLO (listeriolysin O), 48, 120, 129–131, 133, 134, 161, 165, 166, 168–171, 173–175, 178–185, 306
LR (low density lipoprotein receptor class A repeat), 91–94
LUV (large unilamellar vesicles), 280, 281
Lysenin, 122, 129
Lysteriolysin O, 67

M
MAC (membrane attack complex), 8, 15, 32, 83–86, 88, 90–93, 95–99, 101–106
MACPFapi (MACPF domains of apicomplexans), 244, 245, 250
MACPF/CDC proteins, 119, 123, 127–129
MACPF domain, 49, 203, 208
MACPF (membrane attack complex/perforin), 7–16, 18, 19, 21–25, 37, 85–87, 91–96, 97–103, 105, 222, 224, 225
MACPF motif, 11, 12
Macrophage, 200, 203, 213
Macrophage activation syndrome, 221, 226
Malaria, 243, 245
MALS (multi-angle light scattering), 56
MAOP (membrane-attack ookinete protein), 247
MAPK (mitogen-activated protein kinase), 178
MASP (mannan binding protein associated serine protease), 33, 87
MBLs (manose binding lectins), 87
Membrane attack complex, 119–121, 127, 133, 148, 198, 208
Membrane binding, 122, 123, 125, 126
Membrane curvature, 305
Membrane damage, 174, 176, 177
Membrane disruption, 250, 251
Membrane perturbation, 106
Membrane pores, 97
Membrane repair, 132–134, 175–177
Membrane structure, 59, 65
Meningitis, 152–154
MG (macroglobulin domain), 33
MG (macroglobulin), 88
MHC (major histocompatibility complex), 229
Micropores, 151
Midgut, 246–248
MIR (mannosyltransferase, inositol 1,4,5-triphosphate receptor, ryanodine receptor), 259, 261, 262
MIR domain, 15
Mitochondrial damage, 152
Moniliophthora perniciosa, 275
Monomer, 259, 262
Mosquito, 245–248
MPEG1, 200
Mpg-1 (macrophage-expressed gene 1), 10, 16, 17, 23
Mpl (metalloprotease), 170
MTOC (microtubule organizing center), 210

N

NALP3 (NOD-like receptor family pyrin domain containing 3), 150
NBD (7-nitro-2-1,3-benzoxadiazol-4-yl), 300
Nematostella vectensis, 33
Neutrophils, 149–152, 154
NK (natural killer), 221, 222, 226, 227, 229
NLRs (NOD-like receptors), 183

O

Oligomer, 262
Oligomerisation, 63, 65, 67, 72, 73, 75, 119, 121, 130
OlyA (ostreolysin A), 274, 276, 278–280, 283, 286
Ookinetes, 247
Opsonization, 31
Ostreolysin, 274, 276, 277, 279, 283, 284, 286, 287

P

p38-MAPK (p38 mitogen-activated protein kinase), 151
PAI-1 (plasminogen activator-1), 149
PALM (photoactivated localization microscopy), 303
Pancreatic β-cells, 221, 231
Parasites, 242, 243, 245–248, 251
Pathogenesis, 260, 265
Pathogenesis of pneumococcal infection, 145
Pathogen-mediated endocytosis, 260
Pattern recognition receptors (PAMP), 150
PC-PLC (phosphatidylcholine-specific phospholipase), 164, 165, 170, 172
PcPV2, 19
PDMS (poly(dimethyl) siloxane), 305
PERK (double-stranded RNA activated protein kinase (PKR)-like ER kinase), 181
PFN (perforin), 8, 9, 11, 13, 15, 47, 53–56, 84–86, 91, 96, 97, 99, 101, 103, 104, 120, 121, 127–129, 132, 198, 200, 201, 203–209, 211–213, 242, 244, 277, 278
PFO (perfringolysin O), 9, 19, 24, 48–51, 53–57, 63, 120, 128, 131, 134, 166–168, 173, 174, 299
PFP (pore-forming protein), 121, 126, 128, 129, 132, 134, 135
PFT (pore-forming toxin), 120, 121, 130, 134
Phosphatidylcholine, 65, 67
Phospholipase A, 149
Phospholipase D, 258
Phospholipid, 259
Photo bleaching, 296–297, 303
Phylogeny, 20, 21, 23, 24
PI3K (phosphoinositide-3-kinase), 152
PI-PLC (phosphatidylinositol-specific phospholipase), 164, 165, 170, 181
PKC (protein kinase-C), 153
Plasma membrane, 198, 201, 209–212, 257, 260, 261, 263, 264
Plasmodium, 242, 243, 245–247, 250
PLD (phospholipase D), 259, 260, 262
Pleurotolysin, 127, 174
Pleurotolysin A, 15
Pleurotolysin B, 12, 14, 15, 283
Pleurotus eryngii, 274
Pleurotus ostreatus, 274
PLM (planar lipid membranes), 283
PLP (perforin-like protein), 245, 247
PLY (pneumolysin), 48, 56–59, 120, 145–156, 304
PLY (pneumolysin O), 305, 306, 311–313
PlyA (pleurotolysin A), 15
PlyA2 (pleurotolysinA2), 287
PlyB (pleurotolysin B), 272–280, 283–286
PlyB-like proteins, 274–276, 285, 287
PM (polymyositis), 227, 228
Pneumolysin variants, 147, 150
Poly-C9 (polymeric C9), 96, 97
Pore, 47–49, 51, 54, 55, 58–60, 121, 123, 126–128, 130, 134, 135
Pore conductance, 283
Pore formation, 145, 151, 156, 157
Pore forming protein, 121, 126, 128, 129, 132, 134, 135, 176, 180
Pore size, 127, 134, 135
Pore-formation, 123
Pore-forming toxin, 4, 5, 64, 76, 84, 120, 121, 130, 134, 161, 165, 170, 175, 177, 178, 180, 182, 185
Post-translational modifications, 134
PPLP (*Plasmodium perforin*-like protein), 247
Pre-pore, 47, 56–59
Pre-pore complex, 64, 65, 72–76
PRF1, 198, 206, 207, 213
PrfA (positive regulatory factor A), 165
Propidium iodide, 249
Proteolysis, 261, 262
Protostome, 33, 36, 37, 39, 40
Pseudomonas aeruginosa, 279
PSF (point spread function), 302, 307
PsTx-60A, 18, 19
pulmonary permeability edema, 153
PV (parasitophorous vacuole), 243, 245–247, 249

PVM (parasitophorous vacuole membrane), 247, 249, 250
PZP (pregnancy zone protein), 33

Q
Quiescent-like state, 131, 134

R
RB (reticulate body), 257, 258, 261–263
RBC (red blood cell), 245, 246
Receptor binding, 135
Rheumatoid arthritis, 221, 228
RMSD (root mean square deviation), 51, 54
ROCK (rho-associated kinase), 151
ROS (reactive oxygen species), 171

S
Salivary glands, 246, 248
SCR (short consensus repeats), 245
SDM (site-directed mutagenesis), 148
Serglycin, 199, 209, 210
Serine protease, 35, 36
SHRImP (single-molecule high-resolution imaging with photobleaching), 303, 307
SIM (structured illumination microscopy), 307
Single-molecule, 296, 297, 300, 301, 302, 306
Sinusoids, 246, 248–250
SLE (systemic lupus erythematosus), 227
SLEDAI (SLE disease activity index), 227
SLO (streptolysin O), 48–52, 54, 56, 120, 129, 131, 133, 167, 171, 180, 212
SLY (suilysin), 49, 50–52, 54, 56
smFRET (single-molecule FRET), 300
SMT (Single-molecule tracking), 298
SNARE (soluble N-ethylmaleimide-sensitive factor accessory protein receptor), 201, 209, 210
Spatial resolution, 307, 311, 314
SPECT (sporozoite microneme protein essential for cell traversal), 250
SPECT2, 248–250
Sphaerobolus stellatus, 275
Sphingomyelin, 274, 279, 283, 286
SpoC1-C1C, 14
Sporozoites, 246, 248, 249
SR (scavenger receptor Cys-rich), 36
STED (stimulated emission depletion), 307–309
Step-counting, 296
STORM (stochastic optical reconstruction microscopy), 303, 307–309, 311–313
Streptolysin O, 19, 48–52, 54, 56, 67, 120, 129, 131, 133, 167, 171, 180, 212
Structural biology, 85, 86
Structural phylogeny, 51, 53
SUMO (small ubiquitin-like modifier), 179
SUMOylation, 131, 134, 149, 179
Super-resolution, 301, 303, 304, 307–309, 311–313

T
TBX21, 207
Tc1 (type 1 CD8 T cells), 232
Tc2 (type 2 CD8 T cells), 232
TCC (terminal complement component), 37
TCR (T cell receptor), 203, 205
TED (thioester domain), 33
Temporal resolution, 298–301, 309
TEP (thioester-bond containing protein), 33
TgPLP1, 248, 250, 251, 262, 264
TgPLP1 (*Toxoplasma gondii* perforin-Like Protein 1), 123, 244, 248–251, 262, 264
Thioester, 32
TIRF (total internal reflection fluorescence) microscopy, 296, 297, 301, 302
TLO (tetanolysin), 147
TMH (transmembrane hairpin), 48, 49, 57, 58, 75, 85, 92, 93, 95, 99, 100, 101, 102, 105
TMH1 (transmembrane hairpin 1), 74, 75, 85, 99, 100
TMH2 (transmembrane hairpin 2), 74, 75, 85, 99, 100, 105
TNF (tumor necrosis factor), 222, 228, 231
Toll-like receptor 4, 150
Toroidal pore, 102, 128
Toxin, 18, 19, 24
Toxoplasma, 243, 244, 248, 251
Trametes versicolor, 276
Transmembrane pore rich loop, 65, 66, 69
Transmembrane β-barrel, 73
Tryp_SPc (trypsin-like serine protease), 36
TS (thrombospondin type 1 repeat), 91, 92, 96, 103
TSP1 (Thrombospondin type 1 repeats), 37
TSR (Thrombospondin type I repat), 245
Type 1 interferon, 154

U
Undecapeptide loop, 54
UPR (unfolded protein response), 181, 182
Urochordate, 33, 38, 39
UTR (untranslated region), 166

V
Vaccines, 155
Vertebrate, 33, 37, 38
Viral myocarditis, 230
Virulent factors, 120
VWA (von Willebrand factor type A), 36

W
White rotting fungi, 285

X
X-ray crystallography, 56, 60

FSC
www.fsc.org
MIX
Papier aus verantwortungsvollen Quellen
Paper from responsible sources
FSC® C105338

If you have any concerns about our products,
you can contact us on
ProductSafety@springernature.com

In case Publisher is established outside the EU,
the EU authorized representative is:
Springer Nature Customer Service Center GmbH
Europaplatz 3, 69115 Heidelberg, Germany

Printed by Libri Plureos GmbH
in Hamburg, Germany